Sociology and Critical Inquiry

The Work, Tradition, and Purpose

ADVISING EDITOR IN SOCIOLOGY

Charles M. Bonjean
University of Texas—Austin

Sociology and Critical Inquiry

The Work, Tradition, and Purpose

Third Edition

John Walton
University of California, Davis

WADSWORTH PUBLISHING COMPANY
Belmont, California
A Division of Wadsworth, Inc.

Editor: Serina Beauparlant
Senior Editorial Assistant: Marla Nowick
Production: Ruth Cottrell
Print Buyer: Karen Hunt
Designer: Lucy Lesiak
Copy Editor: Betty Duncan
Compositor: Omegatype
Cover: Vargas/Williams/Design
Printer: Malloy Lithographing, Inc.

Printed in the United States of America

1 2 3 4 5 6 7 8 9 10—97 96 95 94 93

Library of Congress Cataloging-in-Publication Data
Walton, John, 1937–
Sociology and critical inquiry : the work, tradition, and purpose / John Walton. — 3rd ed.
p. cm.
Includes bibliographical references and index.
ISBN 0-534-20400-7
1. Sociology I. Title.
HM51.W23 1993 92-41289
301—dc20

Prologue

In the pages that follow, I attempt something unusual by way of an introduction to sociology. I try to present the method and a strong dose of the substance of modern sociology in a manner that is both accessible to the general reader and faithful to the discipline. I want to portray the best work of professional sociologists and carry it forward to matters of public concern, rather than tone it down for indifferent consumption. That is possible, I believe, for two reasons.

First, sociology is a lively subject that engages people of diverse occupations and avocations. It enjoys the distinction among academic disciplines of addressing matters that involve a broad public similarly devoted to understanding pieces of the social order. Sociologists and journalists conduct surveys or opinion polls trying to fathom public moods. Sociologists and lawyers study the development and application of laws. There is a sociology of medicine, welfare, occupations, art, and many more. Sociologists theorize about the roots of social order and conflict, as do others who reflect on, say, the causes of crime or family dissolution. Indeed, many sociological explanations bear a close resemblance to lay theories and vice versa. Many of sociology's central problems coincide with important dilemmas of public policy and social ethics.

Second, sociology is intrinsically lively. Its investigative methods have the intrigue of detective stories. Its results turn up bracing refutations of common sense. Its diagnoses have been incorporated into public policies on social welfare and reform. We all experience society and share in the perplexities of how to explain it. We can see our collective selves in social research about the changing structure of social classes, jobs, leisure, or cities and suburbs.

This book is written for people who are curious about explanations for social phenomena. I address students and, more elusively, the educated public who come to sociology out of their own curiosity. I hope to reach that audience without condescending or holding back important ideas because they defy oversimplification. Yet I have written this for newcomers to sociology, avoiding jargon and the professional idiom in favor of clear language and fertile examples.

This is a book with a purpose. The phrase *critical inquiry* in the title means that the book takes a critical stance on sociology and society. I am critical of social inequality meted out along lines of social class, race, gender, and prestige. Similarly, I criticize varieties of sociology that ignore inequality by celebrating a vaguely alternative "value integration" or justify it according to some "function" it presumably performs in society. More importantly, critical sociology differs from a descriptive, applied, or theoretical sociology by going to the roots of historical and contemporary social practices. Critical sociology is not always grim. It looks for alternatives to pernicious social practices, particularly those alternatives fashioned in the struggles of classes and communities.

Each succeeding chapter develops this central idea from a different point of view. Part I provides the equipment that sociology uses to interpret and investigate society. Chapter 1 begins with a concise history of sociology, suggesting that the discipline enjoys a classical tradition and a distinctive perspective on the world. Sociology is not a new science. The classical tradition began with the inheritors of the French Revolution, was perfected in different ways by Karl Marx and Max Weber, refashioned for North American use by Thorstein Veblen and C. Wright Mills, and is alive in sociological practice today. Chapter 2 gives a description and set of illustrations about how sociologists actually do their work—what the sociological method means in practice and the unity of this method across a range of techniques appropriate to different research problems. The subsequent chapters clarify this perspective by applying it to specific topics. Chapters 3 to 5 of Part II treat the basic dimensions of segmentation in modern society: the rural–urban divide, the crucial concepts social class and status group that are fundamental bases of social organization and inequality. Here, as in most of the chapters, the concepts and explanations are developed in close connection with historical and contemporary society. Chapters 6 to 8 in Part III turn from segmentation to the principal forms of social integration in the modern world: the state, world system, and social control. Chapter 6 treats a key problem too often ignored in sociological analysis—namely, the state that mediates political struggles over class and status, assuming some autonomy in the process. Chapter 7 describes how today's developed and Third World societies are increasingly ensnared in a world system that affects the individual's daily life. Chapter 8 explains how efforts to control society are implemented along lines of privilege, how those efforts change, and how people resist or rebel against them. Finally, Chapter 9 attempts a new synthesis of the foregoing concepts around the idea of civil society and compares the development of civil society in Eastern Europe, the Third World, and the United States.

This book began as an effort to convey a sense of the sociological craft to a wide audience. That proved more demanding and time-consuming than I had anticipated. In the process of revising the original manuscript, I received helpful criticism from a number of colleagues and reviewers including Shelley Coverman, George Kilpatrick, Peter Heller, and Vincent Jeffries. For their comments and suggested revisions of the first and second editions, I am grateful to York Bradshaw of Ohio State University, Michael Kennedy of the University of Michigan, and Boyd Littrell of the University of Nebraska. For their review of the third edition, I thank David Ashley of the University of Wyoming, Gary Green of the University of Georgia, Charles L. Harper of Creighton University, and C. Edwin Vaughan of the

University of Missouri. The third edition was completed during my term as a guest scholar at the Woodrow Wilson International Center for Scholars, Smithsonian Institution, Washington, D.C. I am grateful to the Center and its staff for their generous support.

John Walton
Carmel Valley, California
September 1992

Contents

List of Tables

PART I

THEORY AND METHOD

1

The Sociological Tradition

Thorstein Veblen was the first major North American social theorist but found it difficult to hold academic posts. As John Dos Passos said, "There seemed no place for a masterless man." (Painting by Edwin Burrage Child, with permission of Culver Pictures, Inc.)

THE SOCIOLOGICAL IMAGINATION

Sociology excites a unique set of reactions — it bores some and frightens others. Doubtless, it kindles additional moods, including the passion of a few, but sociology is extraordinary because it can be regarded as both trivial and threatening. This anomaly suggests, at least to a sociologist, that there is more here than meets the eye.

Consider the following loosely connected facts. Before a national television audience in the United States, an eminent broadcast journalist, the late Eric Sevareid, once referred to sociology as slow journalism. Yet only a few years before in 1967, the Greek military junta abolished the teaching of sociology as a menace to the right-wing "Heleno–Christian Revolution." Socialist revolutions have also found reason to outlaw sociology. It virtually disappeared from the Soviet Union under Stalin and was banned for a time after 1952 in China, in both cases owing to its contaminating influence on the "science" of Marxism-Leninism. Despite these credentials, exponents of plain speaking such as the journalist Edwin Newman grumble that "a large part of social science practice consists of taking clear ideas and making them opaque" (1974, 174). Beginning in the 1920s, however, Mussolini's government purged Italian universities of sociologists whose criticism of fascism was all too clear. How is it that the same subject can arouse fears of subversion and state repression in some settings and a smug indifference in others?

An appreciation of the nature of sociology and these diverse perceptions requires some description of the discipline's special perspective and history. That is the purpose of this chapter. Beginning on a necessarily formal note, **sociology** is defined as the study of group life and those aspects of individual lives that are affected by social interaction.

The **sociological perspective** differs from the way we view the world as individuals; this perspective differs in ways that require us to step outside but never abandon personal experience and to gain a new vantage on the world. The reason for doing so is simple: There are many features of social life that cannot be explained otherwise. This principle is demonstrated in the phenomenon of suicide. Individuals understand suicide as a consequence of mental depression, personal or financial loss, or insanity. But none explains the social suicide rate in which Protestants show a much higher incidence than do Jews and educated and single people a higher rate than do the less educated and married. As we will see in discussions that follow, Émile Durkheim could explain these differential rates by unseen but nevertheless potent differences in the value integration of groups. Similarly, people know what it means for their daily life to belong to a particular profession, ethnic group, or religion, but they seldom grasp how those affiliations influence their political choices or how many children they will have. Sociology explores the determinants of individual and collective behavior that are not given in our psychic or biological makeup, but fashioned in the broader arena of social interaction. Once these social influences are revealed they make intuitive sense; they explain something that we did not know before and would not have discovered by searching the contents of individual experience. The question therefore is, How does one gain this perspective?

C. Wright Mills called this viewpoint the "sociological imagination" and suggested that the perspective provides a means for people to interpret their lives and social circumstances:

> [It] is a quality of mind that will help them to use information and to develop reason in order to achieve lucid summations of what is going on in the world and what may be happening within themselves. . . . The sociological imagination enables us to grasp history and biography and the relations between the two within society. (1959, 5, 7)

The sociological perspective is acquired, in the first instance, by asking particular sorts of questions about social life. Illustratively, sociologists do not ask, What makes people do this or that? (which implies that people have individualized dispositions or motives that explain whatever they do); sociologists ask instead, Why does one group of people do this or that in a different way or more often than another group? (which implies social causes of varied group actions). There are many other forms in which proper sociological questions can be asked, but the key is that they address the characteristics of groups and categories of people, that is, **interactive groups** such as families, clubs, and organizations and **distinctive categories** of unrelated persons such as occupations and social classes. Moreover, the sociological perspective assumes that the answers lie in social, as opposed to exclusively psychological or economic, influences. Sociology has no quarrel with the other social sciences. Indeed, it often combines with those in complex explanations. Sociology simply insists that much of social life has its own explanations that cannot be reduced to more elementary compounds from psychology, biology, or some other science.

Mills observed that nowadays people "often feel that their private lives are a series of traps" (1959, 3), that they have little control over the forces that shape their lives. Sociologists beginning with Marx have referred to this feeling as alienation, or the individual's sense of separation and estrangement from useful work, other people, and the self. The sociological perspective should allow people to understand this mood (an aspect of "biography") by relating it to historical and social forces that close in on the individual.

One of the most poignant statements of the alienated mood comes from John Steinbeck's novel *The Grapes of Wrath.* During the depression of the 1930s, an exchange takes place between a tenant farmer who is being evicted from his land and a tractor driver in the employ of the new corporate owners. The farmer begins speaking of his house that is about to fall under the tractor's blade:

> "I built it with my hands . . . it's mine. I built it. You dump it down — I'll pot you like a rabbit."
>
> "It's not me. There's nothing I can do. And look — suppose you kill me? They'll just hang you, but long before you're hung there'll be another guy on the tractor, and he'll bump the house down. You're not killing the right guy."
>
> "That's so," the tenant said. "Who gave you the orders? I'll go after him. He's the one to kill."
>
> "You're wrong. He got his orders from the bank. The bank told him, clear those people out or it's your job."
>
> "Well, there's a president of the bank. There's a board of directors. I'll fill up the magazine of the rifle and go into the bank."
>
> The driver said, "Fellow was telling me the bank get orders from the East. The orders were, 'Make the land show profit or we'll close you up.' "

> "But where does it stop? Who can we shoot? I don't aim to starve to death before I kill the man that's starving me."
>
> "I don't know. Maybe there's nobody to shoot. Maybe the thing isn't men at all. Maybe like you said, property's doing it. Anyway I told you my orders."
>
> "I got to figure," the tenant said. "We all got to figure. There's some way to stop this. It's not like lightning or earthquakes. We've got a bad thing made by men, and by God that's something we can change." (1939, 40).

There are several ways to understand this scene, different explanations of the tenant's frustration and anger. One might say that the tenant has a violent personality, another that he is economically motivated. The sociological perspective would recognize first that the actors in this drama represent broader social categories: hundreds of thousands of tenants who lost their land and livelihood, thousands of tractor drivers who were among the fortunate minority that still had jobs, and hundreds of bankers who foreclosed on their neighbors, perhaps reluctantly on "orders from the East." Why did the tenants want to shoot someone and the tractor drivers counsel restraint? Probably because the tenants felt moral outrage based on the social values of the western United States ("I built it with my own hands"), because they faced an unprecedented situation with no established avenue of redress ("There's some way to stop this"), and, perhaps, because the hard and solitary lives of tenant farmers did not generate politically organized communities which might have found a way to resist ("We all got to figure"). The tractor drivers, on the other hand, came from the same social class backgrounds, but they had the desperately needed jobs and they understood the connection between the evictions and the changing economy ("property's doing it . . . the bank get orders from the East . . . Make the land show a profit or we'll close you up").

In fact the tenants neither shot anyone nor organized politically to save their farms and way of life. Instead, they migrated in response to the destruction of former lives by the combined forces of drought and corporate capital (the banks and eastern owners). And they migrated as social groups. As Steinbeck's extraordinary novel shows, they took to the highways in caravans of families and friends, followed the routes that social networks informed them led to safe passage and California jobs, and helped one another along the way through socially innovated means of mutual support. The migration followed social patterns that reflected their origins in regions of high tenancy, drought, and, in troubled times, a certain cultural solidarity among these "Okies."

In the fifty some-odd years since this fictional incident, the tenant's dawning sense of alienation has diffused to the rest of society. Alienation has become an accepted part of the human condition, only temporarily escaped in leisure, fantasy, mood-altering substances, otherworldly visions, or acts of rebellion. Alienation can be alleviated, too, in effective action or satisfying work, but the difficulty is that that alternative is also harder to find. In 1981 Chicago autoworkers took sledgehammers to an imported Japanese car for the benefit of photographers and policymakers, but their action failed to restore their jobs or halt the steady decline in manufacturing employment. As in the case of the tenant farmer, they failed to find an appropriate or responsible agent on whom their troubles could be pinned. Action to ameliorate social problems is evermore elusive. Does sociology speak to such issues?

Today's alienation is expressed in a set of questions that want a sociological answer: Why is unemployment so high in societies that we are told enjoy prosperity? Will technology eliminate all forms of satisfying work and turn the next generation into computer technicians? Why has the civil rights revolution passed with little change in the income inequality of minorities? Why do problems of crime and drug abuse seem ever on the increase? Why are fundamentalist religion and outer-space movies so popular? Why are developed nations involved in new wars despite the once-vivid lessons of the old? Where are we headed, and what can we do about it?

The sociological perspective is uniquely suited to answer these kinds of questions. It begins by establishing the social facts at issue: rates of unemployment by social category, net effects of technological displacement, actual rates and participants in crime and drug abuse. Next the perspective invents and tries out alternative explanations: whether sloth, production methods, or prejudice explain categorical differences in employment and income. Finding the best available explanation, the perspective asks, What are the implications for related problems and solutions?

When sociology begins to answer these questions, it may also become useful, nettlesome, threatening — in short, controversial. An explanation that helps the tenant to understand may bore the tractor driver and threaten the banker. Sociology is as much a part of the world as the conflicts it examines. Its perspective is necessarily critical to the extent that sociology finds causes for major social changes that are bound to affect some for the better and others for the worse. The best way to describe the perspective and to discover much of the substance of sociology is to review the work of the people who developed it. Similarly, an appreciation of that work depends on understanding the times of the early sociologists and the parts they took in events of their times.

THE CLASSICAL TRADITION

Sociology is not a young science, contrary to some opinion and misguided apologies for the state of its knowledge. Sociology originated in the nineteenth century at a time when many of the modern sciences and humanities appeared. Sociology was a product of the same historical environment that produced these disciplines and, until the turn of the twentieth century, was part of a unified tradition of inquiry that embraced philosophy, law, economics, history, anthropology, and many aspects of the natural sciences. Yet among these pursuits, sociology always had an ambivalent relationship with the "great transformation" from agricultural and commercial societies to the modern world of industrial capitalism, urbanization, bureaucracy, and the nation state. Sociology was both a product and a critical reaction to modernization. This dominant theme created sociology and preoccupied all of its early practitioners.

The Beginnings

Sociology was made possible by the French Revolution. The capture of the Bastille on July 14, 1789, and the Women's March on Versailles introduced a new principle into history by demanding political action to redress mass poverty. The authority of king and church that once justified social inequality was challenged by the philosophical notion of the "general will" among people convinced that their suffering was no longer inevitable. For the first time, public opinion became a legitimate method of expressing social concerns. But the newborn forces of democracy did not at once take institutional shape. The revolution first produced a reign of terror that also justified itself in appeals to the will of the people and later a self-proclaimed emperor, Napoleon. As the political revolution faltered, the Industrial Revolution in England and France gathered momentum, mixing new ingredients to social unrest.

The Count Henri de Saint-Simon (1760–1825) was a product of this age and prophet of a new society. As an aristocratic youth, he fought with the French royalists against the North American revolutionaries, although his sentiments lay with neither side. Later he renounced his title, made and lost a fortune speculating in estates confiscated by the French government, and was imprisoned under the Terror. But he earned a new title: the founder of sociology. Saint-Simon's passion was social reconstruction. He was an eccentric and, in Edmund Wilson's words, "a dilettante on an enormous scale, made restless by an all-trying curiosity" (1940, 81). He studied physics, mathematics, and philosophy; spent his money on dissipation and travel; "and was able to investigate poverty at first hand" (1940, 81).

Saint-Simon's genius was to see what conventional thinkers could not: Society, no less than nature, is governed by scientific laws whose mastery could produce social institutions for the benefit of humanity. History, he surmised, alternated between periods of equilibrium and breakdown. The revolution was a breakdown setting the stage for a new equilibrium based not on the discredited formulas of politics but on the new and scientifically organized industrial system. Political schemes and unrealistic notions of equality (alongside the facts of corruption and inequality) would be replaced by artists and scientists in public service, based on a hierarchy of merit. The aim of industry would be provision of human need and fulfillment. On his deathbed, Saint-Simon remarked "all of my life may be summed up in one idea: to guarantee to all men the free development of their faculties" (Wilson, 1940, 85). His hope was that the new order would be installed, although not governed, by the "party of workers."

The title "founder of sociology" is a somewhat mythical and not altogether enviable status. Saint-Simon had his own mentors in the "science of history" and he was followed by others who claimed parentage of sociology. Auguste Comte (1798–1857), who studied and then repudiated Saint-Simon, first used the term and popularized the notion of a "positive science of sociology." But it was Saint-Simon who first combined eighteenth-century inklings of a social science with an analysis that laid bare the workings of postrevolutionary industrial society in the nineteenth century. Next to that achievement, Comte's hierarchy of the sciences, which placed sociology at the pinnacle (and Comte the father at the head of the new intellectual priesthood), all seemed a bit rarified and derivative. Unlike Saint-Simon, Comte's twin ambition was to create a new science of society and an elite of social experts, not a radically reorganized industrial society based on individual merit.

The modern U.S. sociologist Alvin Gouldner (1970) shows that there were "two roads from Saint-Simon." The first road taken by Comte and the academic sociologists led into the university and compromised commitments to social reconstruction as the price of support by the state — a price that Comte and his followers were eager to pay. The second road taken by the early socialists, Karl Marx just one among them, and by reformers of various political leanings led into politics, the new labor movement, and social action. Because the latter tendency is more consistent with Saint-Simon, academic sociologists have preferred to recognize Comte as their disciplinary founder. Gouldner notes wryly that "if sociologists acknowledge descent from Saint-Simon rather than Comte they are not only acquiring a father, but a blacksheep brother, socialism" (1962, 12).

Whatever these preferences for a founder, that many contributed to the discipline born in response to the trauma of modernization, including political and industrial revolution remains true. With these origins, sociology has always had a dual nature: a disciplined method and field of study and a purposeful activity in society — analysis and action. The manner in which this dual nature is balanced by individual practitioners varies, but the tension between the two is always present.

As sociology evolved from the mid-nineteenth to the early twentieth century, that tension was healthy, cross-fertile, and more often embraced than shunned. Characteristically, works that came to comprise the classical tradition set forth broad interpretations of historical trends and the main drift of modern society. As Mills said of the classical sociologists, "Their intellectual problems [were] relevant to the public issues of their times, and to the private troubles of individual men and women" (1959, 4).

Karl Marx. Karl Marx (1818–1883) was the preeminent social and political thinker of the nineteenth century, and he personifies the classical tradition. Marx and his lifelong collaborator Friedrich Engels (1820–1895) invented a social theory so compelling for their own times, and so provocative for the future, that all subsequent work in the classical mode must first come to terms with it.

Marx was born in Trier located in the western German Rhineland, then under French Napoleonic rule. He came from a middle-class family and grew up in prosperity; his mother was devoted to Judaism, although in an atmosphere of discrimination against Germans and Jews. Marx was an idealistic youth but not a distinguished student. His academic career took direction at the University of Berlin from 1835 to 1841, where he was attracted to the intellectual rebellion brewing in philosophy and completed a thesis on the subject of materialism in early Greek thought.

After the university, Marx hoped to obtain a position teaching philosophy, but the radical ideas of his Berlin training were already unpopular with school officials. So Marx began a career in journalism writing for the *Rhineland Times,* a newspaper financed by liberal industrialists of the region who encouraged his maturing criticism of the Prussian state. Marx proved a witty, relentlessly logical, and devastatingly critical writer. He contributed pieces on Prussian censorship, the oppression of peasants, orthodox religion, and the crisis in the wine industry. Although he was soon promoted to editor and circulation advanced briskly, the paper became embroiled in controversy with a conservative competitor, was accused of Communist leanings, and forcibly closed by the state in 1843. At the time, Marx knew nothing about communism but decided to read about it as he went into exile in Paris.

Paris was a magical world for the twenty-five-year-old journalist: a world teeming with political exiles from all of Europe, French revolutionaries, Saint-Simonians, and partisans of varied socialist sects, all writing and endeavoring to organize the working class. Marx began a new journal, contributed to others, and above all sought to square his philosophical education with the currents of French political thought and society. He knew little about economics and was introduced to its mysteries through an article by Engels on the British political economists. Immediately, Marx began corresponding with his fellow German who was already a published critic of the conditions of the English working class — an odd position because Engels was also the manager of his family's textile-manufacturing firm. They soon met in Paris and began a thoroughly mutual collaboration that outlived Marx by twelve years until Engels's death in 1895. Marx possessed a flashing intelligence and an iron determination to revolutionize social thought. But the partnership was greatly enhanced by Engels's splendid talents as a writer, observer, investigator, linguist, historian, genial human being, and practical businessman who knew industry at first hand and, not least, had an income that helped support the Marx family in frequent hard times.

These were the decisive years for Marx. In 1844 he drafted the *Economic and Philosophical Manuscripts* and, with Engels, completed *The German Ideology* in 1846 — although neither work was published for some time to come. In essence, these writings embodied the social theory that Marx and Engels would elaborate and change over the next fifty years. They claimed to turn on its head earlier philosophy that believed ideals governed history. On the contrary, they said, ideals and social relations arose from the material conditions of production. A new materialist conception of history was fashioned for an eventful future. In Paris Marx embarked on a life of socioeconomic study and political action that never wavered. He decided to change the world.

Marx and Engels built a bridge from the uncertain beginnings of sociology into the twentieth century. They studied

Saint-Simon and agreed that industry, or more properly capitalism, was the driving force of modern society. They agreed that a science of society was both possible and essential for getting to the roots of social ills — but a science of true socialism as distinct from the variety of utopian, ethical, and other naively reformist socialisms that were being proclaimed in Europe at the time. True socialism relied on a materialist understanding of historical development, on an appreciation that social relations were primarily shaped by the conditions of production.

Briefly stated, Marx and Engels believed that societies are governed by basic laws founded in the first instance on property relations. Under capitalism, historically varied forms of property are steadily reduced to just two dominant ones, capital (ownership of the means of production) and labor (ownership of one's own capacity for work). Social classes are defined by these locations in production and therefore tend toward the "two great camps" of capitalists (bourgeoisie) and workers (proletarians). Marx recognized that this was a model summarizing a historical trend and that in any particular society classes were more numerous (including peasants, landowners, small businesses), although moving in the direction of a two-class system. Marx's critics properly dispute this description, but they frequently miss the larger point in which Marx placed great emphasis — the process of changing social relations. As capitalism expanded, everything became a commodity; land, labor, reputation, respect, even love, became things to be bought and sold in the market for the purpose of material gain. Workers, who received much of his attention, no longer produced their own goods as independent artisans but sold their labor for wages to the capitalist. The worker thus was alienated from his or her own labor, which came under the control of the capitalist. As industrial work became more specialized, a laborer produced only a part of the final product — one component on an assembly line, for example — and became alienated now from fellow workers and finally from the self. As classes became more sharply defined, distinctive forms of class consciousness characterized opposing groups.

Consciousness was shaped by the conditions of work but also by opportunities that members of classes enjoyed to share their views with others and form alliances. Capitalists had the advantage here because, with support from government, they could combine while preventing by law the union of workers. The dominant ideologies in society thus reflected the consciousness of the ruling class of owners. The irony, however, was that the very means of capitalist exploitation brought workers closer together — in factories, for example — under harsh conditions leading to the full development of working-class consciousness. In summary, capitalist development is for Marx a process leading to class division at the level of society and alienation for the individual. The result, he predicted, was growing class antagonism, which one day would express itself in a revolution of the system.

The key for Marx and Engels, what distinguished them from the crowd of social reformers, was the premise that labor produced everything of value in society apart from nature's endowments and, under the conditions of capitalist production, labor was exploited — meaning that it received less than the value of what it created. The bulk of the value produced went to the capitalist for reinvestment or for the capitalist's own consumption and wealth. This implied two new and radical interpretations of social conditions: First, poverty, inequality, inhuman factory conditions, and hunger were direct consequences of competitive capitalism; and second, only revolutionary action by the working class would change matters.

In 1848 a series of short-lived revolutions spread across every major European country. Anticipating these events, Marx and Engels prepared a pamphlet for working-class audiences entitled *The Manifesto of the Communist Party* ([1848] 1959), which they hoped would influence the direction of events. No Communist party existed at the time; another purpose of the tract was to coalesce various socialist factions in a new party founded on the insights and proposed strategies of Marx and Engels. Its famous closing line implored "Workingmen of all countries, unite!" By this time, Marx had been expelled from Paris for his "subversive" writing, and he traveled often to London from his new residence in Brussels. Engels introduced him to working-class activists in England (many of them also German exiles) who made up the League of the Just. That organization, a small band by all accounts, changed its name to the Communist League and sponsored the *Manifesto* as its program. The document's publication attracted no attention, and Marx's followers had little influence in the broad socialist movement.

The results of 1848 were nil. Revolutionary governments in Europe were quickly replaced, and reform movements dealt a serious setback. Marx and Engels would not return to political activism until the more successful organization of the International Working Men's Association (the First International) in 1864. In the meantime, urged on by expulsion from both Paris and Brussels, Marx settled in London and began his encyclopedic studies in the British Museum that would result in his magnum opus *Capital* ([1867] 1977).

During Marx's lifetime, only the first of the three volumes of *Capital* appeared in print with his own editing. Engels pulled together the remaining volumes and some controversy surrounds the question of how well they reflect Marx's ever-changing ideas. Nevertheless, several things distinguish this later work. Philosophical concerns are transformed into a more technical economic vocabulary; alienation is replaced by the notion of "surplus value," or basically the mechanics of labor exploitation (see Chapter 2). Revolution and the transition to socialism receive little attention. Marx's single-minded purpose in this work is to reveal the "laws of motion" of the capitalist economic system. Although many of his assumptions are rejected today (especially the labor theory of value), Marx was impressively accurate in his predictions about the direction of capitalist development.

Above all, capitalism is a competitive system that eliminates independent and small producers in a trend toward the concentration and centralization of capital — today's giant corporations. Artisans and small proprietors become "proletarianized," or forced to sell their labor, in Marx's phrase — a trend confirmed today in the growing proportion of wage labor as opposed to independent professionals, shopkeepers, and craftspersons. Technology, Marx believed, would displace labor and create, as we now say, technological unemployment. Capitalism would expand beyond national boundaries to an international division of labor, a development we experience in the multinational corporation and Third World specialization in raw-material exports. And the whole system goes through periodic crises as the profit rate in one branch falls requiring a shift to another, much like today's debt crisis and 1987 stock market crash. To be sure, some of Marx's predictions went wrong, notably expectations about revolution in the most advanced countries. On balance, however, he was probably right more often than he was wrong, which is one test of a useful theory. Another test is the extent to which the theory stimulates new research, and here Marx's work is unrivaled. The reason for that, I believe, is less Marx's specific predictions than his emphasis on class action. He once said that people make their own history, although not under circumstances of their own choosing. Collectively, organized people acting politically can change their circumstances to the extent that they alter such fundamental features of social organization as the system of property ownership. More forcefully than any other classical thinker, Marx gave ordinary working people a decisive role in history.

Émile Durkheim. Sociology was slow in gaining recognition. During the first fifty years since the term's invention in the 1830s, it had actually fallen into disuse in France and was a mere novelty in England and the United States. The survival and growth of the discipline owed much to Émile Durkheim (1858–1917) who decided that the "barbarous neologism" was nevertheless useful for identifying a distinctive and sound method of social study. Yet Durkheim also understood that the new science lacked a persuasive conceptual or practical justification. The indifferent fortunes of sociology over the previous half-century, by the time Durkheim's first writings appeared, would not be reversed until the discipline distinguished itself from history and philosophy by proving that it had practical uses.

Durkheim came from the Lorraine region of eastern France. His birthplace in Epinal lay just 100 miles south of Trier where Marx was born forty years earlier. Durkheim came from a family of modest means and orthodox Judaism. At the age of eighteen, he went to Paris to study philosophy and impressed his teachers as serious to a fault. After completing his first course of study in 1882, Durkheim taught in provincial schools for five years while preparing a doctoral thesis on *The Division of Labor in Society.* Published in 1893 as a book that would become a classic, the thesis helped him land a faculty position at the University of Bordeaux in 1887. In French academic circles, Durkheim's views were controversial, which explains why he remained in the provincial university for fifteen years before the call came from the Sorbonne in Paris; even then, the post was on the faculty of education, and he waited a few more years for a chair in sociology.

In today's language, Durkheim was an academic. Politically, he favored a French Republic in preference to monarchy and a secular state in opposition to the Catholic Church control, and he espoused progressive social reforms. But he was aloof from practical politics, holding to his belief that reform must be based on the most thoughtful and detached analysis. This is far from saying that Durkheim was indifferent to practical action; he simply wanted to place it on a solid foundation. In the preface to *The Division of Labor* he wrote, "we would not judge our research to be worth one hour's trouble if it had only speculative interest. If we carefully separate theoretical problems from practical ones, it is not in order to neglect that latter, but, on the contrary, to become better able to solve them" ([1893] 1960, 33). Durkheim was not a socialist. He doubted the wisdom of its revolutionary wing, despairing of the bitter harvest of the French Revolution, and he had no taste for agitation. But he shared the aims of socialist reform and tried to promote those in his own way.

In Durkheim's lifetime, France suffered chronic crises that added urgency to the need for reform. His youth coincided with the national humiliation of the Franco-Prussian War (1870–1871) and the aborted rebellion of the Paris Commune (1871), which ended in the bloody repression of urban revolutionaries by other French forces. Reflecting on these events in his thesis a decade later, Durkheim concluded that Saint-Simon was right about the deterioration of French society. Sociology, however, had to prove itself with a practical solution for the crisis.

Saint-Simon correctly identified industrialism as the watershed of modern society, the source of social tensions arising from the gap between old customs and modern practices. New institutions that would promote the kind of social solidarity consistent with industrialism were needed. Durkheim believed that, in the absence of new moral rules, social anarchy would prevail in the form of class war and overproduction, much as

Marx had described. Practical reforms could produce the needed solidarity.

For example, class conflict and exploitation stemmed from accumulated privilege (wealth and power) passed along generationally through the institution of inheritance. This inequality invalidated truly free contracts, such as between workers and owners, because the respective parties were not free and independent. One had an enormous unearned advantage over the other, meaning that the free and contractual rules that were supposed to govern industrial society according to its own principles were contradicted in practice. The institution of inheritance was wholly incompatible with individual liberty and opportunity, moral principles appropriate for industrial society. Accordingly, the abolition of inheritance would eliminate a major impediment to social reorganization.

The problem was what to put in place of the exhausted morals and institutions. Durkheim's inspiration on that score is the centerpiece of all his work, running through the thesis to his posthumously published lectures on *Professional Ethics and Civic Morals* (1958). The idea is close to what is called economic democracy today. By the late nineteenth century, Durkheim concluded, the expansion of industry and government had weakened all those "intermediate powers" or "secondary groups" between the family and the state that integrate people into society (civic organizations and voluntary associations). The idea was not new to Durkheim and had been used by Alexis de Tocqueville in his classic *Democracy in America* ([1835] 1959). A decline in associational group life and regulative communities was the source of the social crisis. It fostered the moral confusion, which Durkheim called "anomie," and was related to various social ills including the suicide rate (see Chapter 2). A new basis for group life was required that would capture and express the potential solidarity of an interdependent industrial society.

In *The Division of Labor*, the same problem appeared as a gap between the old, preindustrial ways in which society was organized (which he called "mechanical solidarity") and the new forms that would be compatible with industrial society ("organic solidarity"). Durkheim took the position directly opposed to Marx that modern industry and its cooperative tasks were capable of promoting a closer integration of people around common values. But again that required new forms, ones that reflected people's most basic concepts, of social organization or what Durkheim called the "collective conscience."

The concrete answer that Durkheim proposed was the professional or occupational association as a new focus of group life and as a substitute basis for political representation, superseding the geographical districts that now lacked a common interest or moral integration of citizens. Noteworthy in Durkheim's plan, beyond its continuing appeal to reformers, is a close affinity to Saint-Simon's proposal for a new form of government consistent with the "positive industrial order." Moreover, by returning to the founder, Durkheim developed a synthesis of Comte and Marx. He accomplished this by keeping his promise of a practical sociology because a revitalized sociology was uniquely able to investigate the causes of anomie and evaluate alternative cures. Durkheim was fond of the medical analogy in which sociology diagnosed social ills and prescribed more healthy measures. At bottom it was elegantly simple: Egoism was the bane of modern society, and its amelioration depended on new corporate groups promoting social organization by virtue of their organic connection to the real conditions of contemporary life.

Max Weber. While Durkheim pioneered a sociological renaissance in France, the intellectual world in Germany discouraged such innovation. Universities in the newly united country shared with the state a contentious and conservative bearing, as well as a phobia for socialism and anything associated with their own expatriate Marx. Yet this milieu produced Max Weber (1864–1920), a man whose stamp on the social sciences is still unequalled.

Weber, raised in Berlin, was the son of a prosperous lawyer and National–Liberal Party member of Parliament. He began university at Heidelberg in 1882 as a law student but approached that field, as he would everything in life, from a comprehensive viewpoint that embraced history, economics, philosophy, and a new sociology of his own creation. His dissertation on the history of medieval business organization (1889) and a second work on Roman agrarian history led him to the

University of Berlin as an instructor in law. From 1893 to 1897, he held his first and last regular academic position as professor of economics at the University of Freiberg.

Weber's early years were crowded with professional and civic activity. He held lectures and seminars for nineteen hours a week, participated in state examinations of lawyers, consulted with government agencies, wrote prodigiously, and conducted research whose results he tried to impress on policymakers. Given his own brooding temperament and poor medical history, the load proved crushing. In the fall of 1897, he suffered a breakdown that prevented work for the next four years. For the rest of his life, he held only part-time and visiting, if celebrated, posts, although an inheritance (one Durkheim might have permitted) allowed him to live as a private scholar and political actor. By this time, Weber had achieved some fame and never wanted for offers as a lecturer or public servant. Despite the complexity of his lectures, which are still studied with profit, Weber was a platform attraction, packing halls with students and professors, civil servants, businesspeople, and politicians.

Following his early studies on legal history, Weber accepted the request of a scholarly reform group to analyze parts of a large social survey dealing with German agrarian society. He relished the assignment, and the manner in which he worked his way from the gritty questionnaires to a critical examination of social classes and the state exhibited his energy, values, and extraordinary insight.

Weber's specific task was to analyze data on agricultural production in eastern Germany, the social base of the Prussian state and province of the large landowners, or Junkers. The immediate problem that concerned his reformist sponsors was the migration of German farm workers to western regions of the country and their replacement by Polish and Russian day laborers of a decidedly lower cultural level. From the standpoint of liberal reformers, the social problem was defined by several aspects: Conditions of labor were declining in the East because immigrant workers were paid less and lost an attachment to the land, productivity was low, and German society was being degraded by the infusion of foreign workers and the displacement of a stalwart peasantry from the East to the industrial regions of the West. In the process of exploiting the new labor force, Junker estates were neglecting the agricultural modernization necessary for Germany to maintain its position in the world market of agricultural commodities.

Most of these facts were familiar to public officials and the sponsors of the research, but Weber surprised everyone in his summary of a 900-page report, which he offered in a speech to the reform society's convention in 1893. His analysis laid bare the agrarian problem, explored its roots in the transformation of the state and economy, and ended with some unsettling political implications. Every political ox was gored, and no simple remedies were baited for the reformers. Weber blamed neither the migrating German farm workers, who sought a life free of the drudgery and paternalism of the old estates, nor the large landowners, whose aristocratic society and economic base had enabled political unification of the country. True, Junkers were now becoming an outmoded, even parasitic, class, but in the face of the capitalist transformation of the economy that had overtaken them, they were obeying a certain perverse rationality. As a nationalist, Weber despaired of both developments, although he saw more virtue in the motives of the migrating workers.

The real problem lay in a new alignment of social classes and the state. German unification had benefited the urban and industrial classes of the West to the disadvantage of Junkers who were forced to buy nationally manufactured goods at high prices without compensating stimulants to productive agriculture. Meanwhile, the state was building a new coalition that combined large landowners and the rising industrial capitalists. One of the ways this was done involved granting land trusts that allowed acreage to be taken out of production and rented to reactionary Junkers and status-climbing industrialists who mimicked the style of the old aristocracy. The state assisted in the formation of a group of overindulged capitalists for reasons of its own political support, thereby defying agricultural modernization and corrupting the civil service.

In this work that did so much to establish Weber's reputation as a researcher and political critic, the style is very close to Marx's historical writing. That was no accident. Throughout his life, Weber pursued a spirited dialog with Marxist thought. In the 1890s, his emphasis was on the incompleteness of one-sided materialistic explanations. Weber agreed that economic considerations were potent forces for social change, always present, but seldom the singular causes of social behavior. One must also appreciate the role of ideas that sometimes operate in conjunction with material forces and at other times with a power of their own. The blend and sequence of these complementary influences were more subtle than Marxists allowed. For example, German farm workers who migrated to industry certainly wanted to improve their economic condition, but they were also emboldened by the prospect of greater individual freedom. Weber's insight was the importance of the values that individuals acquired in group settings and how they could be fitted into an analysis that also counted the role of social structure without trivializing either or reducing one to the other.

Ten years later, in his best known book *The Protestant Ethic and the Spirit of Capitalism* (first published in Germany in 1904), Weber ([1946] 1958) elaborated another instance of the point. Why, he asked, did capitalism flourish in western Europe rather than in the Far East, which had comparable material and technological endowments? His answer was that religion, especially Calvinism, inspired an ethic that motivated the aggressive pursuit of gain, not for its own sake but for what it proved in this life about one's place in the next. The rich and pious of this world were God's elect, according to the ethic of the early capitalist entrepreneurs. Ideas not only coincided with the material conditions of capitalism (the resources, technology, and communications) but also had a prior and an autonomous causal effect.

Weber's work lacks the unity of a Marx or Durkheim, not because he lacked a master theme like his predecessors but because he resisted any general theory of history. Weber's broadest position on social change compared the forms of domination that govern societies over time, the bases that support the authority to command. He identified three types: traditional authority based on custom and inherited status; charismatic personal authority based on the heroism or extraordinary, even magical, gifts of a leader; and rational–legal authority founded on a set of rules administered by an organization or bureaucracy. These were pure (or "ideal") types used for analysis, not historical or evolving forms. Instead, Weber analyzed how these types of domination interact and change; how, for example, in their own defense, medieval kingdoms developed courts and princes (royal bureaucracies) that centralized authority and laid the foundations for the modern state. Weber seldom sacrificed the paradoxes of historical change for the sake of some supposed general law.

Yet at other times, particularly in his later years, Weber's interpretation of history emphasized evolving institutional structures with scant reference to subjective meaning or paradox. His analysis of political change, for example, was developed in acknowledged harmony with Marx, and it seems that Weber's apparent aim was to generalize historical materialism from economic to political change.

Two years before his death, in a lecture on "Politics as Vocation," Weber stated that the modern state grows through the expropriation of all private holders of power: "The whole process is a complete parallel to the development of the capitalist enterprise through gradual expropriation of the independent producers" ([1946] 1958, 82). In politics the pervasive consequence that overshadows class struggle is the growth of rational bureaucracy — a society reorganized according to mechanical, depersonalized, and oppressive routine. Although, with characteristic irony, Weber also noted the efficiency and democratizing consequences of bureaucracy, he feared the eventual merger of the state and economy in bureaucratic capitalism. In this he not only extended Marx but showed his own gift for prophecy.

It would be wrong, however, to fix Weber's methodological and political beliefs in any general or timeless mold. He had less interest in and many more reservations about creating a theoretical system than did Marx or Durkheim. His political positions changed, appropriately, with the circumstances that occasioned them and the channels through which they might have some effect. During World War I when he served as an administrator of hospitals, Weber became more embittered and critical of the German state. Yet, as a consultant to the Armistice Commission at Versailles, where the treaty ending the war was drafted, he foresaw the dangers of a punitive peace — the harsh terms of German surrender that subsequent historians would see as the first cause of World War II. Above all, Weber's gift was for intricate and deep inquiry, realism tinged with irony, keen insight held to strict confines, and originality.

The Legacy

The **classical tradition** is a way of thinking and a responsibility for acting. It was built through the efforts of a great many men and women (including among the latter, Harriet Martineau, Beatrice Potter Webb, and Rosa Luxemburg). Marx, Durkheim, and Weber describe its outlines but scarcely exhaust its contents. The classical tradition means engagement with the pain and confusion, as much as with the achievements of the spirit, that exist in society. The classical tradition assesses the character of a social order and asks how it originated, where it is going, and where it could go with the contribution of a clear and humane vision. The classical tradition participates by connecting social issues and individual lives with sociological imagination.

The exemplary classical sociologists differed in their politics and participatory styles: from Marx the intellectual worker and political activist to Durkheim the academic and social planner; from Weber the nationalist and respected critic to Marx the exile and revolutionary; from Durkheim's faith in an organic corporate society to Weber's premonition of bureaucratic state domination. Underlying these differences, however, was a common purpose to beneficially reorganize the prevailing order and participate in its reform.

Classical sociologists spoke to the people. Each would have had something to say to Steinbeck's tenant farmer or today's autoworker: Marx about exploitation by a ruling class of industrial capitalists (the ones to overthrow, if not shoot) and the power of organized workers; Weber about the participation of the state in such injustices and their corrosive effect on legitimate authority; Durkheim about a new social solidarity and nonrevolutionary politics of occupational groups close to the people's interest. The compassionate alternatives are revolution, political reform, and economic democracy — a list that pretty well encompasses our repertoire 100 years later.

THE DEVELOPMENT OF SOCIOLOGY IN NORTH AMERICA

The beginnings of sociology in the United States were tentative and deferential. There was no mad count to propose a reorganization of the industrial system. Indeed, there were no counts, and industry was not yet a system in the 1870s. More important, there was no native tradition of social philosophy on which to draw and only a selective understanding or tolerance for European thought. The first North American sociologists were impatient with the speculative fancies of European intellectuals, some of whom appeared decadent or politically unsavory.

In the early years, the main foreign writers to have a measurable impact on the North Americans were Comte (now seen as the father) and the Englishman Herbert Spencer (1820–1903), a self-educated engineer, self-styled moral philosopher, and enthusiast of Charles Darwin's theory of biological evolution. Spencer developed an elaborate, and certainly speculative, scheme for understanding society much as a biological organism: a creature that evolved, differentiated over time, adapted, and generally edged up the scale of social evolution from simple to modern society. He owed a great deal to Comte and Darwin; although in the case of the latter, Spencer claimed that the influence was mutual — reminding everyone that he coined the phrase "survival of the fittest."

In London's Highgate Cemetery, the modest headstone of Spencer stands opposite Marx's tomb and gigantic bust. These two are closer in rest than ever in life. Spencer had no use for what he saw as unscientific polemics, and he passed that judgment on to his North American readers. The value of Spencer's work is still argued (Mills saw merit in it), but at the turn of the century his impact in the United States was forceful and ironic — Spencer's was the theory the early social scientists read and disagreed with the most.

The Pioneers

Sociology in the United States was begun by two men, contemporaries vigorously opposed on the implications they drew from Spencer: Lester Frank Ward (1841–1913) and William Graham Sumner (1840–1910). Both agreed that a proper sociology would have to be scientific, by which they meant not a historical science in the European fashion but one that was naturalistic and evolutionary. Great appeal lay in the idea that society evolved along the lines of the biological organism and therefore invited the same sort of scientific analysis.

William Sumner was a professor at Yale University and offered the first social science courses in the United States, beginning in the 1870s. Although he differed with Spencer on many points concerning the determinacy of social evolution, Sumner was attracted by the conservative political implications of evolutionary theory. He felt that society evolved progressively, although not mechanically, from adaptive folkways to higher mores in a fashion that was best left alone. Certainly, governments and reformers should not dabble in such verities. Sumner was considered stuffy by his more liberal colleagues.

North American sociology, like populism, was more at home in the Midwest, and Lester Ward was closer to the mood of the new discipline. Born in Joliet, Illinois, his youth was spent on the prairies of Iowa where he acquired a love of nature. After service in the Civil War, Ward settled in Washington, D.C., and accepted a government job as the fruit of his war effort. He obtained a post in the Treasury Department and slowly, with the help of a college degree from night school, moved up the bureaucratic ladder, eventually finding a home in the Geological Survey. Ward mingled in Washington's civic circles dedicated to temperance, women's suffrage, and abolition of capital punishment. He helped establish the National Liberal Reform League in 1869 and edited its paper *The Iconoclast,* contributing articles that praised Lincoln and attacked religious hypocrisy. An unpublished essay on "Washington City" smacked of Mark Twain's criticism of official corruption in *The Gilded Age.*

Critical as he was on some subjects, Ward remained an optimist, reformer, and ambitious public servant. He completed a law degree, also in night school, but became so absorbed in his work that neither the practice of law nor his reform activities could compete with his growing interest in biology and paleobotany. Ward became an authority in the latter field, a scientist of international repute.

Ward's sociological and reform interests, which had been shelved while he pursued his science, revived in later life. He read and enjoyed Comte's *Positive Philosophy* (1853), where he came across the term *sociology*, reporting later, "I fell in love with it." Indeed, Ward decided to retitle one of his manuscripts on social evolution *Dynamic Sociology* (1883), which was published as a book at his own expense. In detached and technical terms, Ward assailed Spencer's (and, later, Sumner's) "gospel of inaction" and mechanistic views of social evolution that ignored the self-conscious and purposeful capacity of societies to improve themselves through law and education. In practical terms, he found their science amateurish and their politics fatalistic.

After a slow start, *Dynamic Sociology* became a standard college text and the first exposure to sociology for the next generation. In the years that followed, Ward wrote more sociology and enjoyed a growing reputation as a bona fide scientist of a new sort. He joined the American Economic Association, helped some of its breakaway members found the American Sociological Society in 1905, and was chosen its first president, much to his own surprise. In the same year, at the age of sixty-five, he retired from the Geological Survey and accepted his first academic position at Brown University. But his liberal views, unorthodox course on "A Survey of All Knowledge," lady admirers (attracted by his advocacy of women's rights), and his geological hikes in the Rhode Island countryside convinced his colleagues to respectfully conclude that "Ward was not of the academic type."

It is odd that Ward, who had no training in sociology or philosophy and only stumbled across a few of the European masters midway through his most influential book, nevertheless became the magnet that drew the next generation of career sociologists to the field: people like Franklin Giddings (1855–1931), Albion W. Small (1854–1926), Charles Horton Cooley (1864–1926), and Edward A. Ross (1866–1951). This second generation of U.S. sociologists was drawn to the idea of a true science of society and led the movement for its professionalization and institutionalization as an academic discipline in U.S. universities.

They had much in common. Like Ward they were aspiring professionals from middle-class and mostly rural backgrounds who wanted to build a respectable science under themselves. In the United States, this meant borrowing the naturalistic premises of the physical sciences, yet mixing those with a soft interpretation of evolution that allowed for human intervention in the improvement of the species. As Small said, Ward saved them all from Spencer's "misconstrued evolutionism." But they also considered historical science a contradiction in terms. Social science had to be built up from basic biological and psychological processes. They were forthright social psychologists, as Ross exemplified when he insisted that sociology and history had to be separated to reduce the unit of analysis and thereby avoid the sins of the speculative Europeans.

They were also progressives in U.S. politics, a happy combination because the root of progressivism was in the idea that society had to be reorganized and managed along scientific lines. They believed, with Ward and against Sumner and English liberalism, that government needed to play an active regulatory role in society to prevent the abuses of unchecked competition. Scientific direction of society was the highest achievement of a purposive evolutionary process. Teddy Roosevelt, symbol of progressivism, wrote a preface to Ross's popular book *Sin and Society* (1907) and Franklin Giddings at Columbia University was a close advisor to the president.

The emphasis on professionalization is reflected in efforts to create an organization that would promote and monopolize the new science. At the American Sociological Society's meeting in 1909, members decided that there should be a standard sociology curriculum developed for the universities:

> In taking rank as a science and in attaining to that dignity and respect which the importance of the subject and the wide interest in it demand, it seems . . . desirable that sociology should standardize its fundamental courses in the same way that the fundamental courses of other sciences are standardized. For illustration, when a student takes Chemistry 1, Physics 1, Biology 1, Economics 1, or Law 1, such course stands for a definite subject matter. . . . (*Footnotes,* American Sociological Association March 1980, 1 & 6)

In 1920 the need for research money was addressed:

> We must also interest rich men in providing money for the prosecution of research until we have shown niggardly boards and legislatures the importance of finding out the facts bearing on social theory and social policy. (*Footnotes*, American Sociological Association March 1980, 6)

Having laid a plan to monopolize control of a standard course of instruction and to find patrons, the last step in professionalization, which dawned on the (renamed) American Sociological Association (ASA) in 1936, was to make itself necessary:

> To get sociological training and field experience recognized as a qualification or substitute qualification for certain Federal and state civil service positions . . . to study ways of getting the graduate training program in sociology to meet the need for equipping students for technical positions in Federal bureaus and administrations in state and local agencies . . . to move wisely and expediently against the practice of hiring persons without any graduate training as teachers of sociology in American universities and colleges. (*Footnotes*, American Sociological Association April 1980, 4)

A new organizational society was emerging from progressivism, and, rather than analyzing or fearing the bureaucratic state as Weber had done, many North American sociologists wanted in on it.

The Reformers

Many others held a wholly different view on the special U.S. problems of urbanization, race, and industrial concentration, topics for which the European tradition was not always a reliable guide. W. I. Thomas (1863–1947) taught at the University of Chicago and (with Florian Znaniecki) studied the city's Polish immigrant population in the early 1900s — their problems of adjustment, occupational lives, patterns of mutual assistance, and neighborhood associations. These were the same workers, streets, and factories that Upton Sinclair illuminated for a shocked and fascinated public in his novel *The Jungle.* The five-volume study of Thomas and Znaniecki, *The Polish Peasant in Europe and America* (1927), is more accurate, equally good reading, and a sociological landmark that rivaled the new social realism in literature.

Thomas was born in rural Virginia and grew up in Knoxville, Tennessee. He had been a professor of English at Oberlin College before his interest in comparative literature led him to study the social origins of various cultural groups, including Eastern European peasants. Reflecting on this period, Thomas said, "My interests . . . were in marginal fields and not sociology as it was taught at the time" (1973, 249). As a result of his writing on "folk psychology," Thomas received a letter from Booker T. Washington in 1910 inviting him to a conference on the American Negro and providing an appreciative criticism of his work. When Thomas arrived at the Tuskegee Institute for the conference, he discovered that his fan and correspondent was not Washington at all but Robert Park. Their meeting began a long friendship and collaboration that developed into a new "Chicago School" of social research.

Robert Park (1864–1944) was from Red Wing, Minnesota, a newspaper reported by trade and a self-described "intellectual vagabond." He attended the University of Michigan and earned a Ph.D. in Germany. Between those engagements, he worked for the *Minneapolis Journal*:

> The yellow journals went in for reform, and I became a reformer. . . . The city editor . . . discovered that I would stay on a story longer than anyone else, so he set me to work hunting down gambling houses and opium dens. This was the beginning of my interest in sociology, although at the time I did not know the word." (1973, 254).

Park practiced what he called scientific reporting; when he learned that it was close to the new methods of inquiry being developed in the social sciences, he returned to graduate school to study the sociology of journalism. This was the way he got into sociology. In 1906 Park worked for the Congo Reform Association and wrote an important magazine exposé on Belgium's exploitation of the African colony. Booker T. Washington then hired him as a public relations officer for the Tuskegee Institute; through this association, Park acquired an intimate knowledge of the condition of southern blacks. He was the only prominent white sociologist in those years to take an interest in the U.S. race problem. Park's black friend W. E. B. Du Bois (1868–1963) held a Ph.D. from Harvard University and, for a time, was a professor of sociology at Atlanta University. Du Bois was ahead of his time; he wrote books on black history and culture such as *John Brown* and *The Souls of Black Folk*. He was forced out of the Atlanta post for criticizing the moderate racial views of Washington, but he went on to head the first militant black political ("Niagara") movement and the NAACP, which he helped establish in 1910. Thomas, Park, and Du Bois began a new sociology focused on urban and racial problems that contrasted sharply with the orientation of many among their academic colleagues — one stressing scientific reform and the other social science professionalism.

Thorstein Veblen and the Classical Tradition in America

There was one bright example of the classical tradition among U.S. social scientists. Thorstein Veblen (1857–1929) was a trained philosopher, self-educated economist, and self-made sociologist. Veblen was the leading theoretician of industrial society, social critic, and academic celebrity of his day. He was a reclusive and, by the standards of polite society, a strange man.

Veblen came from a Norwegian immigrant family that settled in Wisconsin. In 1884 he completed a Ph.D. at Yale University, where Sumner's evolutionary sociology attracted him despite their opposing political views. Veblen found no jobs open to an offbeat philosopher and returned to the family farm to read economics for the next seven years. Later, he found and lost university positions at Cornell, Chicago, Stanford, Missouri, and the New School for Social Research. After the example of Ross, Veblen was fired from Stanford, less for radical political views, which he had, than for living like a bohemian in a rustic cabin with a woman student. Asked whether his dismissal was political, Veblen replied, "The president does not approve of my domestic arrangements, and for that matter neither do I."

Veblen's books attracted a vast academic and public audience from the beginning with *The Theory of the Leisure Class* (1899). Deadly serious and tediously written, his work nevertheless caught the imagination with its verbose criticism of conspicuous consumption and the force and fraud of corporate monopolies built on intangible wealth. Veblen resembled his sociological contemporaries in the strong evolutionary and social psychological aspect to his work, but these were anthropologically and historically anchored. He believed that humans possessed an instinct of workmanship that expressed itself in useful labor, material progress, and self-realization. Social evolution, however, leads to a struggle between instincts and their contamination through conquest, ownership, and the imposition of leisure classes living off productive artisans. Corruption follows once social honor is attached to possessions rather than useful labor. In modern society, business is honored instead of engineering, and consumption instead of production.

These ideas, fully developed in *The Instinct of Workmanship and the State of the Industrial Arts,* which Veblen considered his best book, recall Marx. Veblen knew Marx in depth and wrote critical articles on technical features of the theory. He was the only U.S. scholar intellectually prepared to challenge Marxist theory on its own ground. He agreed that productive labor was the source of value but considered Marx's theory prescientific for its failure to follow evolutionary principles. Moreover, he had little faith in social reformers. Effective social change would come from productive people, engineers and workers. His criticism of Marx and social reform, however, paled in contrast to his views on the academic establishment. In his critical book on U.S. education, *The Higher Learning in America: A Memorandum on the Conduct of Universities by Business Men* (1918), Veblen chided the professors as "captains of erudition" who liked attending "gatherings of the well-to-do for convivial deliberation on the state of mankind at large."

The writer John Dos Passos portrays some of the major figures of the early twentieth century in *The Big Money,* the third novel of his remarkable trilogy *U.S.A.* Veblen is included because

> he established a new diagram for a society dominated by monopoly capital,
> etched in irony
> the sabotage of production by business,
> the sabotage of life by the blind need for money profits,
> pointed out the alternatives: a warlike society strangled by the bureaucracies of the monopolies forced by the law of diminishing
> returns to grind down more and more the common man for profits,
> or a new matter-of-fact commonsense society dominated by the needs of the men and women who did the work and the

> incredibly
> vast possibilities for peace and plenty offered by the progress of technology. ([1933] 1969, 119)

Pathways to the Present

The formative period of North American sociology from 1880 to 1930 was defined by an uneasy alliance between academics who stressed scientific professionalism and progressives who advocated applied science in the service of reform. Sometimes these emphases came together as in the Chicago School of urban sociology. The Chicago sociologists (including Thomas, Park, Small, and, later, Louis Wirth) assisted Jane Addams's settlement work and Hull House devoted to rescuing homeless young people and poor families. The influential philosopher John Dewey was a trustee of Hull House, a natural place for him to practice his pragmatism that so impressed sociology (in part through another Chicago philosopher, George Herbert Mead). While Dewey and the sociologists saw Hull House as a vehicle for reform and for their own research and education, Veblen, who was at Chicago around the turn of the century, considered "the 'solicitude' of settlements . . . in part directed to enhance the industrial efficiency of the poor."

At other times, the two tendencies diverged widely, particularly as academic sociology devoted most of its effort to laying the scientific foundations of the discipline, rather than researching social problems. For every fresh study like *The Polish Peasant in Europe and America,* there were a dozen treatises on the "principles of" (Giddings), "foundations of" (Ross), "general" (Small), and "pure" (Ward) sociology. Doubtless, any new discipline needs to define its philosophical and methodological principles. The distinctive thing about the new North American sociology, however, was that rather than taking these principles from the classical tradition, they were in most cases reproduced in the distinctive U.S. intellectual climate and reinvented at a more abstract level.

The ambiguous result of this history was that academic sociology gained respectability and efficiency. It took its place in the university curriculum and produced record numbers of publications and practitioners, outstripping its European counterpart which was many more years gaining general recognition. At the same time, a schism developed with growing tension between the associated polarities of theory and research, science and reform, or expertise and engagement — a tension that provoked healthy debate but also compromised a clear sense of the sociological purpose. That tension has yet to be resolved.

Illustratively, from about 1930 until 1960, sociology followed two sometimes intersecting paths. Critical research continued. The Swedish economist Gunnar Myrdal was brought to the United States to examine its racial problem; Myrdal's qualification was that he was an objective outsider who explicitly rejected the do-nothing evolutionary theory of Sumner. Yet Myrdal's 1944 classic *An American Dilemma,* which received considerable public attention, did not jar the prevailing social theories with its claim that racial conflict was built into the social order. On the contrary, social theory became more grandly abstract, especially in the much cited and little used 1951 work *The Social System* by Talcott Parsons, which maintained in direct opposition to Myrdal that societies were based on integrated value systems.

This, then, was the setting for the critical and spirited observations of Mills and others — the challenge that sociology somehow bring together its general interpretive ideas and its research. Mills thought the grand theory (of Parsons), as well as mindless fact gathering or abstract empiricism, should be superseded by a publicly relevant sociological imagination. The 1960s began a new era for sociology that continues today, not only in North America but also worldwide. Modern sociology, in the optimistic words of Randall Collins (1987), has come out of the doldrums with renewed interest in historical change, social structure and the individual (or now the "micro–macro connection"), markets, gender, and, above all, the possibility of linking specialties. In the long view, we are on the threshold of a new era, one that moves beyond both the classical and scientific visions described previously. Immanuel Wallerstein persuasively argues that "we are in a third era of social science. It has no obvious name. After philosophical social science and scientific social science we are in the era, or coming into the era, of what might be thought of as social science as interpretation of process" (1987, 1306). Science these days is understood not simply as a formal laboratory method but as an evaluative, systematic, and creative way of reaching controlled interpretations. The new sociology, which is the subject of this book, embraces the responsibility of both objective method and evaluative choice.

SCIENCE, SOCIOLOGY, AND CIVILITY

The paradoxical reputation of sociology as both threatening and trivial, which introduced this chapter, now makes sense. To the extent that academic sociology necessarily spelled out the foundations of the discipline, in part by formalizing commonplace observations, it bored the reader looking for new explanations of social problems. Similarly, to the extent that critical sociology at the hands of a Marx or Myrdal explained poverty or racism as a consequence of the normal operating principles of economy and society, it threatened those with particular stakes in the system. What many of sociology's critics and practitioners fail to see is that these features of the discipline are unavoidable. On one hand, sociology requires its own conceptual language. On the other, if it does the job of analyzing social problems, sociology is bound to antagonize someone. Sociology, after all, is a social activity.

The inevitable question is then, How can something be a social activity and an objective science at the same time? The answer: All sciences share this quality, and they manage, somehow, by producing humanly fallible results that, nevertheless, can be checked by others. In the 1920s, physicists such as Werner Heisenberg had already learned that "science no longer is in the position of observer of nature, but rather recognizes itself as part of the interplay between man and nature" (quoted in Gould, 1981, 21). Later, the historian Thomas Kuhn (1962) would show how scientific communities behave like other societies, changing their theories (or paradigms) in revolutionary jolts rather than by a steady accumulation of fact. The paleontologist and writer Stephen J. Gould (1981) argues that science does not presume to master the enormous variability in nature or to predict in any specific way, as opposed to understanding the general movement of structures. The purported differences between the physical and social sciences, in which the latter are inferior but can emulate the former with more exacting conceptions of causation and powers of prediction, are not such differences after all:

> Science, since people must do it, is a socially embedded activity. It progresses by hunch, vision, and intuition. Much of its change through time does not record a closer approach to absolute truth, but the alteration of cultural contexts that influence it so strongly. Facts are not pure and unsullied bits of information; culture also influences what we see and how we see it. Theories, moreover, are not inexorable inductions from facts. The most creative theories are often imaginative visions imposed upon facts; the source of imagination is also strongly cultural. (Gould 1981, 21–22)

The early North American sociologists, as well as those who doubted the possibility of a social science, were therefore misled by an unrealistic model of what science is. Sociology is a science in the unprepossessing sense that it seeks to discover and codify empirical fact, to organize facts in logically coherent structures (theories) that allow interpretations (explanations) to be drawn that can be verified. Individual sociologists may devote most of their time to one or another of these steps and to the vast technical puzzles each presents. Individuals have their favorite puzzles and regularly plead that either sympathetic discovery or hard-headed verification deserves top priority. But sociology is more than this, as are all the other sciences in that they involve humanly developed standards of what is true (however well-documented technically), what is worth study, and what may be usefully done with the results.

Sociology is a distinctive discipline and social activity that nevertheless struggles with many of the same problems pervading the rest of society. It asks, for example, how to stem the alienation of an increasingly complex society that infuses even its own ranks. Sociologists understand themselves and others as caught somewhere in the rush of uncomprehended changes similar to those faced by Steinbeck's characters: What is this new situation? Where did it come from? Who or what is responsible? How do we change it? Sociology's promise is a means of transcending those circumstances, of providing, in Mills's words, "lucid summations of what is going on in the world" and how it affects each of us — that is the perspective, the unifying element of the tradition, and the civilizing opportunity.

Sociology's tradition is something to envy — nearly 200 years of investigation, reflection, and action coeval with the great transformations of industrialization, urbanization, bureaucratization, and irrepressible human effort to make modern society liveable. The aim of sociology, whatever its divisions, is to foster civility. Civility involves the relations among citizens, one with another and with the body politic, deference or allegiance to the social order befitting a citizen, and solidarity of civil rights and obligations. A social order befitting the citizen is our subject. Sociology is thus a moral matter.

SELECTED BIBLIOGRAPHY

AMERICAN SOCIOLOGICAL ASSOCIATION. 1980. *Footnotes.* March and April: 1, 4, 6.

BENDIX, RINEHARD. 1962. *Max Weber: An Intellectual Portrait.* New York: Anchor Books.

COLLINS, RANDALL. 1987. "Is 1980s Sociology in the Doldrums?" *American Journal of Sociology* 91:1336–55.

DIGGINS, JOHN P. 1978. *The Bard of Savagery: Thorstein Veblen and Modern Social Theory.* New York: Seabury Press.

DOS PASSOS, JOHN. 1933. *The Big Money.* Reprint. New York: New American Library, Signet Classic Edition, 1969.

DURKHEIM, ÉMILE. 1893. *The Division of Labor in Society.* Reprint, translated by George Simpson. Glencoe, IL: Free Press, 1960.

———. 1958. *Professional Ethics and Civil Morals.* Glencoe, IL: Free Press.

GERTH, HANS, and C. WRIGHT MILLS. 1958. *From Max Weber: Essays in Sociology.* New York: Oxford University Press.

GIDDENS, ANTHONY. 1978. *Émile Durkheim.* New York: Viking Press.

GOULD, STEPHEN J. 1981. *The Mismeasurement of Man.* New York: Norton.

GOULDNER, ALVIN. 1962. *Introduction to* Socialism *by Émile Durkheim.* New York: Collier Books.

———. 1970. *The Coming Crisis of Western Sociology.* New York: Basic Books.

HASKELL, THOMAS L. 1977. *The Emergence of Professional Social Science: The American Social Science Association and the Nineteenth-Century Crisis of Authority.* Champaign: University of Illinois Press.

HINKLE, ROSCOE C. 1980. *Founding Theory of American Sociology: 1881–1915.* London: Routledge and Kegan Paul.

KUHN, THOMAS P. 1962. *The Structure of Scientific Revolutions.* Chicago: University of Chicago Press.

MARX, KARL. 1867. *Capital,* Vol. 1. Reprint, translated by Ben Fowkes. New York: Vintage Books, 1977.

MARX, KARL, and FRIEDRICH ENGELS. 1848. *The Manifesto of the Communist Party.* Reprint. *Marx and Engels: Basic Writings on Politics and Philosophy,* edited by Lewis S. Feuer. New York: Anchor Books, 1959.

———. 1932. *The German Ideology.* Reprint. New York: International Publishers, 1947.

MCLELLAN, DAVID. 1975. *Karl Marx.* London: Penguin Books.

MILLS, C. WRIGHT. 1959. *The Sociological Imagination.* New York: Oxford University Press.

NEWMAN, EDWIN. 1974. *Strictly Speaking: Will America Be the Death of English?* New York: Warner Books.

PARK, ROBERT E. 1973. "Life History." *American Journal of Sociology* 79:251–60.

SCOTT, CLIFFORD H. 1976. *Lester Frank Ward.* Boston: Twayne.

STEINBECK, JOHN. 1939. *The Grapes of Wrath.* New York: Viking Press.

THOMAS, W. I. 1973. "Life History." *American Journal of Sociology* 79:246–50.

THOMAS, W. I., and FLORIAN ZNANIECKI. 1927. *The Polish Peasant in Europe and America.* New York: Alfred A. Knopf. Originally published by Badger, Inc. 1918–1919.

TOCQUEVILLE, ALEXIS DE. 1835. *Democracy in America.* Reprint. New York: Vintage Books, 1959.

WALLERSTEIN, IMMANUEL. 1987. "Marxisms as Utopias: Evolving Ideologies." *American Journal of Sociology* 91:1295–1308.

WEBER, MAX. 1946. "Politics as Vocation." Reprint. Pp. 77–128 in *From Max Weber: Essays in Sociology,* translated and edited by Hans Gerth and C. Wright Mills. New York: Oxford University Press, 1958.

WIEBE, ROBERT H. 1967. *The Search for Order: 1877–1920.* New York: Hill and Wang.

WILSON, EDMUND. 1940. *To the Finland Station: A Study in the Writing and Acting of History.* Garden City, NY: Doubleday.

2

Sociological Work

Child labor was common in the early industrialization of textiles. The low wages factories paid women and children offset the costs of expensive new machines and kept profits high. (Culver Pictures, Inc.)

COMMON SENSE

Sociology is many things: a body of knowledge about society to those who study it, a discipline for those who follow it, a practice for those who apply it, a plan for social reorganization for those who revere it, and pretentious restatement of common sense to those who criticize it. For those who do it, sociology is a type of work, a craft like many others. The best introduction to what sociology is, and how each of these descriptions fits into a general appreciation, begins with a look at how sociologists go about their work — the methods and products of their craft.

A common criticism of sociology, one that reveals much about public perceptions and professional aims, is that sociology is "just common sense" — conventional wisdom in pedantic guise. Long ago an ingenious rebuttal was provided by Paul Lazarsfeld, late of Columbia University and a pioneer in the work of its Bureau of Applied Social Research. The Bureau was the first important university social research center in the United States, and it carried out a number of pathbreaking studies of voting, mass communication, and opinion formation, beginning in the 1940s. Under the direction of Samuel Stouffer and Lazarsfeld, this research was largely responsible for the development of public opinion polling, now a major industry and a source of political influence in its own right.

Lazarsfeld was a German immigrant who joined the Information and Education Branch of the U.S. Army, a department headed by Stouffer during World War II. With a team of social scientists, they conducted an enormous study of morale and adjustment in the armed services, which was published in four volumes under the title *The American Soldier: Studies in Social Psychology in World War II* (1949–1950). The study typified what has been criticized as a social science boondoggle: an expensive and long-term binge of fact gathering by a team of so-called experts whose end result was "just common sense." Lazarsfeld's response to the criticism began by openly acknowledging that sociology does sometimes produce findings that sound like common sense, and he went on to list some results, followed by the conventional reasoning that would have led one to expect them:

1. Better educated men showed more psychoneurotic symptoms than those with less education. (The mental instability of the intellectual as compared to the more impassive man-in-the-street has often been commented on.)
2. Men from rural backgrounds were usually in better spirits during their army life than soldiers from city backgrounds. (After all, they are more accustomed to hardships.)
3. Southern soldiers were better able to stand the climate in the hot South Sea Islands than Northern soldiers. (Of course, Southerners were more accustomed to hot weather.)
4. White privates were more eager to become non-coms than Negroes. (The lack of ambition among Negroes is almost proverbial.)
5. Southern Negroes preferred Southern to Northern white officers. (Isn't it well known that Southern whites have a more fatherly attitude toward their "darkies"?)
6. As long as the fighting continued, men were more eager to return to the States than they were after the German surrender. (You cannot blame people for not wanting to get killed.) (1949, 379–80)

Behind this list lurks the question, Was it worth all the time, effort, and money to produce such commonplace results? Or, Didn't we know all this before the experts went out and rediscovered it? Lazarsfeld, expecting that kind of reaction, had a surprise in store: "Every one of these statements is the direct opposite of what was actually found." For example, the better-educated soldiers showed *fewer* signs of neurosis than did the less-educated, blacks were *more* eager for promotion than whites, soldiers were more anxious to go home *after* the fighting was over than while it was going on, and so forth.

The real point of the demonstration, however, is not that the true results just mentioned are great discoveries either. Assuming that we are not being fooled this time, a commonsense rationale can be as easily adduced for each true result. Educated people, one might speculate, are better able to cope with frustration and less likely to become neurotic owing to their experience in dealing with difficult new situations and their knowledge of bureaucratic procedure. An urban background, including exposure to a variety of social groups, might be better preparation for military life than a quiet rural upbringing. Blacks would take to the less-discriminatory armed forces setting and strive for advancement more than in civilian life, where they knew opportunities would be denied them. Lazarsfeld's point, then, is that after the fact common sense is capable of "explaining" or providing a rationale for almost anything, but before the facts are known common sense supports all manner of contradictory conjecture. Common sense, as Bertrand Russell observed, is vague, cocksure, and often wrong.

CONCEPTS

In the interest of uncovering something fundamental about the sociological method, this material can be profitably pursued beyond Lazarsfeld's moving demonstration. For just as Lazarsfeld demolished one myth about sociology, he helped build another.

Lazarsfeld belonged to a generation of sociologists who fancied themselves hard-headed scientists whose primary job was the discovery of social facts. As a methodologist, Lazarsfeld was convinced that research provided the bricks out of which sociological theory is constructed. The first order of sociological work, accordingly, is the discovery of those facts. Many like-minded sociological "positivists" believe that the discipline will not achieve its scientific potential until a solid inventory of fact is created to provide the elements from which any generalization may be cautiously assembled and evaluated.

It is no surprise that popular conceptions of sociology have assimilated this view and, having heard that sociologists are in the business of discovering "unknown" facts about society, respond with understandable criticism when elaborate studies turn up nothing new. The trouble is that this simple fact-gathering view of the sociological enterprise is mistaken in the first place. Sociological reasoning and research is not a matter of discovering the unknown or of fabricating theoretical chimneys from factual bricks. Rather, it is a matter of making interpretive sense of facts, new and old. Sociological work essentially involves producing ideas.

Returning to the example, the facts about the morale of soldiers in World War II have very little general interest, particularly these days when we might safely assume that they are no longer true. Perhaps people enjoy military service today because it provides secure jobs or training in occupational skills that may be used elsewhere after a brief enlistment period. The early research would have some general sociological significance only if it helped us understand other situations,including current ones, about the military and, better yet, about how people adjust to membership in groups. By asking for a demonstration of such general significance, we are asking for a **concept**, a shorthand summary for a large number of similar observations. Concepts,not facts, are the building blocks of sociological interpretation.

Following the publication of *The American Soldier,* Lazarsfeld's colleagues at Columbia, especially Robert Merton and Alice Rossi, began to reflect on the more general meaning of these facts. They were drawn to the study's repeated demonstration that the amount of deprivation soldiers felt for having been inducted into the army (as most were) depended on their social status and to whom they compared themselves. Older and married men, for example, felt more resentment about their induction than the younger and unmarried. The reason was that the older and married soldiers knew more people who had been granted deferments and compared themselves to those exempted from service. Similarly, soldiers who remained in the States felt less deprivation than those sent overseas, who in turn considered themselves better off than overseas combat soldiers. Black soldiers stationed in the South felt themselves much better off than the black civilians who lived in the South. In short, deprivation was a relative matter, depending on individual comparisons with "like-statused" people (other married men, other blacks, etc.).

This was well understood by the authors of *The American Soldier* who used the notion "relative deprivation" to explain many of their results. But Merton and Rossi (1957) were more ambitious: "By provisionally generalizing these concepts, we may be in a position to explore the wider implications of the materials for social theory." The first step toward generalizing these results consisted of replacing "relative deprivation" with a broader concept. The concept **reference group** was chosen for several reasons. Merton and Rossi noted that relative deprivation was a rather narrow psychological notion referring mainly to the individual and not really exploring the question — relative to whom is deprivation felt? — other than the blanket allusion to others of like status. In a meticulous review of the original study, they discovered that individual soldiers compared themselves and their social status (e.g., their education or marital status) with a variety of social groups, some in which they held "membership" (other married men) and others in which they placed themselves vicariously (southern black civilians), all for the purpose of deciding how well off they were relative to others. Deprivation therefore depended on which of many possible groups individuals turned to as reference points or pertinent standards of comparison.

The more general question was not so much that soldiers felt deprivation, but how much they felt as a function of which reference groups they selected as most pertinent to their circumstance (itself a composite of their social status and present situation). The reference group provided a sociological construct with more explanatory potential than the psychological idea of relative deprivation because it captured the many kinds of comparisons that soldiers actually made and it linked individual judgments to the social context.

Finally, Merton and Rossi come to the important question of the general significance of reference groups: Does the concept apply to a variety of situations? They claim that it does and propose a general theory of reference groups based on the rule that "insofar as subordinate or prospective group members are motivated to affiliate themselves with a group, they will tend to assimilate the sentiments and conform with the values of the authoritative and prestigeful stratum of that group" (1957, 308). So stated, the principle is intended to apply to a wide variety of situations, to constitute therefore a noncommonsense generalization that can be drawn from the original study.

In a less wordy rendition, Merton and Rossi are saying that newcomers motivated to join new groups assume the attitudes of prestigious members of the group rather than those of persons in the same objective circumstance as themselves. How good is their generalization? Can it be applied to other situations in a manner that explains something that we did not already know? Let's try it. It is often noted with some surprise that new immigrants to the United States who arrive poor and unskilled nevertheless identify with higher-status groups in the population. New Hispanic groups, for example, identify with the Republican Party and conservative values, not unanimously by any means, but far more frequently than would be expected based on their modest economic position and Roman Catholic religious preference, statuses characteristically associated with liberal politics. The Merton and Rossi proposition explains this fact that has puzzled others who ignore reference group theory and assume, wrongly in some cases, that people naturally identify with others in their economic (or class) situation.

Returning to the matter of common sense, does this generalization have practical implications? Does the new proposition say anything that we did not already know about social policy? I think it does. Illustratively, none of us would be surprised by a pronouncement from public school officials to the effect that black children are pessimistic about the value of education for their occupational futures — a "fact" justifying special programs in the schools to encourage optimism in black children. But suppose, as Merton and Rossi imply, that black children in integrated schools turn out to be nearly as optimistic as all the other children and that pessimism is less a matter of being black than of attending segregated schools. The second "fact" (which we have reason to believe is true) would be revealing by the standards of common sense and importantly would carry very different policy implications for programs of educational reform.

DISCOVERY

The larger point in this discussion of reference groups is that the distinctive **sociological contribution** comes not with the uncovering of certain facts (a valuable task that journalists, pollsters, and others also do very well), but with tracing the *implications* of facts to a new and fruitful interpretive idea. If there is a discovery involved in this, it is the discovery of a new idea. The reference group is one such idea, and one that has come in for much debate and critical research, which is appropriate if it is doing its job. The essential work of sociology, far from simply discovering facts or coining terms, lies in the invention of concepts and in reasoning and research about their implications. The point is gracefully stated by Howard Becker:

> I think it generally true that sociology does not discover what no one ever knew before, in this differing from the natural sciences. Rather, good social science produces a deeper understanding of things that many people are already pretty much aware of. . . . [The virtue of any analysis] does not come from the discovery of any hitherto unknown facts or relations. Instead, it comes from systematically exploring the implications of [a] concept. (1982, x)

Social research operates in the large space between unsystematic common sense and unlikely discovery of things hitherto unknown. And within that space, the most distinctive contributions of research reveal facts and relations that are *unexpected.* Unexpected research results then lead to new ways of understanding with the development of concepts and their implications.

A classical example of this proposition is provided in Merton's discussion of the "serendipity pattern [that] involves the unanticipated, anomalous and strategic datum which exerts pressure upon the investigator for a new direction of inquiry which extends theory" (1957, 105). Researchers studying a new working-class suburb called Craftown observed that a large proportion of the residents belonged to more civic and political organizations than had been the case in their previous places of residence and that this was true for parents of young children. The finding was inconsistent with common sense and much previous research about the low rates of political participation of young, working-class families. When questioned about this, residents replied that it was easier to find babysitters in Craftown because of the greater number of teenagers living there. Yet, when the researchers looked into this observation, they discovered that it was not true. Actually fewer teenagers lived in Craftown than in the urban neighborhoods from which the residents had moved. As investigators probed this question, Craftown respondents provided the clue when they told interviewers that they felt more confident about leaving their children with teenage babysitters when they knew their neighbors. Merton (p. 107) concludes, "It is not that there are objectively more adolescents in Craftown, but more who are *intimately known* and who, therefore, *exist socially* for parents seeking aid in child supervision" (1957, 107). More generally, the researchers learned that participation in local organizations is a function of perceptions of community and confidence in one's neighbors. Merton's researchers made their "discovery," or learned something new, by pursuing the implications of an observation in light of other known facts and some theoretical reasoning aimed at explaining their results.

Strictly speaking then, sociology does not, in Becker's words, "discover what no one ever knew before," but sometimes it does produce unexpected results. Sociology uncovers certain facts whose importance is mainly in the refutation of complacent common sense. Debunking myths is a useful result of social research. Sociology also reveals relationships that no one had thought much about previously. Both sources of revelation account for some of the adventure in doing and reading about social research. Research may also produce expected results, which is reassuring once we have become skeptical about common sense. At its characteristic best, however, sociology fashions concepts that organize all these facts and generates from them implications that in sometimes quite unexpected ways produce new explanations of social phenomena. Sociologists do not haphazardly blunder into great finds or pile up pure facts until the pile sprouts the flower of an idea. On the contrary, they labor over closely focused problems and search for that just-right concept that will catch the essence of a process while suggesting a set of implications about its operation. This is the fundamental objective of sociological inquiry and the goal that research methods must serve if they are to be anything other than technical exercises. In place of fact gathering or discovery, Arthur Stinchcombe (1968) provides a more appropriate phrase and image for sociological work. Sociological accounts of the world, "theory" in a practical sense, "*ought to create the capacity to invent explanations.*"

METHOD

Beyond inventing concepts and explanations for social phenomena, sociologists spend a good deal of their time determining whether their interpretations are true. Ideally, the sociological method generates or assembles factual evidence in a purposeful manner; it evaluates concepts, follows implications from the arena of reasoning to results, and tests explanations. When such conceptual tools are not yet available, social research may perforce undertake exploratory missions with the aim of finding evidence that will inform new concepts. In any case, **method** involves a series of logical and procedural steps that link concepts and explanations to facts. Method is a bridge between the two realms of reasoned interpretation and fact.

There is an important difference between method in this sense and research procedures or techniques. There is also a regrettable, even misleading, tendency to equate the sociological method with a series of techniques — an unfortunate habit in textbooks and specialized writing to separate methodology from other (substantive) sociological work and to treat method exclusively as a statistical or technical problem. In a proper sense, however, there is one sociological method and a number of useful techniques or research strategies that, in a given study, appropriately implement that method. Some of these strategies are illustrated in the sections that follow and need only be characterized here.

The **sample survey** is the most conventional research strategy and involves obtaining responses to a standard questionnaire from a systematically chosen sample of people representing some population. In situations where greater intimacy or flexibility are desired, researchers employing **participant observation** enter social group settings, thoroughly recording events and conversations in their own research journals. Sociologists do research with previously, perhaps officially, collected information such as the census of population using **archival sources**; and they move into the recesses of archives, into museums, records, and diaries, to do **historical research.**

Practical strategies of sociological method are as varied as its problems and imagination. Another adventure in research lies in inventing ways to obtain inaccessible information. Some sociologists put college sophomores through laboratory experiments, whereas others interview heads of state; some live in urban ghettos and others at computer centers. In any case, the value of the research strategy depends on whether its evidence informatively and critically addresses the concepts and explanations invented in sociological work. A good bridge is firmly anchored on two sides of some otherwise unnavigable expanse, and good method ensures brisk traffic between explanations and evidence.

In general then, there is only one sociological method, although there are many strategies for implementing that method. The philosopher Abraham Kaplan (1964) has done us a service by distinguishing among different meanings of methodology. One of those is "technique" in the sense of the social survey or statistical procedure just mentioned. Another Kaplan calls "honorifics," or statements about what constitutes *the* scientific method. Third, "methods" per se are more precisely construed as

> logical or philosophical principles sufficiently specific to relate especially to science as distinguished from other human enterprises and interests . . . such procedures as forming concepts and hypotheses, making observations and measurements, performing experiments, building models and theories, providing explanations, and making predictions. (1964, 23)

Methodology unavoidably involves all these things, but a practical emphasis on the last enables appreciation of method as a general way of proceeding with social science research that distinguishes it from other kinds of inquiry and provides a standard for evaluating the fruits of that inquiry. "In sum," Kaplan says with refreshing clarity, "the aim of methodology is to help us *understand,* in the broadest possible terms, not the products of scientific inquiry, but the process itself" (1964, 23).

Why Questions

The sociological method is, first, a general process of inquiry, rather than an ensemble of techniques, procedural loyalties, or honorific pieties about science. The process conforms to scientific tenets in the sense that it invents concepts and hypotheses, refines and tests these, and aims at more general explanations and theories. Focusing only on this aspect, however, can turn methodology into a rather dreary topic, which is unfortunate because the core problem of method — how to find and document the answer to a problem — suggests some excitement. A second and distinctive thing about sociological method, the point at which it becomes exciting, is that it begins with genuine puzzlement about what is happening or has happened in society — with a why question or problem consciousness. In *The American Soldier,* investigators (and their official sponsors) wanted to know why there were problems of morale in the armed services. Merton wanted to know why Craftown residents perceived their neighbors as trustworthy.

This goes back to the question of fact gathering and the difference between sociological research and other useful forms of inquiry. Undisciplined research seldom has an explicit why question in mind before setting about its work, and the result is bad if it does not find one as it goes along. The work may be aimlessly descriptive or abstractly speculative without such a question, just as it may be argumentative and self-serving if the question and the answer are selected prior to the research — the latter by no means a rare occurrence. The German sociologist Ralf Dahrendorf developed this contrast thirty years ago in words that have lost none of their bite:

> "The Social Structure of a Hospital," "The Role of the Professional Football Player," and "Family Relations in a New York Suburb" . . . chosen because nobody has studied them before or for some other random reason, are not problems. What I mean is that at the outset of every scientific investigation there has to be a fact or set of facts that is puzzling to the investigator: children of businessmen prefer professional to business occupations; workers in the automobile industry of Detroit go on strike; there is a higher incidence of suicides among upwardly mobile persons than among others; Socialist parties in predominantly Catholic countries of Europe seem unable to get more than 30 percent of the popular vote [then]; Hungarian people revolt against the Communist regime. There is no need to enumerate more of such facts; what matters is that every one of them invites the question "Why?" and it is this question, after all, which has always inspired that noble human activity in which we are engaged — science. (1958, 123)

Dahrendorf makes two observations that effectively summarize this discussion of method. First, sociology begins its effort to interpret and explain social life with "the simple impulse of curiosity, the desire to solve riddles of experience, the concern with problems" (p. 123). Second, once that curiosity is formulated in researchable (why) questions, the issues of method present themselves in a manageable form: "Problems require explanation; explanations require assumptions or models and hypotheses derived from such models; hypotheses, which are always, by implication, predictions as well as explanatory propositions, require testing by further facts; testing often generates new problems" (p. 124). In short, sociology will go a long distance toward its substantive and scientific goals by becoming more problem consciousness.

SOCIOLOGICAL PROBLEMS

Dahrendorf is right, but that leaves us with the question, perhaps the real question of method: How does sociology generate a fertile problem consciousness? Where do the why questions come from? I suspect that there are many answers and none are fully satisfying because sociology, like any other creative activity, has many intuitive wellsprings. Nevertheless, we can make a lot of headway by showing some important ways in which problem consciousness arises. It comes from at least three sources: the work of our predecessors, important activities and events in the society around us, and reflection on our own social lives.

In the second category, for example, a rich source is provided by the cultural and political worlds that are full of purported explanations for daily events, explanations, ironically, that are often drawn from antiquated or misunderstood social science. Unemployment, we are told, is high because people do not want to work or prefer receiving welfare benefits for not working; people riot in the cities because they are "outside the mainstream" of organized society; revolution brews in Central America because of Communist subversion; children do not perform well in schools because educational fundamentals are ignored. Although we do not customarily think of it this way, and it may glorify ordinary prejudice, there is a sense in which all of these answers are also popular theories of social behavior. Moreover, they are sometimes consequential theories to the extent that they inform public policy (e.g., to eradicate unemployment by cutting off welfare benefits or to snuff out revolution by military containment of subversion). As theories, they invite problem-consciousness evaluation.

Theoretical Problems

Some illustrations will help clarify these three sources of problem consciousness, while showing their interconnectedness. Among our sociological predecessors, none is better known or more controversial than Karl Marx. Marx, as we will see later in this chapter, endeavored to explain the process of capitalist development and, particularly, what it meant for people engaged in various branches of economic production. One of his central hypotheses was that the growth of capitalism destroys small business and the independent crafts and professions (the petty bourgeois stratum), turning those occupations into wage labor (or proletarians). This experience of degradation, Marx inferred, would help to generate a revolutionary mood (or class consciousness) in the new working class. C. Wright Mills (1951) explored this question in his book *White Collar: The American Middle Classes* and documented Marx's prediction insofar as the destruction of the old middle classes was concerned. On the other hand, Mills found that far from sinking into the working class, members of the former craft and professional occupations were reabsorbed into a decidedly unrevolutionary new middle class of white-collar, salaried employees in large public and private bureaucracies. Mills regretted this development and chided the new white-collar workers in his own vigorous style:

> They are worried and distrustful but, like so many others, they have no targets on which to focus their worry and distrust. They may be politically irritable, but they have no political passion. They are a chorus, too afraid to grumble, too hysterical in their applause. They are rearguarders. In the short run, they will follow the panicky ways of prestige; in the longer run, they will follow the ways of power, for, in the end, prestige is determined by power. In the meantime, on the political market-place of American society, the new middle classes are up for sale; whoever seems respectable enough, strong enough, can probably have them. So far, nobody has made a serious bid. (1951, 353–54).

Mills pruned one theory and advanced another that still has a decided critical appeal. As a Texas-born populist, Mills revered the ethic of Jacksonian democracy and small-town society, which seemed to him lost in modern bureaucratic society, and he clearly mixed his personal values with his research interests.

Problems from Society

The second source of sociological questions, events and activities surrounding us in society, is especially rich when those events come packaged in popular theories. These are expressed everywhere — in the newspapers, television, public deliberations, official judgments, and so forth. Broken homes produce juvenile delinquents. Television makes children violent, or nuclear war scares them. Some of these ideas influence society and deserve evaluation of just how true they are.

In late April 1992, a riot broke out in South Central Los Angeles just hours after an all-white jury acquitted four police officers accused of beating Rodney King, a black man suspected of drunk driving. When it was all over, the Los Angeles riot of 1992 resulted in fifty-eight deaths and $785 million in property damage (much of that from over 5000 arsons), thereby surpassing an earlier record of thirty-four deaths and $40 million in damage established in the Los Angeles riot of 1965. Although separated by twenty-seven years in which Los Angeles had suffered no such major civil disorder, the two riots arose under similar conditions. Charges of police brutality were the precipitant in each case, and members of the black community claimed general neglect that is reflected in unemployment and inadequate public services.

The initial Los Angeles, or "Watts" (after the neighborhood of origin), riot prompted a new type of public policy and sociological research. Watts was far from the first race riot or massive civil disorder in U.S. history, but it was, perhaps, the most noteworthy in a series that occurred in the late 1960s: It was the most destructive to that date; it inaugurated a new type of public disorder in which issues of civil rights were intertwined with ordinary mayhem in the eyes of participants and observers; it happened in Los Angeles where no one expected it, given a less discernible ghetto, few slums or tenements, and more home ownership in contrast to older cities of the East and Midwest; and it prompted the first in a new series of quasi-social science, riot commission studies.

Beginning with the mishandled arrest of a black youth by white police officers and police brutality in the eyes of a crowd that gathered at the tense encounter, the disturbance raged for the next six days over the forty-six square-mile area of South Central Los Angeles. When it was over, 10,000 people had joined in rebellion, 34 were killed (3 white law enforcement officers and 31 black residents), more than 1000 injured, 4000 arrested, and some $40 million worth of property damaged in the looting and burning of 600 buildings and stores. One week after order had been restored, the McCone Commission was empaneled to unearth the causes of riot and provide, in the short 100 days allotted for its inquiry, a "deep and probing" analysis of what troubled Los Angeles.

The McCone Commission found three basic causes for the riot: unemployment and "idleness," low cultural and educational levels among black youth, and poor police–community relations. In each case, this conventional diagnosis, with much liberal handwringing, laid the blame on the black community. The absence of employment opportunity in South Los Angeles was noted forthrightly, but little attention given to why this was the case (e.g., urban disinvestment or discrimination in hiring). The report emphasized problems of black youth on the unexamined assumption that most rioters were young. Educationally, young people were seen as caught in a "dull devastating spiral of failure" — ill-prepared for school in the first place, leading to poor performance and a high dropout rate. The matter of police–community relations was treated obliquely, indeed delicately, by a recitation of those purported characteristics of the black community that made it difficult to police (such as the assumed high number of recent migrants from the South with unrealistic expectations about California living). In sum, the report leaned on what Anthony Oberschall (1968), a sociologist who lived in Los Angeles, called the "criminal riff-raff theory" of civil disorder. Oberschall demonstrated this interpretation in the report and then asked, Is it true? Are there facts that allow for testing the theory or inventing a better one?

One striking fact was that the McCone Commission, with all its experts and $250,000 to support its research, had never asked itself Oberschall's questions. The commission used some capable sociologists as consultants, including Edward Ransford,

Melvin Seeman, and Robert Blauner (who later wrote a scathing analysis of its "White Wash over Watts"), but, as Blauner noted, the report was mute on the question of who participated in the riot and why. Oberschall did the sensible thing; he went to the record of police arrests of participants in the riot. The facts exploded the riff-raff theory. Arrested riot participants were not seasoned criminals or the dregs of society. Seventy-five percent had lived in Los Angeles for more than five years, refuting the recent migrant assumption. Their average educational level matched the population of South Los Angeles as a whole. They were not especially young: 41 percent were between twenty-five and thirty-nine years of age, and another 17 percent were over forty, meaning that 58 percent were far from teenage gangsters. And they were representative of the area-wide lower-class background, not "scum." From all this, Oberschall concluded,

> What strikes one is the extent to which the riot drew participants from all social strata within the predominantly lower-class Negro area in which it took place. The riot cannot be attributed to the lawless and rootless minority which inhabits the ghetto, though, no doubt, these were active as well. The riot is best seen as a large scale collective action, with a wide, representative base in the lower-class Negro communities. . . . (1968, 329)

Contrary to the official commission's judgment, Oberschall showed that Watts was a rebellion and that people rebelled because the conditions with which they lived came to outrage their sense of justice. People do not engage in massive civil disorder mainly, in Edward Banfield's calloused phrase, "for fun and profit" (1968, 185). The rebellion in Watts came from the grass roots and attacked particular representations of perceived injustice: policemen and their patrol cars, the stores of exploitative white merchants. Violence was the anguished expression of insistence on simple civil rights. It had little to do with the malintegration of these citizens into U.S. urban culture. On the contrary, it was energized by that culture, the civil rights movement and the esteem attached to social mobility.

The fact that the riot commission failed to see these simple truths merely reflects the public mood of denial, scapegoating, and resentment. It was more convenient, even reassuring, to locate the problem within the black community itself. The fresh sociological analyses that exposed this prejudice were no doubt motivated by individually held values of equality and a passion for debunking, but their analytic cunning followed from a simple question, Why?.

Although we have not yet had time for a careful analysis of the 1992 riot, observers have noted differences from 1965. Los Angeles journalist Tim Rutten says that we have seen "something new, what might be called the nation's first multi-ethnic urban riot" (1992, 52). Participants in the protest and looting included some of the area's Hispanic population, in addition to African Americans; the principal targets attracting looters and arsonists were businesses owned by Koreans who have settled in neighboring communities, especially in the last decade. Yet, some contemporary analysts also share the McCone Commission point of view. Writing in *Commentary,* Charles Murray (1992) explains the 1992 riot as a result of the paternalistic reforms of the 1960s that undermine community responsibility through misguided antipoverty programs — an argument that leaves similar protest events of the 1960s, which predated such programs, unexplained.

There is no way to completely eliminate the kinds of myths perpetuated by, among others, public agencies and the mass media. They can, however, be minimized and rooted out in two ways: through the imperfect process of self-correcting tests and reformulations and by reconceptualization of a problem and its causes — much as Oberschall reversed the question of riot causation by shifting it from deficiencies within the black community to the reasons for rebellion against the larger society. The first tack is imperfect, slow, and sometimes the source of new myths. The second is bold and often unpopular. For all its methodological blunders, the commission report and the prejudices behind it had a stronger impress on public policy than did the work of a few critical sociologists. The problem therefore is how to introduce the results of social research into the dialog of public policy. A third example deals with that general issue.

Problems of Experience

Reflection on experiences of daily life can be a valuable source of sociological questions, provided that personal impressions are framed in broader issues — provided, in the words of Mills, that research questions seek to establish the connection between "personal troubles of milieu" and "public issues of social structure" (1959, 8). An enormously successful example of this approach is the book by Robert Bellah and his associates, *Habits of the Heart: Individualism and Commitment in American Life* (1985).Drawing on their own experience, this research group began with the unsettling and widely shared belief that white, middle-class Americans are a troubled people — relatively affluent in material terms, yet unfulfilled in any deeper meaning of their lives. In the most general forms, the researchers' questions were, "Who are we as Americans? How do we think we ought to live? How to preserve or create a morally coherent life?" (p. vii). At this stage, the very general question needed specification, some link between the private troubles of experience and public issues in society. The researchers found this link in the classical examination of *Democracy in America* written by the French historian and diplomat Alexis de Tocqueville in the 1830s. Tocqueville, perhaps the most acute observer of American society, praised its tradition of equality but wondered whether that virtue engendered a stress on individualism that undermined the very institutions of public life that guarantee equality. Following Tocqueville, "We therefore decided to concentrate our research on how private and public life work in the U.S.: the extent to which private life either prepares people to take part in the public world or encourages them to find meaning exclusively in the private sphere" (Bellah et al. 1985, viii).

The research team actually conducted four studies integrated in one interpretation. In the private realm, they investigated how Americans find meaning in love and marriage and in therapy; in the public sphere, they focused on community voluntary associations and political organizations. Their interviews portray both the strengths and limitations of the American character. One woman in her early thirties credited hard work for her successful career as a therapist and her good marriage to an engineer. She regarded personal fulfillment as life's highest aim, one that could be attained only through maturity and self-reliance. Yet, her stress on individual well-being ruled out any sense of social responsibility. She lacked the principles or the language that would describe any responsibility to others, even to her own children and husband. "In the end, you're really alone and you really have to answer to yourself" (p. 15). In a sensitive reading of this testimony, the authors note that "there is no wider framework within which to justify common values" (p. 16), no inkling of a standard for the public good that individuals should observe for their mutual welfare. Although this attitude might be expected in therapists who work with individual clients, the researchers soon learned that it was pervasive. Another respondent had deserted from the Marine Corps during the Vietnam War and spent years underground before surrendering to a military court and brief imprisonment. Today, he works as a community organizer helping tenants, including the elderly, maintain their affordable homes against efforts by housing developers to drive them out with condominium conversions or demolition making way for expensive high-rise apartments. He wants to liberate people, give them power over their own lives. "But what specific *kinds* of things should these newly liberated people create in society?" (p. 19); what is social justice and what principles should it work for? The community organizer who devotes his life to creating a better society is "strangely inarticulate" about what that society would look like.

Yet, the people interviewed by Bellah and his associates vaguely sense their own isolation and struggle to put the feeling into words. Forthright individualists, for example, express a desire to "reconnect" with others. Many interviewees imaginatively long for community, the homes and churches they felt obliged to leave as a requirement of success in a competitive world. A striking result of the research was the number who nostalgically yearned for small-town life. Ironically, just as they idealized these "communities of memory," their current activities were increasingly centered in what the authors called "life style enclaves" — places of residence, leisure, and consumption activities that segregate similar kinds of people in pursuit of private pleasures (e.g., retirement or singles' housing developments, tennis clubs, or private gyms). As a result, the vague if genuine yearning for "community degenerates into life style enclave" (p. 154). People who then live their lives in segregated spaces devoted to self and consumption never learn the "practices of commitment" that define responsibilities to a community. The whole pattern is reflected, finally, in the language available to Americans for the description of their search for meaning. "The language of the self-reliant individual is the first language of American moral life, the language of tradition and commitment to communities of memory are 'second languages' that most Americans know as well, and which they use when the language of the radically separate self does not seem adequate" (p. 154). But the term *second language* is used deliberately to suggest one that is less developed, less instinctive as a set of ideas that comes to mind for interpreting experience. At bottom, the modern problem of Americans is the absence of a moral discourse for expressing or debating civic virtue.

Long ago the United States was founded on complementary traditions that stressed both individualism in the pursuit of happiness and republicanism in the civic participation that would guarantee everyone's liberty. But the second, public tradition, was best sustained in the local society of the nineteenth century, whereas modern technological society dominated by large corporations and government reinforces the first language of private individuals. The authors conclude that moral coherence will be restored to our lives only to the extent that we are able to reinvigorate common public values. Although they are realistic about the enormous task we face, they are also constructive and responsible in the suggestion that social science has a role to play in this process by documenting and explaining our present condition.

CRITICAL SOCIOLOGY

The foregoing examples show the varied ways in which sociologists locate problems and translate them into why questions. They also show how research on important problems leads directly to critical evaluations of social conditions — racial inequalities that lead to protest or weakening civic traditions that deprive people of meaningful ways of engaging social life, for example. All of this means that research is a social activity with unavoidable implications for how society is managed — that is, a political reality. Sociology does not lack public recognition or political relevance. On the contrary, when riot commissions and other deep-and-probing official studies of social problems posture at sociology's diagnostic authority, it is an ironic compliment to the discipline. More directly, *Habits of the Heart* has reached a wide public and sold more copies than any recent social science book. If anything menaces sociology's independent, demythologizing role, it is an easy familiarity with conventional ideas and the temptations or compromises that come with celebrated attention from those in power. Sociologists need to strain in a critical direction simply to ensure evenhandedness in a society where established ways of doing things are already privileged. Critical sociology therefore is radical in the sense of radical surgery that goes to the root of a problem. Howard Becker and Irving Louis Horowitz argue, convincingly in my judgment, that good and radical sociology are highly coincident, though not identical:

> Good sociology is sociological work that produces meaningful descriptions of organizations and events, valid explanations of how they come about and persist, and realistic proposals for their improvement or removal. Radical sociology also rests on a desire to change society in a way that will increase equality and maximize freedom, and it makes a distinctive contribution to the struggle for change. On the one hand, it provides the knowledge of how society operates, on the basis of which a radical critique of inequality and lack of freedom can be made. On the other hand, it provides the basis for implementing radical goals, constructing blueprints for freer, more egalitarian social arrangements. . . . (1972, 50, 52–53)

Critical sociology is summarized in three characteristics: interpretation, evaluation, and proposed alternatives. Critical sociology (1) provides a revealing description of social reality and interpretively explains the forces that brought it about, (2) evaluates the (intended and unintended) consequences of that reality against the explicit value standards of freedom and equality, and (3) proposes alternative policies or social arrangements that, based on arguments from research, would more fully realize the values. The second and third tasks require that the sociologist has clearly in mind an alternative, preferable state of affairs that is not present in the immediate reality. In fact, we all operate with such implicit ideas, but the critical sociologist must make them explicit alternatives — "utopias," in the phrase of Karl Mannheim (1936), that project a standard against which we can compare the present reality. I should stress that urging critical sociology is itself a value reflecting my belief that it is the most coherent integration of the useful roles that social science can play. Yet, it is also true that not all sociology fits the description of critical sociology and some very good sociology may focus on only one or two of the three defining criteria. The work by Bellah and his associates on the mood of America addresses the three standards of critical sociology. Mills's study of white-collar work, however, says little about alternatives to the occupational structure, and Oberschall concentrates his research on an explanation of the urban riots. Critical sociology, in short, is a special type of work that, in my judgment, shows off the best of what sociology can contribute.

EXEMPLARS OF THE SOCIOLOGICAL METHOD

The three previous examples of good sociology show the varied sources of research questions and make a single point. Good research and method derive from good why questions, irrespective of their source, and good questions assume a critical stance by asking for explanations of the origins, persistence, and change of social arrangements. In the passage from this kind of question to the many inventive strategies for doing research work, one general method is the sociological method. Having stated that case, it remains to support it with a closer look at social research.

Method comes alive in practice as sociologists go about their work. The preceding points are illustrated here in a series of research biographies about important and engaging exemplars of sociological work. The following set of studies spans nineteenth-century classical and recent work, a variety of methodological strategies and substantive problems, and what will appear as some unlikely bedfellows. This is deliberate, as we will see in the end.

Karl Marx on Profit and Labor-Time

A good place to begin is with the work of Karl Marx because it shows the continuities between classical and recent work, while demonstrating that the classical sociologists were not just speculators but energetic fact mongers. Throughout his fifty years of research and writing, Marx pursued one master question: What are the "laws" of socioeconomic and, particularly, of capitalist development? He wisely refined this with a series of more limited questions that would yield to conceptual and factual answers: What is the source of value in the production process, where does the capitalist's profit come from, and what share of all this does the worker receive?

Marx rested his interpretation of capitalist development on the concept of "surplus value," the result of a process in which workers are exploited by factory owners who pay workers less than the full value of their labor. The new concept reversed earlier accounts of where value came from, explanations by the "bourgeois political economists" who held that the value of a commodity — say, a pair of shoes — was somehow in the shoes themselves or in the process whereby the shoes were exchanged for money. These were not explanations at all but pure mysticism according to Marx. What sense did it make to say that there is value in a pair of shoes? How did it get in there? If one follows the career of the shoes, it is clear that they have a "use value" for their buyers and an "exchange value" for their sellers, but neither of these explains how value is created in the first place or where the capitalist seller's profit comes from. The key for Marx was that the origin of value could be found only in the process of production and, specifically, in the amount of "labor-time"required to produce the shoes (because the additional requirements for production such as equipment and raw material came either free from nature or from past labor).

This, it should be stressed, is Marx's premise, advanced for the purpose of making other facts intelligible and not itself a factual assertion. The utility of the concept and the basis for its factual evaluation come with its ability to explain the origin of profit. Profit is now understood concretely as the amount of "unpaid labor" that the worker involuntarily donates to the capitalist — the difference between the amount of labor-time "embodied" in the commodity and the value at which the capitalist exchanges it. Where else, Marx asks rhetorically, could value come from? The essential source of profit for the capitalist, whose livelihood depends on being as competitive or more so than the next capitalist, lies in extorting from the worker more value in labor-time than is returned in the purchasing power of wages — or simply to get more unpaid labor-time from the worker.

The capitalist can generate profit in a variety of ways such as increasing work to get more produced for the same time and wage, lowering wages for the same amount of time and work, mechanizing production to get more produced with the same or a lesser amount of labor (and even with higher wages, provided that the additional value of the wages does not exceed that of the additional commodities produced), or simply extending the length of the working day at the same wage.

Marx researched the operation of all these mechanisms in the English industrial setting. Moreover, as a sociologist rather than an abstract economic theorist, he researched the social and political conditions through which the drive for profit worked. His analysis of "the working day" (Chapter 10 of *Capital*, Vol. 1 [1867] 1977) is a fine example of how his invented explanation and its implications make sense of a complex history: "Nothing characterizes the spirit of capital better than the history of English factory legislation from 1833 to 1864" (p. 390).

Marx traced this history beginning with the first "Statute of Labourers" enacted in 1349 after the Great Plague as a means of attracting scarce workers and holding down their wages. By the early nineteenth century, the length of the working day had extended well beyond the initial twelve-hour "natural day" for adult males to fifteen hours (5:30 AM to 8:30 PM) for all workers including women and children. The lives of the factory workers were menial, bleak, and cut short by disease and hazardous conditions. Hardship and a growing working-class culture produced a political movement dedicated to a new People's Charter and a Ten Hours' Bill.

Here the story becomes intriguing. The workers mounted their struggle for ten hours against the factory owners and sought the support of Parliament for passage of the bill. Parliament at the time was still dominated by aristocrats and landowners, with whom the industrialists were struggling for repeal of the Corn Laws and other protective tariffs. The industrialists wanted free trade and, particularly, elimination of duties on grains imported into the country from Europe (which protected English agricultural interests) because that would result in cheaper food for workers and less pressure for wage increases.

Marx saw the situation as rife with *interclass* conflicts (between workers and industrial owners) and *intraclass* conflicts (between industrialists and landowners). In the maneuvering that followed, the industrialists were most clever, taking advantage of the situation in an unexpected way:

> The spokesmen and political leaders of the manufacturing class ordered a change in attitude and in language towards the workers. They had started their campaign to repeal the Corn Laws, and they needed the workers to help them to victory! They promised, therefore, not only that the loaf of bread would be twice its size, but also that the Ten Hours' Bill would be enacted in the free trade millennium. ([1867] 1977, Vol. 1, 393)

The new class alliance worked. Factory owners and workers joined forces, and the Corn Laws were repealed along with duties on cotton and other raw materials. The new Factory Act of 1847, which promised a reduction of the legal working day, came as a concession to the working class for their support — helped along by revengeful landowning interests in Parliament that still smarted over their defeat by the industrialists. However, as soon as the industrialists had what they wanted, they began to sabotage the new Factory Act by reducing wages. Thus, although the workers now had the ten-hour day, it netted them much less money, and they were required to return to the longer day for the same wage.

The manufacturers claimed in Parliament and the press that workers voluntarily took on additional hours. This falsehood was exposed by the factory inspectors, for whom Marx had great respect. Marx cited the work of one Leonard Horner, an aggressive inspector and investigator who helped interview 10,270 workers in 181 factories, showing that 70 percent preferred the ten-hour day even for some loss in wages, if they had a choice in the matter, which they did not. Nevertheless, with all their maneuvers and power, the industrialists were still having trouble nullifying the Ten Hours' Bill. Providence intervened on their behalf. In June 1848, a revolution in France sent shock waves through Europe and across the channel:

> Insurrection in Paris and its bloody suppression united, in England as on the Continent, all factions of the ruling class, landlords and capitalists, stock-exchange sharks and small-time shopkeepers, Protectionists and Free Traders, government and opposition, priests and free-thinkers, young whores and old nuns, under the common slogan of the salvation of property, religion, the family, and society. Everywhere the working class was outlawed. . . . The manufacturers no longer needed to restrain themselves. They broke into open revolt, not only against the Ten Hours' Act, but against all the legislation since 1833 that aimed at restricting to some extent the "free" exploitation of labour-power. ([1867] 1977, Vol. 1, 397–98)

Through political power, guile, and luck, the industrial–owner class won this engagement in the class struggle and went on to enhance their profits by reextending the working day and adding new forms of exploitation. In effect, while they were engaged in a Parliamentary struggle with the landowning aristocrats, they made a temporary political alliance with the working class, based on a promise of better working conditions. Once the industrialists had beaten the old aristocrats, they took back the concessions to labor and added new methods for extracting more labor-time at no additional cost to their profit margins. On the contrary, industry emerged from the struggles richer and more politically powerful.

Marx's research method involved two key features. First, it was the concept of surplus value that got to the heart of events by showing the connection between profit and labor — just as it helped to explain later why the extended day could be profitably abandoned with further mechanization. Second, Marx unraveled the intricate ways in which the drive for surplus value worked itself out through meticulous research. Studying in the British Museum, he poured over hundreds of volumes of *Reports of the Inspectors of Factories,* Parliamentary inquiries by the *Children's Employment Commission,* the *Factories Inquiry Commission* and the vast *Parliamentary Papers.* The amount of factual evidence that he synthesized is amazing, although it is also true that his own preference was for research with historical and archival documents. For "original data," he relied mainly on people like Horner, although in 1880 he did draft an enormous, 100-item questionnaire for distribution among French workers. Evidently, very few of the 25,000 printed questionnaires were ever completed, and it is no wonder. Inspection of the questionnaire, reprinted by Tom Bottomore and Maximilien Rubel (1956) shows that it would have required a great amount of time and sophistication to fill out. Marx was no pollster, but he was an exceptional researcher.

Émile Durkheim on Suicide

Émile Durkheim, like Marx, studied the "great transformation" from feudal to industrial society. Durkheim's research in the late nineteenth-century began with questions and operative concepts much different from those of Marx. The Frenchman wanted to know why late nineteenth-century society was disintegrating and what could be done about it. The concept he developed to summarize the disjuncture between past and potential future forms of social solidarity was "anomie" — the absence of normative regulation or binding commitments to a stable moral order. The consequences of anomie included class struggle and inequality. But Durkheim, in a book first published in 1897, wanted to show how a uniquely sociological method could explain pathological behavior. The book was *Suicide: A Study in Sociology.*

Choosing suicide as a problem to be explained sociologically was strategic. What could be a more individual and appropriately psychological question than why people take their own lives? What could sociology — the study of "social facts" that are "exterior and constraining" to the individual according to Durkheim — explain about the ultimate act of self-renunciation? Naturally, Durkheim had an answer. Whatever the individual factors in suicide might be, they scarcely exhaust the problem because they cannot explain variations in the social suicide rate across time in a single society or across societies and social groups at a given time: "At each moment of its history, therefore, each society has a definite aptitude for suicide" (Durkheim [1897] 1951, 48).

As Durkheim began to analyze the European data on suicide rates, he became convinced that psychological, geographical, and other nonsocial factors explained very little. There was, for example, no systematic difference in the suicide rates of the sane versus the insane, no unique connection with alcoholism, no greater incidence in the gloomy climes of northern Europe than in the sunny South, no differences by race and "national character." Indeed, as Durkheim tracked down some fairly preposterous theories about the effects of climate and season, the only regularities he found were higher rates of suicide in daytime, summer, and cities — all of which he interpreted as occasions of more intense social activity. This was a clue.

Having excluded many false leads, the question narrowed to which social groups might have especially high (or low) suicide rates and, if those could be found, what would explain the differences. Now a wealth of meaty results began to appear: Suicide rates were higher for Protestants than for Catholics, although Catholics, in turn, had a higher rate than Jews; highest for single people, next for widowed and divorced, less for married, and least for the married with children; rates rose positively with education; they were higher among members of the army than for civilians; males were four times more likely to kill themselves than females; and the urban exceeded the rural rate. Durkheim identified many more correlations and examined them in detail looking for some underlying coherence. In trying to make sense of the results, the overall pattern, rather than particular results, requires attention. Illustratively, one could make a good deal of the differences in religions by searching for theological injunctions against suicide, but if religion were the principal factor preventing or disposing to suicide, why would education or military status, neither highly correlated with religion, also show patterned differences?

As Durkheim reflected on the question of a pattern, he reasoned that there was a common denominator in all these statuses. Being Protestant, single, and educated, for example, all involved greater individualism (or "egoism"); Protestants enjoyed greater freedom from religious dogma and greater latitude of individual interpretation, single people had fewer responsibilities to a family, educated persons were more accustomed to "think for themselves," and so on. Jews, married persons with children, and the less-educated were all less individualistic and more bound to social responsibilities or group expectations. In short, and this should come as no surprise, suicide rose as normative regulation declined. Suicide was a positive function of anomie — just as Durkheim had suspected or "hypothesized."

At this point, one might conjecture that Durkheim has been engaged in an elaborate demonstration of what he already knew or wanted to prove. In a way, that is true. He certainly had an inkling that the moral disintegration he had worried about all his life was connected with the kind of social pathology represented by suicide. But here is where the unexpected came in, owing to the implications of the invented explanation. That is, if suicide varies with anomie, there should be other sorts of "anomic" circumstances that bear a relationship to suicide.

Durkheim reasoned this way and checked out the implied expectations. Periods of economic collapse, depression, and recession, for example, all clearly disrupt the normative order of society and should produce a higher suicide rate according to the argument. Durkheim then discovered that the suicide rate does indeed rise with downturns in the business cycle. Common sense!, one cries. People jump out of skyscrapers or into the moats of French châteaux during depressions. That is figuratively true also. But Durkheim is ahead of us. He reasoned that periods of rapid economic prosperity should also disrupt the normative order, and he went on to show, quite contrary to common sense, that the suicide rate rises during economic boom times.

What would common sense say about periods of political upheaval? What should happen to the suicide rate during wars, revolutions, or electoral crises? Conventional thinking might fudge here and predict higher suicide rates as in the case of economic crises. That would be wrong. Durkheim, focusing conceptually on anomie rather than the unexamined events of social crisis, reasoned that war and revolution would reassert the moral order, call for greater social regulation, and thereby lower the suicide rate. His prediction was borne out by the evidence on suicide during political upheavals throughout Europe.

The concluding chapter of *Suicide* is entitled "Practical Consequences," and it asks, in effect, What is the remedy? The simplistic "social engineering" answer might be that because suicide is correlated with high education, Protestantism, and being single, to solve it you give people less education, convert them to Catholicism, or better yet to Judaism, and marry them off — or, failing these, you live with it. That is nonsense to Durkheim who, as a radical sociologist in every sense of the word, went to the root of the problem and proposed a solution based on social reorganization.

Religion, education, and the rest were different symptoms of the anomie that had overtaken society with the atrophy of institutions: "Thus, the only remedy for the ill is to restore enough consistency to social groups for them to obtain a firmer grip on the individual, and for him to feel himself bound to them" (pp. 373–74). The next question is, But what groups? Durkheim argued that they should not be the traditional religious, political, or family groups, nor the "far removed state." They should be decentralized groups that have an "organic" connection to modern life: "The only decentralization which would make possible the multiplication of the centers of community life without weakening national unity is what might be called *occupational decentralization*" (p. 390). So, Durkheim's most rigorous study and the one classic that continues, properly if exclusively, to be read as a primer in sociological method, came home to his reform passion.

Seymour Lipset, Martin Trow, and James Coleman on Union Democracy

Occupations have been the focus of some fascinating sociological research. For many years at the University of Chicago, Everett Hughes encouraged his students to go into the community and learn how people do their work. Monographs resulted on such diverse jobs as taxi drivers, funeral directors, and dance musicians. The occupational association that interested Durkheim has attracted modern researchers, although few of those were inspired by Durkheim's theory or method. The new issues that concerned researchers by mid-twentieth century are reflected in the study *Union Democracy: The Internal Politics of the International Typographical Union* by Seymour Martin Lipset, Martin Trow, and James Coleman (1956). Lipset, who initiated the collaborative project, has written its research biography in a way that reveals much about the personal and situational influences on sociological work.

Lipset began graduate school in 1943 at Columbia University, studying with Merton and Lazarsfeld. His interest in the International Typographical Union (ITU) began with his father who was a printer and took him to union meetings. Later, as a graduate student, Lipset decided to write about the union because it had a unique internal, two-party democratic system. It might have been an illustration of Durkheim's decentralized occupational association, but Lipset was viewing it in a different light appropriate to his own circumstance. As a youth, he had participated in socialist politics, but like so many of the young U.S. intellectuals of the 1930s and 1940s, Lipset had become skeptical about the fate of democracy:

> The experience of the left and labor movements in various countries indicated that the building of a large socialist or labor movement, or even its coming to power, was not sufficient to democratize a society. In the Soviet Union, a supposed Marxist revolution has resulted in the creation of a new, even more exploitative, form of class rule than existed in capitalist countries. The social-democratic parties and trade-unions in other countries seemed to exhibit more concern with organizational stability and survival than with the need to advance radical solutions to social problems. . . . The question was why. (1964, 112–13).

Lipset found the best available answer to that question in the book *Political Parties* by Robert Michels (1959), who proposed that combative organizations such as unions and parties follow an "iron law of oligarchy" that eschews internal democracy for the sake of doing battle with industry and the state. But the ITU belied that law, providing a fertile exception to the undemocratic process in most large organizations. Lipset's question could now be refined and focused on an explanation for the uniqueness of the ITU. What was different about printers?

The answer to this question was a long time coming. Lipset drew on his personal knowledge, study of the union's history, particularly the development of its two-party system, and interviews with members of the New York locals. With a proper question in mind, the clues began to appear in certain extraordinary features of the printers' trade. On the job, they took a good deal of pride in their work and thought of themselves, if not as middle class, then certainly as a highly skilled pinnacle of the working class. Off the job, they engaged in a variety of leisure activities with one another: social clubs, veterans' organizations, athletic teams. This made further sense because the printing trade was relatively homogeneous (status distinctions between printers were small), and it involved odd working hours. The leisure-time activities of printers took place when most people were working, meaning that printers spent most of their time together. From this kind of theoretical reflection and factual material, Lipset reached his conceptual invention: Printers had an "occupational community" rooted in the conditions of their work, which maintained democratic practices in their union.

Just as Durkheim had a pretty good idea of the consequences of anomie before he collected his data on suicide rates, Lipset had discovered the occupational community before he, Trow, and Coleman conducted the survey research on union members that makes up the bulk of *Union Democracy*. Indeed, it was this refined understanding of the union that made an informative social survey possible. Lipset found that

> The methodological innovations evidenced in our sample design did not stem from any special concern with creative methodology. Rather, it and the questions reflected the fact that the survey was planned to test a highly complex analytic model; it was a sophisticated survey design precisely because years of prior investigation of the attributes of a complex system had preceded it. Having the hypotheses formulated well in advance meant that the methodology almost created itself. (1964, 125)

The parallel with Durkheim's research style goes further. Lipset's colleagues could ask what was the use of the elaborate survey (in three stages, including a sample of 434 regular members selected by the size of shop in which they worked, a subsample of union stewards, and another of leaders of the internal political parties) given what he knew before getting to this phase of the research. Was this just quantitative window dressing? Lipset notes that the question "challenges the principal assumptions on which most sociologists work." His answer was that not only did the invented explanation about democracy in the union need to be evaluated but also many of its implications had to be followed through to issues that the researchers did not presume to know beforehand.

The survey research supported the explanation, but it added many new and unanticipated insights: The size of the union shop influenced a flourishing democracy because small shops were unable to support two-party differences as amicably as large shops; special hiring practices encouraged the occupational community even before printers became regular union members; the structural effect of shop size revealed and helped explain why only the larger locals had a competitive party system; yet at the individual level, political participation was explained by the extent to which printers had working-class reference groups.

Having shown why democratic unionism survives in this atypical setting, the book concludes on a pessimistic note. The implications for organizational democracy are sobering precisely because the conditions that supported it in the ITU, the pride and skill of the craft, are being destroyed by industrial rationalization, business unionism, and technological advances in printing. Michels's oligarchy was more grimly prophetic than Durkheim's occupational associations. In their conclusion, Lipset, Trow, and Coleman anguish over their research because it

> does not offer many positive suggestions for those who would seek consciously to manipulate the structure of such organizations so as to make the institutionalization of democratic procedures within them more probable . . . [but they also precaution vaguely that it] would be foolhardy to reject the possibility that major changes in the social structure will increase the potential for democracy within the labor movement. (1956, 404–405, 406)

Although they provide no hint about what those changes might be, they honestly confront the unhappy facts. If nothing else, the results suggest that we need to look elsewhere for the conditions under which communities struggle for autonomy and moral order.

Erving Goffman on Asylums

The social order that Durkheim saw changing with industrialization, Erving Goffman discovered in the unlikely confines of a mental hospital. This is not surprising because Goffman chose to examine the society of mental patients from the standpoint of the inmates themselves, rather than through the eyes of their psychiatrists and custodians. His book *Asylums: Essays on the Social Situation of Mental Patients and Other Inmates* (1961) poses a question about mental illness that departs radically from conventional views by proposing to study "the moral career" of a mental patient.

Goffman chose the career concept precisely because of its two-sidedness, its capacity to integrate matters internal to the individual and a public or official aspect of that person's life. The idea that people have a moral career allows study of the individual in a social setting by linking public and private sides of the person in ways that are lost in psychological studies — particularly those that assume that individuals suffer diseases of the mind. Goffman believed that more could be learned about mental illness through an "institutional approach to the study of the self." To follow the career of mental patients, Goffman got as close to them as he could by working in a minor staff position at a large public hospital (he ruled out actually becoming a patient as too restrictive of his freedom to study the institution). Once Goffman had entered this unique but sensible world, he found he was "participating in a community not significantly different from any other" (p. 130).

The unique question in this study was, What are the effects on a person of being treated as a mental patient? Goffman notes that people get to the mental hospital for all sorts of reasons besides their mental condition. Hospitals admit people that certainly appear crazy, but they admit others who seem quite normal. Psychiatry, an unreliable science at best, cannot yet distinguish between the more and less sick. What really distinguishes arriving mental patients from the general public is less a common condition than the fact that the individual "somehow gets caught up in the heavy machinery of mental-hospital servicing." In Edwin Lemert's phrase, patients are less victims of some reliably diagnosed disease than of "career contingencies," such as the energy with which their relatives pursue their commitment. "But once started on the way, they are confronted by some importantly similar circumstances and respond to these in some importantly similar ways" (Goffman 1961, 129). That is, the career of a mental patient begins, and this has relatively uniform effects on individuals apart from their mental condition on entering the hospital.

Much of Goffman's fascinating book describes a socialization process in which people learn to behave as mental patients. On admission they are alienated from friends and from conspiring hospital staff. They undergo a degradation ceremony in which their possessions, clothes, and other symbols of self are taken away. Staff supervision robs them of freedom, invades their personal space, and conveys a judgment of incompetence. The hospital's ward system manipulates small privileges in a way that constrains the person to accept official views of his or her craziness. The main thing that the person learns is how to make out under these circumstances by ingenuously playing the system, by learning "to practice before all groups the amoral acts of shamelessness" (Goffman 1961, 169).

But mental patients are not crazy in the specific sense that they submit to all this compliantly. Playing the system creates an "underlife of a public institution," all those adjustments and subversions by which people maintain some privacy, sense of efficacy, and self-respect. They break the rules in small ways: smuggle liquor into the hospital, pursue romantic liaisons with other patients or community volunteers, ridicule therapy sessions, keep stashes of personal or illicit possessions, find safe and unsupervised places to meet, and develop an underground system of communication. Thus, rebellion in all these ways is actually what is taught in the mental hospital. Unwittingly, the staff denies the patient many of the perquisites of self-respect that are then regained in subversive underlife activities.

The important implication of this moral conflict is that the subversion of institutional rules is caused by other rules. In their hospital career, patients are deprived of many features of their selves. To the extent that they grudgingly cooperate with this process, patients enable the institution to perform its official functions (conducting the therapy sessions, or therapeutic "dances"). But in exchange for patient cooperation, the staff lays itself open to reciprocal demands; it bargains cooperation for illicit privileges such as reduced supervision or a blind eye to contraband. As a result, the institutional system corrupts itself, and the self that results from all this is an institutional product. Goffman states that

> each moral career, and behind this, each self, occurs within the confines of an institutional system, whether a social establishment such as a mental hospital or a complex of personal and professional relationships. The self, then, can be seen as something that resides in the arrangements prevailing in a social system for its members. The self in this sense is not a property of the person to whom it is attributed, but dwells rather in the pattern of social control that is exerted in connection with the person by himself and those around him. This special kind of institutional arrangement does not so much support the self as constitute it. (1961, 168)

With his concepts of moral career and the institutional self, Goffman takes us a long way toward a new way of understanding and dealing with mental illness. Goffman's work is read and used by people interested in reforming psychiatric treatment and by advocates of "deinstitutionalization," or the elimination of hospitals and reformatories in preference to community-based treatment.

Goffman reflected on what his research implied for other kinds of institutions. Mental hospitals resemble in some respects other institutions where people live, work, and pass their leisure all in the same place. He calls these "total institutions" and cites other examples such as the military, prisons, boarding schools, and monasteries. Now, if this much is valid, Goffman reasons, then the institutional self produced in those institutions should bear essential similarities to the moral career of the mental patient. A good deal of evidence supports his inference. Army recruits begin their military careers with degradation ceremonies in bootcamp. Novitiate monks are stripped of their personal possessions along with their mortal corruptions. Prisons generate an underlife of illicit activities, a special argot, and open rebellion. The total institution is another conceptual invention that leads research to unexpected factual results.

THE SOCIOLOGICAL METHOD

Interpretation of the four studies unabashedly joins Marx and Goffman, or Durkheim and Lipset, because they are all skillfully engaged in a similar enterprise. If the four research projects were classified by the usual categories of method, one would be considered historical research (Marx), one participant observation (Goffman), and two survey or quantitative work (Lipset and Durkheim). That classification would miss the more interesting similarities, which also seem to capture the purpose and flavor of the work.

Let us recall the similarities that run through all these exemplars:

1. Each began with a clear question and important problem: Why does the suicide rate vary with social factors? Why does democracy survive in one exceptional union?
2. Each investigator invented an explanation for the problem and summarized that in a key concept (surplus value, occupational community, moral career).
3. Each study pursued a critical approach by challenging conventional ideas, reversing the usual assumptions, and going to the root of the question rather than treating its superficial correlates.
4. Each study employed multiple research strategies in complementary ways. Marx consulted the surveys by factory inspectors to test his historical predictions; Durkheim statistically analyzed official records but also investigated the literature on religion; Lipset engaged in years of historical work before his survey; Goffman participated in the mental hospital routines and then read widely about organizations of other total institutions.
5. Finally, each study is explicitly or implicitly critical and thereby sympathetic to social reform — the investigators had clear value commitments. Marx and Durkheim are explicit about the changes they propose to advance equality and freedom. Lipset is concerned about the fate of organizational democracy. Goffman is bothered by the amoral experiences of a mental patient's career, and reformers have supported deinstitutionalization with his

> results. These investigators vary in the extent to which they carry the lessons of their research into social action, a choice they have every right to make. All, however, are concerned about social ills and the practical applications of their work.

Sociological work at its best involves inventing explanations for important problems and finding out whether they are true. Concepts summarize observations of social life in distinctive ways, and they embody implications about other things that should be observed if the concepts are fertile. The sociological method, subsuming a variety of techniques, is a process of investigating whether explanations and implications fit the known facts and generate new ones. This is the purpose and the beauty of method, what makes it at once inseparable from theory and a challenge to imaginative detectives in search of critical facts. When methodology is lifted out of the sociological craft and turned into a specialized technique, both theory and method suffer — the first because it cannot move into the world of experience and the second because it wanders there aimlessly.

The work that sociologists do testifies to the unity of the sociological method — inquisitive and critical work that, in the end, is useful because it convinces us that we now understand something that was once baffling. Edmund Wilson's description of the passion that drives writers to reinterpret their times applies as well to sociologists:

> With each such victory of the human intellect, whether in history, in philosophy, or in poetry, we experience a deep satisfaction: we have been cured of some ache of disorder, relieved of some oppressive burden of uncomprehended events. ([1940] 1983, 353)

SELECTED BIBLIOGRAPHY

BANFIELD, EDWARD C. 1968. *The Unheavenly City.* Boston: Little, Brown.

BECKER, HOWARD S. 1982. *Art Worlds.* Berkeley: University of California Press.

BECKER, HOWARD S., and IRVING LOUIS HOROWITZ. 1972. "Radical Politics and Sociological Research: Observations on Methodology and Ideology." *American Journal of Sociology* 78:48–66.

BELLAH, ROBERT N., RICHARD MADSEN, WILLIAM M. SULLIVAN, ANN SWIDLER, and STEPHEN M. TIPTON. 1985. *Habits of the Heart: Individualism and Commitment in American Life.* Berkeley: University of California Press.

BOTTOMORE, T. B., and MAXIMILIEN RUBEL. 1956. *Karl Marx: Selected Writings in Sociology and Social Philosophy.* New York: McGraw-Hill.

DAHRENDORF, RALF. 1958. "Out of Utopia: Toward a Reorientation of Sociological Analysis." *American Journal of Sociology* 64:115–27.

DURKHEIM, ÉMILE. 1897. *Suicide: A Study in Sociology.* Reprint. Translated by John A Spaulding and George Simpson. Glencoe, IL: Free Press, 1951.

GOFFMAN, ERVING. 1961. *Asylums: Essays on the Social Situation of Mental Patients and Other Inmates.* Garden City, NY: Anchor Books.

KAPLAN, ABRAHAM. 1964. *The Conduct of Inquiry: Methodology for Behavioral Science.* San Francisco: Chandler.

LAZARSFELD, PAUL F. 1949. "The American Soldier: An Expository Review." *Public Opinion Quarterly* 13(3):378–80.

LIPSET, SEYMOUR MARTIN. 1964. "The Biography of a Research Project: Union Democracy." Pp. 111–39 in *Sociologists at Work: The Craft of Social Research,* edited by Phillip E. Hammond. Garden City, NY: Anchor Books.

LIPSET, SEYMOUR MARTIN, MARTIN A. TROW, and JAMES A. COLEMAN. 1956. *Union Democracy: The Internal Politics of the International Typographical Union.* Glencoe, IL: Free Press.

MANNHEIM, KARL. 1936. *Ideology and Utopia: An Introduction to the Sociology of Knowledge.* New York: Harvest Book.

MARX, KARL. 1867. *Capital,* Vol. 1. Reprint. Translated by Ben Fowkes. New York: Vintage Books, 1977.

MERTON, ROBERT. 1957. *Social Theory and Social Structure,* rev. ed. Glencoe, IL: Free Press.

MERTON, ROBERT, and ALICE ROSSI. 1957. "Contributions to the Theory of Reference Group Behavior." Pp. 279–334 in *Social Theory and Social Structure,* by Robert Merton. Glencoe, IL: Free Press.

MILLS, C. WRIGHT. 1951. *White Collar: The American Middle Classes.* New York: Oxford University Press.

———. 1959. *The Sociological Imagination.* New York: Oxford University Press.

MURRAY, CHARLES. 1992. "The Lagacy of the 60's." *Commentary* 94:23–30.

OBERSCHALL, ANTHONY. 1968. "The Los Angeles Riot of August 1965." *Social Problems* 15:322–41.

RUTTEN, TIM. 1992. "A New Kind of Riot." *The New York Review of Books* June 11:52–54.

STINCHCOMBE, ARTHUR L. 1968. *Constructing Social Theories.* New York: Harcourt Brace.

TOCQUEVILLE, ALEXIS DE. 1835. *Democracy in America.* New York: Vintage Books Edition, 1959.

WILSON, EDMUND. 1940. "The Historical Interpretation of Literature." Reprint. Pp. 338–53 in *The Portable Edmund Wilson,* edited by Lewis M. Dabney. New York: Penguin Books, 1983.

PART II

SEGMENTATION

3

Country and City

Strip mining brought jobs and a temporary boom to Appalachia but destroyed farms and pushed once independent mountain people off the land and into the "little Kentucky" slums of Cincinnati, Baltimore, Chicago, and Detroit. (© Arthur Tress)

THE PEOPLE

Hillbillies tell one of the diagnostic stories of urbanization. They tell it literally in oral histories of their experience and figuratively in what they represent as a social category. If we hope to understand the process that transforms rural societies into urban–industrial worlds, we must be able to explain the fate of hillbillies and similar people.

Urbanization, the growing concentration of a population in cities, is a story of the displacement and social reorganization of people: farmers, immigrants, small-town folk, and the colorful but disappearing breeds of lumberjacks, miners, and cowboys. The story of hillbillies is a revealing cross section of this process — an experience of traditional rural independence, regional despoliation, struggle for a self-sufficient way of life, and defeat at the hands of a centralizing corporate society. In many ways, the history of hillbillies parallels the transformation of North America summarized in the master process of urbanization. An imaginative line connecting the first British explorations of the Carolina coast in the 1580s and urban life in the 1980s would run straight through the society of southern Appalachia.

A sociological understanding of this experience begins by reflecting on the term *hillbilly*. In its present general usage, the term refers derisively to people of the southern mountain regions — their alleged rascality, lack of culture and education, loose morals, fondness for alcohol, clannishness, and penchant for living in the trash-infested "little Kentuckys" of northern cities. With the exception of a slumming indulgence of hillbilly music, the popular image is pejorative. That stereotype, however, is recent and at odds with the regional origins of the term.

As late as 1900, the *New York Journal* felt obliged, with some indifference to geography, to inform its readers that "a Hill-Billie is a free and untrammeled white citizen of Alabama, who lives in the hills, has no means to speak of, dresses as he can, talks as he pleases, drinks whiskey when he gets it, and fires off his revolver as the fancy takes him." Yet the term has a longer and more complimentary provenance. Southern Appalachia (the western Virginias and Carolinas, eastern Kentucky and Tennessee, and northern Georgia) was populated beginning in the late seventeenth century by British escapees and freed bondsmen from the eastern seaboard plantations. The hardy survived in this wilderness by adopting the ways of their Choctaw and Shawnee Indian neighbors — clothes made from the skins of game they hunted, the tomahawk, and Indian corn, which they grew and distilled with the Scottish whiskey brewer's skill. Like Daniel Boone, they lived in family clans no closer than sighting distance of their neighbors' chimney smoke. British fondness for the name William, Bill, and the diminutive Billy provided the eponym for the society of hill people — "hillbillies" in local parlance was the affectionate counterpart of the "mountainmen" and women who earned the nation's esteem in the Revolutionary War.

As the rustic and insular society prospered, the more successful river-bottom agriculturalists acquired a few slaves and became merchants. Incipient classes and neighbors were divided by the Civil War: yeomen farmers and frontiersmen siding with the union and prosperous farmer-merchants with the Confederacy. Legendary feuds began with the depredations and vendettas of the war, imposed as they were on an austere and clan-centered society. Whiskey production was the only source of cash income and commerce with the external world. Indeed, the isolation and feuding of kinship clans led to inbreeding and a small factual basis for the image of mental retardation. As the *New York Journal* piece suggests, however, the ridiculing hillbilly stereotype was still ill-formed in 1900.

Apparently, the shift in the nation's esteem from mountainman to hillbilly occurred with the growing commercial contact between Appalachian society and the industrializing midwestern and northern cities. In the 1870s, national demand for hardwoods brought timber buyers into the mountains who dealt the guileless frontier people out of their forests. Later, particularly from 1912 to 1927, rapacious coal companies bought mineral rights for trifling sums and converted hillbillies into rent-paying, commissary-indebted miners living in company towns. No wonder that mountain people, lured by the opportunity for a regular cash income exceeding anything they had known, came to be called dumb hillbillies by the industrial North. Now the stereotype flourished with the lovable morons of the "Li'l Abner" comic strip, the querulous Hatfields and

McCoys, the fatuously clever "Beverly Hillbillies" of television, and the barbarians of James Dickey's novel and film *Deliverance.*

The southern Appalachian coal boom was stilled by the Great Depression, but out-migration was small simply because there was no place to go for a job. Instead, "the relief" (welfare payments) came to the mountains and stayed, converting the once independent frontier people into an enclave of welfare dependents. The mines revived during World War II, but afterward they were increasingly mechanized. Successful struggles by the United Mine Workers from the 1930s onward encouraged the non-conglomerate coal companies to eliminate as many miners and company towns as possible. So began the mass exodus to the "little Kentuckys" in Baltimore, Cincinnati, Chicago, Detroit, and points beyond. Harry Caudill, in his masterful *Night Comes to the Cumberlands: A Biography of a Depressed Area,* observes, "The same industry which required seven hundred thousand men to provide the nation's coal in 1910 was able to provide all of the same fuel required by a vastly larger nation in 1958 with fewer than two hundred thousand men" (1962, 263). With the introduction of strip mining in the 1950s, most of the remnants of a culture were bulldozed away with the coal, soil, towns, and people.

Southern Appalachian migrants found jobs but not better lives in the cities to the north. Families and women were strained by the confinement of working-class neighborhoods, as Harriet Arnow's novel *The Dollmaker* (1954) about a Kentucky family in Detroit showed. Jobs were often scarce, temporary, and low paying for those equipped, if at all, with a miner's skills. In a more recent collection of interviews by Todd Gitlin and Nanci Hollander, *Uptown: Poor Whites in Chicago,* hillbillies tell their own stories. John Dawson, who found work through "labor [contracting] companies" describes employment conditions:

> They's a poor class of people that come to Chicago and come to different cities like Detroit, Chicago, New York, Los Angeles, and places like this that they thought that when they got there that they could just walk in and get a job of work that would pay 'em big money, but places like these slave-labor markets hurt a lot of those people trying to make a decent living for their families. There's a lot of people that comes here with a family and they find out they

> can't get a job where they can make a living and don't have enough money to pay for theirselves. All right, they hop over here and they think, "Well, I can make a living in this place here," and they don't know that the place is only paying one-twenty-five, one-thirty an hour till they get out there. Well, in fact they wind up with about a dollar an hour when they get off the job. They cain't feed their family. Well, they get the attitude, the hell with it, and that's the cause of a lot of people from the South that just don't give a damn. . . .
>
> It wasn't like it is back home. Most of the class back home, you walk in a plant whether they know you or not and if they give you a job you can explain to the personnel manager that you had to have some money at the end of the week or had to have some money the next day after you work. Well, he would make arrangements for you to get money to live on till you drawed a payday. Well, these plants here won't do that, for they've got this cutthroat with these slave-labor markets here, they get their labor to them much cheaper. (1970, 90–91)

Dawson lived in an uptown apartment where the landlord failed to provide heat and where Dawson's child contracted lead poisoning from paint peelings. After a fight with the landlord, he moved to another apartment.

> At that time we had a nice manager there, when we moved in. We stayed there about four months, very nice manager. This woman she was from Kentucky, had some kind of sickness and the doctors advised her to go back to a different climate, you know. And she left and went to a different climate. So they got one of these damn smart-asses in to manage the buildin. And she had the attitude that southern people were all dogs. So I had the attitude that she was just a sonofabitch. And I just told her there in the hall one afternoon, she was runnin her damn mouth, and I just told her in the hall, I said, "Well, what in hell you think you're talking to?" I said "I ain't no dog." I said, "I'm makin more money right now than you make." I said, "I ain't askin for a damn thing." I said, "I ain't got to rent in this buildin." (1970, 84)

Hillbilly migration was just one stream that joined a current of blacks, Hispanics, motley immigrants, and other poor whites. The United States became a rapidly urbanizing society during the decades around the turn of the century. In *The Grapes of Wrath,* John Steinbeck describes the migratory flight from Oklahoma's dust bowl in the depression years. Okie tenant farmers, evicted "under the tractor's blade," took to the highways: "Highway 66 is the main migrant road . . . the path of a people in flight, refugees from dust and shrinking land, from the thunder of tractors and shrinking ownership" (1939, 127–28).

Hillbillies and okies describe one important side of the urbanization story but certainly not the only side. Other people from comfortable circumstances were drawn to the cities by hopes of social mobility. Flights from oppression by Russian Jews, or from famine by Irish peasants, combined with the English weavers' pursuit of opportunities to ply their acquired industrial skills or, in the phrase of U.S. sojourners from small towns, to "get into something."

This illustration based on the hillbilly experience demonstrates the central theme of this chapter: Urbanization is a social process, a basic transformation in the lives of individual migrants and in the character of society. Whatever the group motives for migration, urbanization reorganizes individual and social life. The seemingly neutral urbanization process generalizes a great many forces that break up rural societies and shape the emerging cities in decisive ways. Understanding that process requires a close examination of the conditions that produce urbanization rather than exclusive attention to what happens to people once they reach the city. City and countryside help explain one another — there is no adequate urban sociology apart from a rural sociology. The hillbilly experience, for example, encapsulates an important part of the meaning of urbanization and urban life. Urbanization is a process, a global process and the hallmark in many ways of the modern world. But it is also an abstraction built from the uprooting experience of hillbillies, Okies, peasants, workers, and sojourners around the world.

URBANIZATION

The first great social division is between country and city. Marx once remarked that "the foundation of every division of labor which has attained a certain degree of development . . . is the separation of town from country" ([1867] 1977, 472). The American anthropologist Robert Redfield (1954) believes that the appearance of cities marks a "revolutionary change" in human history. "In fact the story of civilization may then be told as the story of cities" ([1954] 1969, 206), a task made possible, he thinks, by contrasting folk society and modern urbanized society as two ideal models. About one-half of the world's population and three-quarters of the United States's live in urban areas today; just 100 years ago, the figure was about 3 percent for each. Few changes in the modern world have been so dramatic for their speed and effect on the way we live.

Urbanization refers to the **redistribution** of a population, a shift in the proportion living in cities rather than in villages and rural areas. Urbanization therefore involves a distributional change in the population of some geopolitical unit — whether that unit is a country, a state, or the world as a whole. Because urbanization is defined by the shifting percentage of a population residing in cities versus rural areas, its measurement requires specification of what constitutes a city. This is no simple matter because the notion of a city conveys something physical (the size and density of a population) and something social (an urban culture or way of life). The issue needs treatment in two parts, beginning here with a description of urbanization as a worldwide population phenomenon. The next section focuses on how sociology conceives of urbanism as a social form.

Two facts about the modern world, and particularly the twentieth century, distinguish it from much of recorded history: It is predominantly an urban world, and it got that way rapidly. Demographers, who study populations and their change, have used two standards for measuring world urbanization. The first is a fixed standard applied to all countries. Defining urban places as settlements with 20,000 or more inhabitants, in 1800 just 3 percent of the world's population was urbanized. During the nineteenth century, partly as a result of industrialization, this began to change, affecting first the more developed parts of the globe in Europe and North America. In the twentieth century, this change accelerated. The demographer Kingsley Davis puts in pointedly:

> Compared to most other aspects of society — e.g., language, religion, stratification, or the family — cities appeared only yesterday, and urbanization, meaning that a sizable proportion of the popu-

lation lives in cities, has developed only in the last few moments of man's existence. (1955, 429)

Table 3.1 shows that in 1920 there were 266 million people (of the then-total world population of 1.86 billion) living in cities of 20,000 or more. Most of these were in the more developed or industrialized regions. This absolute number represented 14.3 percent of the world's population. The more developed regions had 29.8 percent of the populations in cities and the less developed regions 6.9 percent.

From 1920 to 1985, enormous changes took place. In just sixty-five years, the total world population more than doubled (from 1.86 to 4.89 billion, an increase of 263 percent), while the urban population increased eightfold (from 266 to 2034 million). This combination meant that in percentage terms the world's urban population increased from 14.3 to 41.6 percent. The more developed areas, led by North America, passed the 70 percent mark in urbanization, while the less developed

TABLE 3.1 World Urban Population and Population Growth Based on the Fixed Standard of Urban Places Comprising 20,000 or More Inhabitants, by Region: 1920–1985.

	1920	*1930*	*1940*	*1950*	*1960*	*1970*	*1985*
World population total (absolute numbers in billions)	1.86	2.07	2.30	2.52	2.99	3.58	4.89
World urban total (absolute numbers in millions)	266	338	432	533	760	1010	2034
World total (%)	14.3	16.3	18.8	21.2	25.4	28.2	41.6
More developed							
Major areas (%)	29.8	32.8	36.7	39.9	45.6	49.9	71.5
Europe	34.7	37.2	39.5	40.7	44.2	47.1	71.6
North America	41.4	46.5	46.2	50.8	58.0	62.6	74.1
Soviet Union	10.3	13.4	24.1	27.8	36.4	42.7	65.6
Oceania	36.5	38.0	40.9	45.7	52.9	57.9	71.1
Less developed							
Major areas	6.9	8.4	10.4	13.2	17.3	20.4	31.2
East Asia	7.2	9.1	11.6	13.8	18.5	21.7	28.6
South Asia	5.7	6.5	8.3	11.1	13.7	16.0	27.7
Latin America	14.4	16.8	19.6	25.1	32.8	37.8	69.0
Africa	4.8	5.9	7.2	9.7	13.4	16.5	29.7

SOURCE: W. Parker Frisbie. 1977. "The Scale and Growth of World Urbanization." In *Cities in Change*, 2d ed., edited by John Walton and Donald E. Carns. Boston: Allyn and Bacon; and *The Prospects of World Urbanization Revised as of 1984–85,* 1987. Population Studies No. 101. New York: United Nations.

areas, led by Latin America, urbanized even faster — although they started from and reached lower overall levels (from 6.9 to 31.2 percent). From 1800 to 1970, world population increased four times, but the number living in cities increased fifty times. As a result, the urban population of 1970 was larger than the entire world population of the early nineteenth century, and that number doubled again by 1985.

Table 3.2 introduces the second, variable standard that demographers use to measure world urban population. Here urban means those places that nations themselves consider cities. The evident result of this self-defined standard is to raise the percentages (e.g., from 14.3 to 19 percent urban in 1920). Another difference in Table 3.2 is that past trends are used to estimate urban population growth through the year 2000. The results are bracing but not incredible. Although the percentages are higher than those in Table 3.1, the trend is much the same: a steady increase in worldwide urbanization, led by the developed areas, and a narrowing gap between the more and less developed. In both tables, the 1920 figures show the developed regions roughly five times more urbanized. By 2000 the difference will be just double. If the estimate is right, by the year

TABLE 3.2 Estimated Percentages of Urban Population as Nationally Defined, in the Total Population of the World and Major Areas: 1920–2000

Major Area	*1920*	*1940*	*1960*	*1980*	*2000*
World total	19	25	33	46	51
More developed areas	40	48	59	70	80
Europe	46	53	58	65	71
North America	52	59	70	81	87
Soviet Union	15	32	49	68	85
Oceania	47	53	64	75	80
Less developed major areas	10	14	23	32	43
East Asia	9	13	23	31	40
South Asia	9	12	18	25	35
Latin America	22	31	49	60	80
Africa	7	11	18	28	39
More developed regions	39	47	60	71	81
Less developed regions	8	12	20	30	41

SOURCE: *Growth of the World's Urban and Rural Population, 1920–2000,* 1969. Population Studies No. 44. New York: United Nations, Department of Economic and Social Affairs.

2000 the less developed areas will be as urbanized as the industrialized world was in the 1950s.

Urbanization therefore is a massive, world process that uproots people and reorganizes societies. It suggests that an ensemble of new and adaptive experiences, including the hillbilly's, characterizes millions of people in recent history and at present. Urbanization also refers broadly to the advent of a new kind of society, economy, and culture. But what kind?

THE CITY

The nature of the city and the social meaning of urbanism have always engaged sociologists. Lively debates focus on whether urbanism represents a society coercively organized according to the needs of capitalist economic development — and therefore an experience of personal alienation from traditional bonds — or a new culture of liberation from rural and small town parochialism — the touchstone of modernity. For some, urbanism is a segmented and impersonal way of life, for others a blessed toleration of diversity. From debates and research prompted by such questions, a coherent urban sociology has emerged, based on the classical tradition and recent accomplishments in urban political economy.

Defining Features

What is a city? The meaning of that term is far from obvious. Cities have existed for perhaps 5000 years, or so the legends and archaeological remains of the Middle East suggest. Cities were the centers of great civilizations in Egypt, India, and Mayan Central America before the rise of Athens, Rome, and the other European latecomers to the urban world. Is there a conception of the city that embraces these historical forms and the modern metropolis?

Max Weber posed this question in a collection of essays on *The City* ([1921] 1958). He noted the existence of many definitions that "have only one element in common: namely that the city consists simply of a collection of one or more separate dwellings but is a relatively closed settlement" (p. 23). Most definitions center on the size and density of a settlement, but

Weber rightly observed that "in terms of what it would include and what it would exclude size alone can hardly be sufficient to define the city" (p. 23). For example, small but densely settled commercial centers could be excluded and large sprawling settlements included by this standard. Instead, Weber looked to the activities of cities as the hallmark of their distinctive characteristics. Of these he discovered three.

First, cities are settlements where there is a regular exchange of goods. "The market becomes an essential component in the livelihood of the settlers." No settlement could be called a city unless its inhabitants possessed a local market that met their daily needs with locally produced or acquired goods. "In the meaning employed here the 'city' is a market place" (pp. 23–26).

Although economic activity is a necessary component of the definition, alone it is incomplete. "The mere fact that merchants and tradesmen live crowded together carrying on a regular satisfaction of daily needs in the market does not exhaust the concept of the city." Second, and historically, cities have also been the locus of political and administrative authority over the urban hinterland, and that authority is used in ways that ensure the prosperity of the local market. In Weber's words, the "politico-administrative aspects are conceptually fused with pure economic aspects and conceived as forming one whole" (p. 31). Indeed, the "fusion" of market and political authority is essential for the city's survival, growth, and prosperity. Cities characteristically develop practices, which include "regulations in the interests of permanently and cheaply feeding the masses and standardizing the economic opportunities of tradesmen and merchants" (p. 31), for dealing with the countryside.

With reference to the first two activities, Weber spoke of "the city as the fusion of fortress and market . . . the politically oriented castle and economically oriented market" (pp. 35–36). The third element is a population of citizens who enjoy rights and status of their association, what Weber called the "urban community." The urban citizenry are united in a corporate unit administered by authorities whom they elect.

In sum, a city has three defining features: market, administrative center, and community. The virtue of Weber's definition is the fusion of demographic (population) and economic conditions with a special form of association, or community, and the recognition that these become fused in political activity. Cities do not simply well up on the land. They are created by groups of people with political and economic aims. This should be the central proposition of any urban sociology, and that it has sometimes been forgotten has only meant that it is periodically rediscovered, most recently in a renewed emphasis on urban political economy.

If Weber was right, the rapid, even revolutionary, changes in world urbanization must have produced far-reaching changes in economy, politics, and social organization. Specifically, political and administrative practices must have been devised to regulate the urban marketplace and to organize the urban community. Urban social research in the early twentieth century, however, understood this only vaguely. The sociological study of cities did not begin with any clear or singular conception of its subject. Much of that research was produced in the United States where Weber's ideas were slow in gaining recognition.

Urban Ecology

More important, the pragmatic North Americans focused on urban social problems, which had a special urgency about them in the political and intellectual climate of progressivism (Chapter 1). A rich and energetic urban sociology developed in this setting, particularly in Chicago where reform-minded researchers gathered at Hull House, the University of Chicago, and in the city's motley ethnic neighborhoods. Chicago sociologists had a natural laboratory and a purpose. Some of their work was scientifically pretentious or conceptually misplaced, but it began to provide the raw material from which our current understanding is fashioned.

One wing of the Chicago School pursued a scientific description of basic urban forms and processes. In search of a proper metaphor from the physical sciences, Robert Park, Ernest Burgess, and Roderick McKenzie (1925) hit on the notion of ecology — the science of relationships between organisms and their environment. They proposed an urban ecology in which the city landscape, perhaps even its institutions and ways of life, could be understood as adaptations to the environment. In this naturalistic metaphor, the city is an agglomeration of people in space governed by ecological processes. Human communities perform in a spatial setting the essential functions for survival such as production, distribution, and consumption. The form and life of the city arise from processes of dominance and competition that govern the functions.

Illustratively, Burgess and his colleagues depicted the city as a set of concentric circles and zones: the innermost devoted to business and commerce, industry and worker housing in the next zones of transition, followed by a ring of middle-class neighborhoods, and upper-class residences at the periphery (Figure 3.1). Whether the picture fitted precisely every city, the point was that urban life could be explained by ecological competition over land use and function.

After unsuccessful efforts to find a similar land-use pattern in cities other than Chicago, new models were proposed. A sector theory said that the concentric rings were sometimes cut by wedges that form as cities grow and, particularly, as the lower class moves into areas abandoned by the upper class fleeing the city's bustle to the outer rings or suburbs. A multiple-nuclei theory countered that urban growth leads to a patchwork of specialized zones rather than a geometric design; some activities benefit from contiguity, and others want insularity. Different models seem to fit different cities and histories of urban growth, casting doubt on the explanations given for the generality of each. Underneath all these portraits, however, was the basic (some even said the subsocial) idea that urban form obeyed certain physical or economic *forces* — that it followed a general process governed by dominance and competition.

With its appearance of rigor and tangible results, urban ecology continued to have a strong impact until research began attacking its conceptual foundations. While some research offered more refined land-use models, other work claimed to uncover specific causes for variety in the urban landscape. For example, an important study by Everett C. Hughes (1928), a Burgess student, observed that Chicago did, indeed, have a clear-cut land-use pattern, but one did not have to resort to mysterious, subsocial forces to explain it. The pattern reflected

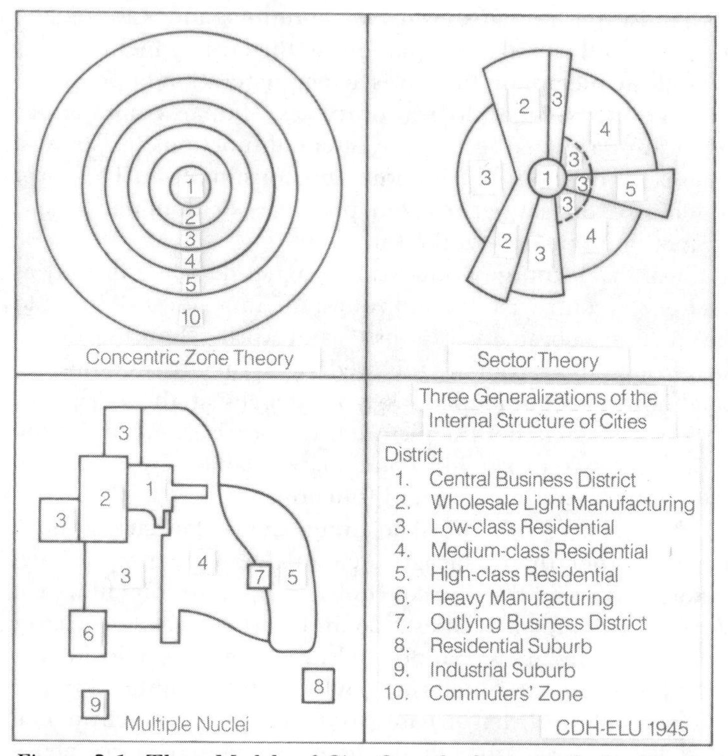

Figure 3.1 Three Models of City Growth. SOURCE: C. D. Harris, and Edward L. Ullman. 1945. "The Nature of Cities." The Annals *242:7–17.*

instead the interests and political muscle of the Chicago Real Estate Board. Business influence and realtor practices determined which zones were set aside and promoted for specific uses, including where the lower classes and minorities should live and where they were prevented from living. To the surprise of no informed citizen, city politics had more to do with urban structure than ecological competition. Another study of Boston by Walter Fiery (1947) noted that "sentiment and symbolism" were the reasons for maintaining the Boston Commons in the middle of prime urban real estate and an exclusive residential area in the heart of the city. The urban form is determined in a market of property values, but that market is decisively affected by politics, civic values, and, no doubt, by conflicts between those two.

In short, research along the lines of Weber's ideas about the marketplace, political control, and the urban community appears more discerning than other accounts based on population growth and naturalistic metaphors. The notion of an "urban ecology" was appealingly simple and stimulated a great deal of informative descriptive work. Its drawback was a failure to *explain* anything, and that stemmed from its retreat into the vagaries of some supposed "master process" (here, dominance and competition). This is not to say that the research of the early, ecologically minded urban sociologists was wrong. It only needed to be fitted to a different explanatory model — one that was only beginning to emerge at the time.

Urbanism

Another related wing of the Chicago School of urban studies was advanced by W. I. Thomas, Park himself, and preeminently by Louis Wirth. Wirth was part of the generation of progressive reformers in Chicago during the early twentieth century, and his special interest was in the social psychology of city life — "urbanism as a way of life" in his words. Wirth took one leaf from Weber's book by defining "the city as a particular form of human association." His influential view of what human association involved in the urban setting was that

> characteristically, urbanites meet one another in highly segmented roles. They are, to be sure, dependent upon more people for the satisfaction of their life-needs than are rural people and thus are associated with a greater number of organized groups, but they are less dependent upon particular persons, and their dependence upon others is confined to a highly fractionalized aspect of the other's round of activity. This is essentially what is meant by saying that the city is characterized by secondary rather than primary contacts. The contacts of the city may indeed be face to face, but they are nevertheless impersonal, superficial, transitory, and segmental. The reserve, the indifference, the blasé outlook which urbanites manifest in their relationships may thus be regarded as devices for immunizing themselves against the personal claims of others. ([1938] 1969, 152–53)

Wirth equated this condition with Émile Durkheim's description of anomie. Yet, as he went on to describe the effects of urban density and heterogeneity, he noted that "different parts of the city acquire specialized functions, and the city consequently comes to resemble a mosaic of social worlds in which the transition from one to the other is abrupt. The juxtaposition of divergent personalities and modes of life tends to produce a relativistic perspective and a sense of toleration . . . " (p. 155). There is a deliberate irony in Wirth's description of urban life. Urbanites are involved with more people and groups than rural folk, although that involvement is divided into many specialized relationships, none especially deep. Such, however, is the cost of toleration or freedom from the oppressive parochialism of rural life.

When Wirth generalized his social psychology to the urban landscape, the result was a similar variety of segmented communities. The many segmented roles of the individual urbanite had a direct parallel in the specialized communities that comprised the city. And the parallel extended further with the inference that segmentation produced ill effects. Specialized but fragile institutions took over industrial, educational, and recreational activities, causing "the weakening of bonds of kinship, the disappearance of the neighborhood, and the undermining of the traditional basis of social solidarity" (p. 160). Wirth envisioned what later writers would call a mass society, including a strong connection between "virtual impotence as an individual" and the appeal of mass movements.

Wirth stimulated a great deal of research with his provocative ideas and confident assertion that "all these phenomena can be substantially verified through objective indices" (p. 160). The overwhelming result of that research, however, was that urbanites are not the segmented, anomic beings he described. On the contrary, they have ties to local community institutions, a neighborhood life mirrored by the community press, and geographically extended networks of "primary contacts." Urbanism, in short, is a much more engaging and social way of life than the pessimistic theory suggested. Yet, if Wirth was refuted on that score, his more lasting influence was to encourage countless chronicles of the "mosaic of social worlds." Many of the classics in urban sociology were produced in this milieu — studies of the texture of city life as it appeared in the social world of hobos, the organization of streetcorner gangs, or the recreation afforded by taxi dance halls.

Wirth's contribution is not tainted by his pessimism or anti-urban views. Ironically, he characterized the city as anomic for generations of followers who dug up vivid evidence of community bonds. It is also true that urban life is harshly impersonal for those poor and black who are not welcome in the mainstream of occupational and civic life. But that is not what Wirth and other students of urbanism intended to explain. The problem with urban social psychology is not that it fails to observe the social disorganization of communities or the pains of underprivileged life, but that it falls back on oddly mechanical explanations for those facts, such as the purported effects of population density, size, and heterogeneity. Political and economic exploitation, for example, play no part in individual or community troubles.

The two wings of the Chicago School, with all their valuable research contributions, came to grief in the 1960s. An urban crisis suddenly appeared, which the earlier theories did not expect and could not explain. Urban riots spread from Watts to Washington, D.C., because of anger rather than anomie. Central cities were deteriorating into sprawling ghettos rather than zones of transition. Social problems had less to do with population heterogeneity than with race and class inequality.

The New Urban Sociology

The crisis generated a different approach — which writers began to call urban political economy (Logan and Molotch 1987), or simply the new urban sociology (Zukin 1980). The new approach not only builds on the classical tradition and the Chicago School but also attempts to explain urbanization itself — its timing, rate, location, causes, and consequences — as a historical process rather than taking it for granted as some sort of natural or evolutionary fact. The new urban sociology argues, with Weber, that cities are created by particular economic and political conditions. Among other implications, this means that the physical form and social organization of cities are different in socialist (Szlenyi 1983) and capitalist (Harvey 1985) societies. As a special case of capitalist urbanization, the United States has seen its cities develop according to the interests of local "growth machines" composed of investors, developers, and government officials (Molotch 1976). Yet the approach stresses *political* economy by noting that although "the substance of urban phenomena [are found] in the actual operations of markets . . . the nature of human settlement, including its market organization, is a product of social arrangements" (Logan and Molotch 1987, 17, 49). Markets function according to certain rules (including laws); those rules are created and governed by political choices. People experience urban life in ways that are shaped by these forces — in suburban communities by residents who organize politically to defend their property values (against minority buyers, drug rehabilitation centers, or toxic dump sites), for example, or in central-city neighborhoods by residents who demonstrate for jobs, better schools, and health-care services. Consciousness of urban life is fashioned by these concrete economic and political conflicts rather than by some vague notion of urbanism as a "way of life."

Specific urban processes such as residential segregation, community organization, or class and ethnic politics are understood in connection with the economy and patterns of political control. The new urban sociology deals with change, particularly as it grows out of conflicts between and among classes and status groups. These conflicts are the basis of the political process and, increasingly, the business of the state. Changes in the economy are politically generated and mediated. Political and social changes are in no sense independent of the economy. Studies of land use and the spatial distribution of productive or residential areas have a place in urban political economy, provided they are understood as a result of historically changing economic strategy and political influence rather than dubious ecological processes. There *is* an urban way of life, one situated in group action and urban social structure rather than flowing from population size, density, or heterogeneity. Community studies of the slum or suburb are the "blood" of urban structure, but they make something more than anecdotal sense only when we can explain why such divergent neighborhoods exist in the first place and how they are segregated.

Chapter 2 of this book suggested that the sociological method has less to do with discovering previously unknown facts than with developing the implications of facts from an interpretive standpoint. Accordingly, the definition and framework proposed for analyzing cities suggest that the process of urbanization will work itself out differently under historically different political and economic conditions. In a particular country such as the United States, the pattern of urbanization and the experience of urban life should change over time. Similarly, in broad comparisons of the developed and underdeveloped countries, which live under wholly different economic and political circumstances, cities should reflect a different structure. If the proposed framework does its job, we should be able to identify and explain these differences. The next section and another in Chapter 7 devoted to urbanization in the Third World explore that proposition.

URBANIZATION IN THE UNITED STATES

By world standards, urbanization in the United States is unique. In the historically brief span of 200 years since the first population census, the number of people living in cities has risen from virtually none to almost all — from 5 to almost 75 percent urban. These figures vary with the particular definition of urban, and the U.S. census offers several.

Dating from an earlier era, the census defines the "urban population [as] all persons living in urbanized areas and in places of 2500 or more inhabitants outside urbanized areas." Although this is a minimal definition more appropriate for the early nineteenth century, the census retains the category for purposes of comparison. Closer to present standards, an "urbanized area" is defined as a central city with a population of 50,000 or more and its surrounding closely settled territory or urban fringe. Yet even this is outmoded with the modern megalopolis that includes millions spread over areas that sometimes cross state lines. As a result, the U.S. Bureau of the Census developed the category "metropolitan statistical area," or MSA (formerly SMSA), corresponding to "a metropolitan area [having] a large population nucleus [the urbanized area], together with adjacent communities which have a high degree of economic and social integration with that nucleus." In 1990 there were 284 MSAs in the country (less than the 318 SMSAs enumerated in 1980, owing to changes in definition of the categories).

Table 3.3 traces the growth in "urban population" (the first definition above) from 1790 when the nation included just under 4 million people, of whom only 5 percent lived in towns or cities larger than 2500 people. The figures show a rate of urban growth and distinct periods of accelerated increase. Two cautions are necessary for reading the table. The precise definition of *urban* shifted with each census, and the total land area of the country kept expanding with annexation of frontier

TABLE 3.3 Urban Population of the United States: 1790–1990

	Total U.S. Population	*Percentage Urban*
Current Urban Definition		
1990	248,709,873	75.2
1980	226,545,805	73.7
1970	203,211,926	73.5
1960	179,323,175	69.9
1950	151,325,798	64.0
Previous Urban Definitions		
1960	179,323,175	63.0
1950	151,325,798	59.6
1940	132,164,569	56.5
1930	123,202,624	56.1
1920	106,021,537	51.2
1910	92,228,496	45.6
1900	76,212,168	39.6
1890	62,979,766	35.1
1880	50,189,209	28.2
1870	38,558,371	25.7
1860	31,443,321	19.8
1850	23,191,876	15.3
1840	17,069,453	10.8
1830	12,866,020	8.8
1820	9,638,453	7.2
1810	7,239,881	7.3
1800	5,308,483	6.1
1790	3,929,214	5.1

SOURCE: U.S. Department of Commerce, Bureau of the Census. 1991. 1990 *Census of Population, Characteristics of the General Population.* Vol. 1. Washington, D.C.: Government Printing Office.

territories. The latter had the effect of reducing the urban average because it added new people but seldom any towns of important size to the U.S. population total. With that in mind, the increasing urban population is all the more impressive, particularly in the period from 1840 onward. Generally, there are two major periods of urbanization: the first from 1850 to 1930 (especially the middle years of 1880–1910) and the second from 1950 to 1970. In the last decade or two, urbanization has slowed perceptibly, perhaps as a "ceiling effect" is reached. It is noteworthy, however, that in 1990 the urbanization in the United States topped the 75 percent figure.

The data for MSAs tell the same story, with 77.5 percent of the nation's population within these units. This indicates that the vast majority of urbanites live in the very large cities or, more precisely, in the cities and suburbs that together comprise metropolitan areas. The number of MSAs has nearly doubled in the last forty years, much of that growth resulting from suburban expansion. The Census Bureau estimates that by 1992 just over one-half of the nation's population will reside in the suburbs of metropolitan areas. The pattern requires historical explanation.

THE HISTORICAL DEVELOPMENT OF U.S. CITIES

The historical and contemporary experience of urbanization in the United States is usefully divided into four periods, each with its distinctive urban form, function, and conflicts: the **commercial city** (colonial times to 1850), the **industrial city** (1850–1930), the **corporate city** (1930–1960), and the **world city** (1960 to the present). Each period and urban form is described as it appears chronologically in a current of historically changing influences on the city. Distinct conflicts mark each period and its evolution to the next. The first three periods are adapted from a lucid article by David Gordon (1978) and used to summarize a rich history. The fourth period is contemporary, even futuristic, and is reserved for the following separate section.

In keeping with the framework, the U.S. city in each period is a result of the interacting forces of the economy, polity, and community. Changes in the economy in some ways demand reorganization of the urban system, but alone they do not determine it. Change is implemented in the political process that responds to economic demands and community conflicts. It is the resolution of these forces in distinct social and spatial forms that defines each period. Moreover, the transition from one period to another is marked by conflict among these spheres. The changing character of urban life and the nature of the evolved present require treatment from the standpoint of historical sociology.

During the first 200 years of U.S. history, four commercial cities dominated: New York, Philadelphia, Boston, and Baltimore. In 1790 the three port cities (New York, Boston, and Baltimore) included 67 percent of the country's urban population. The early cities were devoted principally to foreign trade and the commercial marketplace. Artisan crafts existed but no significant industry. Later, when the textile industry began to grow in the 1820s, it located in New England towns with available water power — in Lowell, Lynn, Waltham, Lawrence, and around Philadelphia.

The Commercial City

The commercial city centered around the port itself, the business of the docks and shipping. The occupational structure included merchants and their retainers, artisans, and laborers in transportation work (seamen, dockworkers). Most business establishments were small and combined work and the extended family in one place. Social classes and economic activities were closely integrated in a maze of narrow streets that crossed at irregular angles. Street life flourished and the city gave, in the words of historian Sam Bass Warner (1972), a feeling of the "unity of everyday life." The only obvious segregation placed the itinerant poor in shanties and rooming houses on the outskirts.

Under these conditions, the urban population grew slowly. Indeed, from 1690 to 1790, it actually dropped (from 9 to 5 percent) as a result of westward movement and new agricultural settlement. Until independence the number of cities held fairly constant because England granted charters for the legal establishment of towns and followed a policy of minimizing the number of commercial centers that might compete with their trade.

Social conflict in commercial cities focused on customs, trade restrictions, and the impressment of seamen into British maritime service. Pauline Maier's description of popular uprisings in the eighteenth century indicates that "when the press was of long duration, moreover, or when it took place during a normally busy season, it could mean serious shortages of food and firewood for winter, and a general attrition of commercial life that sustained all strata of society in trading towns" (1970, 12). In cities such as Boston, the urban mob frequently served as an extralegal arm of the community, preventing grain exports that might result in food shortages. An accumulation of such conflicts led to the American Revolution in which the commercial cities played a leading role.

After independence the cities expanded rapidly without any fundamental change in economic activity. Among the important commercial cities, only Philadelphia converted to a center of industry. In the absence of English control, big cities competed for trade dominance with one another and with newer commercial towns that began to flourish. The result was a growing prominence of New York, which enjoyed the special advantages of connections to inland markets through the Erie Canal. By 1860 New York controlled most foreign trade and took its present place as the country's financial capital.

The transition from commercial to industrial city in the mid-nineteenth century was produced by a number of intersecting forces. Until about 1830, U.S. industry was concentrated in the smaller New England towns and devoted mainly to textiles. As the country grew, industry diversified under tariff protections from European imports. A key problem for industrial development was the shortage of a skilled and disciplined labor force. The United States had not yet produced its own industrial working class. New England textile mills relied on British immigrants and, later, young women from rural areas who lived in company dormitories and operated spinning machines (e.g., the Lowell girls). The labor shortage was made worse by free public land in the West and what seemed an attractive alternative to tedious factory work.

Yet the methods of industrial production and the demands of the internal market were also developing. Steam-powered machines made the urban factory possible, and a protected, expanding home market promised successful diversification of manufacturing. Equally important, the state began building an urban infrastructure of roads, canals, and the all-important railway for the profit of private firms. Governments at all levels intervened in the economy with enormous subsidies to the railroads. Historian Carl Degler (1959) estimates that between 1850 and 1870 the state made cash loans to the railroads totaling $150 million and awarded them 180 million acres of public land as right-of-way — an area the size of most of western Europe.

At the same time, the first massive wave of immigrants was arriving. From 1820 to 1880, roughly 15 million people came to the United States, mainly from the British Isles and northwestern Europe. As Table 3.1 reflects, this period witnessed a sharp increase in the populations of the older commercial cities as well as the rapid development of new industrial market hub cities such as Chicago, Detroit, Minneapolis, Kansas City, St. Louis, and Omaha. Immigrants and railroads were essential ingredients of the industrial city, but the encouragement of both was a political policy choice.

Vast as these changes were, they neglect the most traumatic social and political reorganization that laid the basis for urban–industrial society, namely, the Civil War. The Civil War was fought over slavery, but slavery was only one manifestation of the conflicting economies of North and South. Northern industrial development by 1860 had reduced its dependence on textiles and, consequently, on southern cotton. As it diversified and moved toward the Midwest, industry looked for new markets and a political alliance with western agriculturalists, themselves fearful of competition from slave-based plantation agriculture. North and West struck a bargain based on mutual support for free public land and the tariff — an agreement that joined the prime interests, respectively, of the West and North in opposition to the South. Barrington Moore concludes that

> the focal point of the difference was slavery . . . [but] the more fundamental issue became more and more whether the machinery

of the federal government should be used to support one society or the other. . . . The American Civil War was the last revolutionary offensive on the part of what one may legitimately call urban or bourgeois capitalist democracy. (1966, 136, 112)

The Industrial City

With the end of the Civil War, the industrial city began to boom and to suffer periodic recessions. From 1870 to 1930, the U.S. population tripled, and the number of workers in manufacturing increased five times. The urban population increased from 20 to 56 percent. A major determinant of population and urban growth was the new or second wave of immigration drawn mainly from eastern and southern Europe (Italians, Greeks, Poles, Russians, Slavs in general, and many Jews). In just fourteen years, from 1902 to 1916, the number of new immigrants exceeded all that had arrived in the sixty years from 1820 to 1860.

The new industrial city relied on the controllable immigrant labor force. Industries preferred the big cities where labor was abundant, cheap, and easily disciplined. The U.S. working class was formed at this time largely from immigrant stock. And it was at this stage that the urban ecology described by the Chicago sociologists developed, through no accident, because Chicago was the prototypical new industrial city. Factories were centrally located along railroad lines and shipping canals. A business and shopping district formed the city center. Wedged among these two were the working-class neighborhoods as close as possible to the factory gate or trolley line. The middle and upper classes now preferred living at a peaceful distance from the noise, stench, and crowds in the center. Unlike the commercial city, the industrial city was segregated by social class and economic activity.

In *The Jungle* (1906), Upton Sinclair provided a memorable portrait of urban life in the industrial city and, specifically, of Chicago at the turn of the century. An extended family of Lithuanian immigrants settles in an ethnic community near the stockyards, buys (and later loses) an overpriced house on contract. Then a son drowns in a huge pothole in the street, and the family finds and loses jobs until its few survivors learn the corrupt ways of union and machine politics. The validity of Sinclair's grim picture is supported in the writings of Jane Addams, who created the first settlement house in the same neighborhood to deal with urban social problems. The work of Hull House, in turn, figured prominently in the first volume of the *American Journal of Sociology* published in Chicago in 1895. As the research of Thomas, Park, Wirth, and others filled in the factual foundations for novelists and reformers, a general and critical understanding of the industrial city grew out of the Chicago experience. Never were professional sociology and crucial developments in American history so closely connected.

Like the commercial city and its historical period, the industrial city reflected a characteristic urban form and a pattern of social conflict. The important conflicts, moreover, occurred in both urban and rural areas — the latter's fortunes being severely affected by urbanization. The reorganization brought by urban industrial society produced a generation of social conflict unprecedented before or since. From the late 1870s until World War I, a mortal struggle was fought in the countryside and the cities, the stakes of which were nothing less than the kind of society the United States was going to be.

In the Old South, Texas, and the Great Plains, an agrarian revolt rose against concentrated capital. Following the debt-financed Civil War, the federal government adopted a harsh deflationary monetary policy — the results of which benefited bankers and financial capital but severely contracted farm prices, credit, and income. Farmers were forced into a crop-lien system, pledging anything from 20 to 50 percent of their harvest in exchange for seed, fertilizer, and provisions from the "furnishing man" (or, to black farmers, simply "the man"). When the crop came in, many farmers were scarcely able to settle accounts, and others went further behind as farm prices continued falling. Large numbers lost their farms and were forced to migrate to the cities or to remain as tenant farmers in the service of local lenders who consolidated landownership. Farm tenantry and sharecropping, which had been infrequent and largely the fate of black farmers prior to 1870, rose to 25 percent of all farmers in 1880 and 28 percent by 1910. In selected regions such as the Old South, Iowa, and Kansas, half the farmers were reduced to tenants by the 1930s depression.

In the late 1870s, rural society responded with The Farmer's Alliance, a cooperative mass movement opposed to the railroads, credit merchants, and the money trusts. Alliance men and women set up cooperative stores, warehouses, credit bureaus, and procurement and marketing arrangements. A series of Alliance newspapers like the *Rural Citizen, American Nonconformist,* and *Progressive Farmer* argued their case to the general public, while hundreds of traveling lecturers carried to the small towns and farms the details of an Alliance Plan to reform the U.S. Treasury and save the family farm. In their political demands, the Alliance sought "such legislation as shall secure to our people freedom from the onerous and shameful abuses that the industrial classes are now suffering at the hands of arrogant capitalists and powerful corporations" (quoted in Goodwyn 1978, 53). They endeavored to unite with the Knights of Labor and the new labor union movement in a general strike against the railroad. With their own cooperatives, they boycotted local merchants and lenders. At its peak in 1890, the Alliance included 400,000 members in chapters from Georgia to the Dakotas.

In the end The Farmer's Alliance was defeated by government indifference, the opposition of finance capital, and the movement's own recourse to a People's Party that was absorbed into the Democratic party. Far from assisting the innovative cooperative or ameliorating the plight of the family farmer, government programs favored the more prosperous commercial farms and, later, the emergence of agribusiness. As Lawrence Goodwyn's study concludes,

> The collective effort of twentieth-century agricultural legislation . . . was to assist in the centralization of American agriculture at the expense of the great mass of the nation's farmers. The process of extending credit, first to the nation's most affluent large-scale farming interests, and then in the 1920s to sectors of the agricultural middle class — while at the same time denying it to a "whole class" of Americans who worked the land — had the effect of assisting large-unit farming interests to acquire title to still more land at the expanse of smallholders. Purely in terns of land-ownership patterns, "agri-business" began to emerge in rural America as early as the 1920s, not, as some have suggested, because large-scale corporate farming proved its "efficiency" in the period 1940 to

> 1970. In essence, "agri-business" came into existence before it even had the opportunity to prove or disprove its "efficiency." In many ways, land centralization in American agriculture was a decades-long product of farm credit policies acceptable to the American banking community. (1978, 268–69)

Small towns suffered with the decline of the family farm and the rise of mechanized corporate agriculture linked more directly to national and urban markets. Many of the towns simply disappeared as younger generations out of necessity migrated to urban educational and career opportunities. Others persisted but without a solid economic foundation that could maintain local investment and the amenities of small-town life. A little-known study by C. Wright Mills and Melville Ulmer on "Small Business and Civic Welfare" (completed in 1949 for the U.S. Senate Special Committee to Study Problems of American Small Business) concluded that "in small-business cities the environment was favorable to the development and growth of civic spirit" (p. 126) and to local welfare as measured by health, housing, church membership, and public expenditures on cultural facilities. For the same committee in 1946, the anthropologist Walter Goldschmidt (1978) compared two agricultural towns in California's Great Central Valley. One based on large-scale farming had a more unequal structure of income and social class, and it performed poorly on civic welfare.

The consistent conclusion of such research is that, whatever amenities the cities themselves generated, urbanization and agribusiness marked the eclipse of a small-town way of life conducive to the public good. Many people besides the radical populists of the 1880s realize this. In *The Resisted Revolution* (1979), David Danborn describes how the Country Life Movement fought to preserve farm towns but lost to federal policies that promoted industrialized agriculture.

In U.S. history, there is no better example of how a changing industrial economy, with zealous political support, produced a sharp increase in urbanization and a new kind of city. Many of the costs of this accomplishment were visited on rural society, indicating once more that urban and rural changes are necessary to explain one another. The costs of urbanization are poorly understood because the city is mistakenly studied as independent from the countryside.

Political defeat of the agrarian revolt, in tandem with agricultural mechanization and commercialization, contributed to a huge internal migration. The labor force engaged in agriculture fell from 60 percent at the outset of the industrial city to just under 40 percent at the turn of the century, continuing steadily downward to the present level of 3 percent. Farmers joined hillbillies and immigrants in the explosive urban growth of 1890–1920. In the cities, however, the new industrial society generated a related set of conflicts centered on the labor question.

Until the mid-1880s, labor unrest was confined mainly to the mines and railroads. In 1886 something new happened. The Farmer's Alliance joined the Knights of Labor in the Great Southwestern Strike against the Missouri–Pacific Railroad. Indeed, 1886 became known as "the year of the great labor uprising." Previous years had averaged 500 strikes involving 150,000 workers nationwide. In 1886 the figures climbed to 1400 strikes joined by 500,000 people. The labor economist John Rogers Commons called it "a great movement by the class of the unskilled which had finally risen in rebellion. . . . The movement bore in every way the the aspect of a social war" (quoted in Zinn 1980, 267). On May Day, the recently formed American Federation of Labor called a general strike that attracted 350,000 sympathizers in support of the eight-hour day.

During a large demonstration two days later at the McCormick Harvester Works outside Chicago, police fired into the crowd of strikers, killing four. A protest rally was called for the following evening of May 4 in downtown Chicago's Haymarket Square. When 180 police began to move on the crowd, a bomb exploded, killing 7 officers and several bystanders. The Haymarket massacre is legendary for what it told about the forces of reaction to labor unrest. At the urging of a citizen's committee, eight anarchists, seven of whom had not even attended the rally, were arrested as instigators of the bombing. Although no evidence about who threw the bomb was produced at their trial, four anarchist leaders of the Working People's Association were hanged for the deed. The labor wars continued with such celebrated instances as the Homestead, Pennsylvania, strike at the Carnegie Steel Plant in 1892 and the Pullman Strike near Chicago in 1894, where Pinkerton detectives and the Illinois National Guard attacked and killed strikers. The Haymarket, Homestead, and Pullman cases only highlight a thirty-year labor struggle that lasted with no resolution until World War I.

Although organized workers eventually won their demands for the eight-hour day, an adequate wage, and Social Security, most of those gains came in the 1930s as a consequence of government action following the depression. The principal result of the great labor uprising at the turn of the century was a new stage of social reorganization — the transition from industrial to corporate city. With the first major corporate mergers of the late 1890s, new conglomerate firms began to decentralize industrial production. Although suburbs had formed in the late nineteenth century as middle- and upper-class residential areas, the movement of industry escaping labor unrest gave the greatest impetus to suburban growth. David Gordon quotes a revealing illustration from Graham Taylor's 1915 book, *Satellite Cities: A Study of Industrial Suburbs*:

> In an eastern city which recently experienced the throes of a turbulent street-car strike, the superintendent of a large industrial establishment frankly said that every time the strikers paraded past his plant a veritable fever seemed to spread among the employees in all the work rooms. He thought that if the plants were moved out to the suburbs, the workingmen would not be so frequently inoculated with infection. (quoted in Gordon 1978, 49–50)

The Corporate City

The corporate city gained its characteristic features as a result of industrial decentralization. As large corporations diversified their production and came to dominate larger shares of the market, they integrated specialized activities on a nationwide basis. Management, production, and marketing could take place in different locations with some efficiency. Corporate headquarters might remain in the central city, but new plants in outlying areas could be linked into national transport networks. Decentralization was all the more convenient when suburban governments cooperated with low taxes, land acquisition, and noninterference in labor politics and when skilled workers followed to the new industrializing suburbs. No doubt, the workers who could live and be employed in the suburbs found the move a pleasant change and a valued symbol of their own social mobility. But it was not their preferences or the appeal of green grass that created the suburbs. The corporate city assumed its decisive and familiar shape as a result of depression era and postwar state policies. Confronted with the depression crisis and supported by unprecedented congressional majorities, the New Deal administration began to rebuild the economy in a manner that also built the corporate city.

The Democratic Party was rebuilt in the same process, particularly as it appealed for political support from the urban working classes and status groups: second-generation immigrants, blacks, Catholics, and union members. But Roosevelt's New Deal programs successfully reached out for a broader coalition that included business and suburban interests as well as farmers and the Old South. New Deal (and Truman's subsequent Fair Deal) policies for rebuilding the cities included federal mortgage insurance, which stimulated single-family home construction; urban renewal, which eliminated blighted inner-city areas of working-class housing to make room for downtown commercial development; public works — such as roads, bridges, waterfront projects, libraries, and municipal buildings — which employed many of the depression's jobless; and limited amounts of public housing. Democratic administrations were never hostile to the corporate interests that continued to make their headquarters in the cities. Although public housing was promoted for the urban poor, it was made palatable in exchange for programs that channeled far more costly government subsidies to single-family housing through the private mortgage market and construction industry.

Despite the inclusiveness of this progrowth coalition, political interests were acquiring new and higher stakes in what John Mollenkopf calls *The Contested City* (1983). Republican administrations returning in 1952 had their own metropolitan constituents: suburbanites, real estate developers, and corporations. Ironically, many of these were precisely the beneficiaries of depression era policies and their contribution to economic recovery. Typical among those beneficiaries were second- and third-generation ethnic families who had moved to the suburbs and enjoyed some social mobility. In the process, however, their political allegiance changed. Splits developed in the old Democratic progrowth coalition. The Republican policy alternatives emphasized urban renewal with little compensatory low-cost housing, freeways and the interstate highway system, and unalloyed suburban development.

At bottom, the urban policies that created the corporate city were manifestations of political party struggles as each party addressed social and economic problems in different ways. The New Deal worked through reform-minded local officials, bypassing wherever possible the old machine politicians. Eisenhower Republicans built their base in the suburbs and in the newer western cities. Kennedy's New Frontier and Johnson's Great Society appealed once more to old and big cities, now dominated by less mobile working-class ethnics, blacks, the new Hispanic immigrants, and the elderly — minorities in general who were willing to join as citizen participants in inviting Community Action Programs. Nixon's New Federalism was a characteristic switch back to "impartial" support of the cities in the sense of abandoning priority for the inner city and the "urban crisis" in favor of formula funding to state and local governments for projects of their own choosing (revenue sharing rather than antipoverty programs). And so the pendulum swung in narrowing arcs during the Carter and Reagan administrations as urban policy languished in the face of recession. In a capsule summary of the progrowth coalitions that built the corporate city, Mollenkopf concludes that

> over the five decades since the New Deal, the web of federal urban development programs and the intricate national-local political coalitions woven largely by Democratic political entrepreneurs succeeded in reshaping American cities and establishing the Democrats as the nominal majority party. With federal program tools, local progrowth political coalitions . . . transformed old industrial cities into modern administrative centers populated by corporate headquarters, [and] advanced corporate service firms, hospitals, universities, and other service institutions. The same programs hastened the displacement of industrial investment to the suburbs and the newly developing regions of the country. These successes dissolved the urban conflicts of the industrial city and helped build

> the modern Democratic party, much as Franklin Roosevelt and his New Deal entrepreneurs had hoped.
>
> It is hardly surprising that changes of this magnitude also generated new kinds of urban conflict which the founders of the New Deal could not anticipate. Seeing Democratic successes, Republican national political entrepreneurs and the conservative coalition in Congress sought to blunt the political efficacy of Democratic programs while substituting alternatives designed to reinforce their own constituencies. Suspended between this conservative opposition and the divisions of interest among their own urban supporters and within their party, Democrats could preserve only programs which reinforced market trends. (1983, 254)

The contemporary urban landscape is dominated by the mature corporate city. It appears as an aging and generally neglected core, save for the commercial central business district. The outer city and expanding suburbs flourished with new housing, sources of employment, and the characteristic shopping centers linked by freeways. Paradoxically, city savings and loan associations have redlined local neighborhoods and reinvested (or "disinvested") the money of their urban depositors in suburban development projects, while city taxpayers pay for the urban services that support downtown business establishments and commuters' jobs. Metropolitan regions are now segregated by class, ethnicity, age, and economic function.

Yet the social life of this segregated metropolis seldom conforms to popular images of a pathologically disorganized inner-city slum and a blandly homogeneous suburb. Beginning in the 1950s, a new folklore grew up in the mass media about the pleasures and restrictions of suburban life. In favorable light, they were pictured as conducive to a family-centered life-style, church-going, neighborly, and favored with leisure time — the "American Dream" for a deserving postdepression and postwar generation. But critics, such as William H. Whyte, Jr., in *The Organization Man* (1956), argued that suburbs encouraged standardization, status anxiety, conformity, and cultural degradation — the "split-level trap" of commuters and unliberated women who pilot station wagons between schools, shopping centers, and the veterinarian.

A series of sociological monographs challenged both kinds of description, albeit with little effect on popular stereotypes. In

The Levittowners: Ways of Life and Politics in a New Suburban Community (1967), Herbert Gans reported on the ways of life and politics in a new suburban community in New Jersey. The residents and their styles of life were no more homogeneous than their urban counterparts. In fact, most of the new suburbanites came from city neighborhoods and did not significantly alter their lives in the move. As suburbanites they maintained quasi-primary relations with their neighbors, generally enjoyed their new homes, and conformed mainly to standards such as keeping a well-groomed lawn and exterior because they agreed with them. People who reported important changes in their lives, such as more involvement in child rearing, also said that the changes were intended and among their reasons for moving to the suburbs. Bennett Berger's study, *Working Class Suburb: A Study of Auto Workers in Suburbia* (1960), in California reached similar conclusions. New suburbanites maintained their (mostly Democratic) political preferences, went to church as seldom as ever, socialized mainly with kin, and saw their future as pretty much like their past and their homes as the fulfillment of a dream. The evidence from Gans, Berger, and others showed that the life-styles of the people living in suburbs were more a consequence of social class position and age than of any attributes of the suburbs themselves.

Stereotypes of urban social disorganization were proven to be equally distorted when Gerald Suttles published *The Social Order of the Slum: Ethnicity and Territory in the Inner City* (1968). By becoming a sociological resident of Chicago's multiethnic Westside community and exploring how poor people managed their lives, Suttles revived the old Chicago School. In the slum, a provincial morality thrived that was "not so much the heartfelt sentiments of people as a set of defensive guarantees demanded by various minority group members" (p. 4). Slum dwellers have the same desire for security and self-respect as do members of the larger society, although they live in a more dangerous and economically precarious environment. Consequently, their moral order is different but not independent of the larger society. Lacking conventional bases for mutual trust, slum dwellers restrict their social relations to close associates and inquire closely into the personal character of people outside those provincial circles. As a result,

> the major lineaments of the area's internal structure are such commonplace distinctions as age, sex, territoriality, ethnicity, and personal identity. Some ethnic customs have been preserved, and numerous localisms have been developed. Many of the residents are so thoroughly acquainted with one another that only personalistic standards are relevant. Frequently the residents emphasize these structural components to the exclusion of educational, legal, and occupational considerations. Taken out of context, many of the social arrangements of the . . . area may seem an illusory denial of the beliefs and values of the wider society. Seen in more holistic terms, the residents are bent on ordering local relations where the beliefs and evaluations of the wider society do not provide adequate guidelines for conduct. (Suttles 1968, 3–4)

Another study and modern classic by Herbert Gans, suggestively entitled *The Urban Villagers: Group and Class in the Life of Italian-Americans* (1962), found similar patterns in Boston's West End. The neighborhood included a plurality of first- and second-generation Italians, along with a mixture of other ethnic groups, employees of a major hospital, and sundry artists and bohemians: "Various ethnic groups, the bohemians, transients, and others could live together side by side without much difficulty, because each was responsive to totally different reference groups. . . . Life in the area resembled that found in the village or small town, and even in the suburb" (p. 15). Italians lived in fairly close-knit kin and friendship groups and valued their community, not least for quality of housing it provided at reasonable rents. Gans described the West End as a "peer group society" because most relationships were with people of the same sex, age, and life-cycle status. As in the case of his Levittowners, the distinctive features of the urban village derived from social class and life-cycle considerations rather than any purported unique influence of ethnicity or the urban neighborhood by themselves.

The social conflicts generated by the corporate city and signaling the start of its present transformation began in the 1960s. Essentially, they revolved around the consequences of metropolitan segregation and urban decline. The main similarity they bore to their predecessors of agrarian revolt and labor struggle was that change penalized some while rewarding others. The first victims were the inner-city poor and minorities who saw their neighborhoods cannibalized by urban renewal and abandoned by public and private investment. Urban riots, as we saw in Chapter 2, were protests against unemployment, exploitative merchants, police, and the lack of public services such as transportation, affordable housing, and hospitals.

But the conflict went far beyond destructive rage. Neighborhoods also organized using the Great Society's palliative Community Action Programs as vehicles for what Daniel Bell and Virginia Held (1969) called a community revolution. The community revolution was precisely an effort by blacks and the urban poor to wrest power from the political system, although it was articulated in the more acceptable language of participation in a crazy quilt of organizations: local planning boards, committees for local control of schools and police, community development corporations, employment task forces, youth service agencies, welfare rights groups, and many more.

Emboldened by the participatory blandishments of the Great Society and by contemporaneous Vietnam war protests, the community revolution expressed itself in direct action. Marches for jobs and housing took place. Neighborhood defenders camped on proposed urban renewal sites to prevent housing destruction or unwanted new uses. Welfare and renewal agencies were occupied, even trashed. And small victories were won because liberal Democrats in Washington and in city hall seeking new bases of urban political support were also part of the revolution. Bell and Held capture the complexities of this process in a passage about the contestants and their varied strategies:

> What some of the social scientists only dimly sensed quickly became a reality: that for the indigenous leaders of the poor, particularly the militant blacks, the movement for "participation" became a drive for "power." This challenge was aimed at the local political machines, and was correctly understood as such by the local politicos. In San Francisco, black leaders succeeded early in 1965 in taking over control of the poverty program, and they sought to use the program to build a political base for black control. Plans were laid to have a part-time community organizer for almost every block and a full-time organizer for every eight blocks. Little of the monies went into services for the residents. In the political struggle within the black community, and between it and city hall, the

> original black leaders lost out. In Philadelphia, a program dominated by independent blacks was pulled back to city hall influence, though not with total city hall control. In New York City, the situation was reversed. The Poverty Program was used by the Lindsay administration to build a new political base by putting the black militants on the community action payrolls, tying them in with the Urban Action Task Forces, and using them as a battering ram against the older political machines. (1969, 152–53)

The community revolution died a slow death after 1968 as the Nixon administration withdrew political support from the inner cities in favor of another coalition with the suburbs, states, and western Sunbelt cities. Added to the recession of the 1970s, this change marked not only the defeat of the inner cities but also of entire cities in the Northeast and Midwest whose economies were based on heavy industry. Detroit, Cleveland, Pittsburgh, Akron, Gary, Buffalo, and many more became known as the rust bowl in a fitting analogy to the Oklahoma dust bowl and the 1930s depression.

The recessionary damper on industrial production and growing imports of foreign steel and automobiles combined in stagnation for the older cities. The changing economy encouraged new growth in the Sunbelt, which benefited from petroleum and high-technology industries. Local officials in the declining cities joined autoworkers and steelworkers in protest of plant closings. Without national political support for protection against the closings, however, they had little chance of saving their communities — except in rare instances where employee buy-outs of industry could be financed and made profitable. Youngstown, Ohio, is typical of these dying towns:

> A sense of gloom pervades this city. Youths moping on street corners call it a ghost town. One of every eight people in surrounding Mahoning County is on welfare, and, government officials add, almost 60,000 people are unemployed in the metropolitan area of 500,000.
>
> It wasn't always this way. Mahoning Valley steel mills around this eastern Ohio town helped build America and win two world wars. At one time, more steel was made in Youngstown than in any other city in the United States.
>
> The plants are dead or ill for for a number of reasons. Workers blame industrialists for not investing money in archaic plants. Steel

executives concede they made mistakes but also point to salary demands, foreign imports brought here at below cost and environmental regulations.

[An ex-union leader observes] "These are people who did everything right. . . . They went to college. They went to church. They bought into the system. Now what? We're wasting people. . . . " [And an unemployed steelworker] "They say high tech will come here. What do I know about computers? How are they going to teach someone like me something like that. A 20-pound sledgehammer is what I know. I know gears. I don't even know how to play Atari. And maybe 30 percent of the people I worked with in the mill couldn't read or write. I guess we're down." ("Portrait of a Dying Steel Town," *Sacramento Bee,* September 11, 1983)

What one observer called the "manufacturing Appalachia of the Midwest" is part of a broader national phenomenon — the latest transition from the corporate to the world city.

The World City

The world city is presently in the making and something to be observed less as a mature physical form than in its emerging apex. The term comes from John Friedmann and Goetz Wolff, who provide a description and some examples:

> As cities go, they are large in size, typically ranging from five to fifteen million, and they are expanding rapidly. In space, they may extend outward by as much as 60 miles from an original center. These vast, highly urbanized — and urbanizing — regions play a vital part in the great capitalist undertaking to organize the world for the efficient extraction of surplus. . . . [They reflect] the mode of their integration into the world economy . . . [and] include such metropolises as Tokyo, Los Angeles, San Francisco, Miami, New York, London, Paris, Randstadt, Frankfurt, Zurich, Cairo, Bangkok, Singapore, Hong Kong, Mexico City, and São Paulo. (1982, 309–310)

Just as the industrial and corporate cities came to constitute a new apex at the top of an urban hierarchy that retained smaller towns with specialized, realigned, or residual functions, so the world city overarches and reorganizes persistent national cities. The effects of world-city formation can be seen locally in economic shifts that move production to the Sunbelt or abroad at the convenience of multinational corporations. The deindustrializing of the United States expresses in a critical way the rise of a new urban system. Older cities like Youngstown may wither or, like Boston, attract a piece of worldwide production and marketing for high technology. World cities such as Los Angeles become a magnet for domestic migrants, multinational corporate headquarters, and the "new ethnics": Asians, Middle Easterners, and Hispanics. Armenians and Salvadorans, for example, have their own communities in Los Angeles.

One study of the new immigration by Saskia Sassen (1988) draws some striking contrasts with older migrations and shows how cities are increasingly linked in a global network of capital and labor flows. Today's immigrants are principally Asians and, secondly, Hispanics. Unlike earlier sojourners, they come from the more, not the less, developed countries of the Third World (e.g., Korea, both Chinas, Mexico, and Caribbean states such as the Dominican Republic). Labor and capital flows influence one another and explain the new pattern. That is, U.S. capital and corporations move to these countries and disrupt traditional labor practices by creating a few manufacturing jobs in export industries, a new female work force preferred by the multinational firms, and much unemployment as jobs in the Third World country are lost to competition or the young female workers are replaced after a few years with the new youth. In the United States, manufacturing jobs are lost (exported to the Third World), but new low-wage jobs are created in services for the large corporations and their affluent employees (in cleaning, building maintenance, restaurants, and specialty shops). The pattern is most noticeable in the large world cities such as New York and Los Angeles, which receive the greatest proportion of immigrants. "The same set of basic processes that has promoted emigration from several rapidly industrializing countries has promoted immigration into several booming global cities" (Sassen 1988, 22).

Based on these discoveries, Sassen went on to ask the question of how the forces causing global population shifts were simultaneously altering the shape of world urbanization. Sassen's book *The Global City: New York, London, Tokyo* (1991) compares these cities. Her argument is that those cities epitomize a process of globalization in which the functions of finance and control become more concentrated in a few great cities while production is increasingly decentralized. The two trends are causally related: Greater concentration of financial and business services is required to coordinate the fragmentation and dispersal of manufacturing to nonmetropolitan subsidiaries, offshore platforms, and low-wage enclaves around the world (see Chapter 7). One familiar result of this process is deindustrialization in developed countries and the decline of old manufacturing cities such as Manchester, England, and Detroit. Another is the growing polarization of the global cities themselves as the metropolitan labor force is divided between the well-paid professionals in law, trade, and finance and the low-paid service workers who cook, clean, and babysit for the professionals or who labor in the new electronic and apparel sweatshops. "The fundamental dynamic posited here is that the more globalized the economy becomes, the higher the agglomeration of central functions in a relatively few sites, the global cities" (Sassen 1991, 5) leading to cities that produce mainly specialized services, a truncated national urban system, and polarization within the global city. Interestingly, Sassen's comparison shows that the "fundamental dynamic" works as suggested in New York and London, but not in Tokyo, suggesting that cultural factors temporize the influence of capital on urban social organization.

New ethnic communities grow up in world cities of the United States such as Los Angeles where a fascinating study by Ivan Light and Edna Bonacich (1988) shows that Korean immigrants not only form a distinctive part of the city but also create their own ethnic enterprises and employment sources. The immigrant entrepreneurs establish a variety of new businesses such as franchise operations (from fast food to service stations), subcontracting in the assembly of garments or electronics, groceries, liquor stores, restaurants, cleaners, and many more. Most of these businesses employ family labor, work long hours, and operate with small profit margins. The bracing and persuasive conclusion that Light and Bonacich reach is that small ethnic business actually provides a form of "cheap labor" for large U.S. companies. Small business is interpreted as labor here in the sense that the Korean immigrants exploit themselves and their family labor, making a meager living by distributing the products of U.S. corporations and opening markets that would otherwise be closed. Illustratively, the Korean proprietor who, in order to provide for a large extended family, keeps a liquor store or service station open to all hours of the night in a high-crime area is risking life and leisure for a small share of the profit that goes to corporate distributors.

As the discussion of the 1992 Los Angeles riots in Chapter 2 shows, the global city is vulnerable to ethnic and economic strife. If Korean merchants may be viewed as self-exploited from the standpoint of large U.S. corporations that profit from their small businesses, African Americans may see the same merchants as exploiters of the black community, merchants who charge high prices, employ few blacks, speak English roughly, and behave clannishly and fearfully. Both perceptions may be correct, describing from different vantage points and with different interests the ways in which the Korean community has adapted to their migratory and economic circumstances.

There are ample parallels between the formation of industrial cities in the early 1900s and world cities in the late 1900s. The key difference, however, is that today's world city follows the dictates of a global political economy and grows at the expense of the older industrial and corporate cities. Because we are just coming into this fourth period of U.S. urban development, the conflicts and consequences of its restructuring effects are not fully known. Certainly, deindustrialization and the struggle over jobs and plant closings are major issues (see Chapters 4 and 7). A number of interest groups and the U.S. Congress are pressing for the restriction of foreign imports and workers. A perceptive study by David Bensman and Roberta Lynch, *Rusted Dreams: Hard Times in a Steel Community* (1987), of southeast Chicago's plant closings finds former steelworkers struggling to rebuild the community through churches, small business, and job-creation organizations, even while recognizing that the mills were closed in response to corporate strategies for adjusting to the world economy. Much is yet to happen, but we see our major cities changing in response to a new era, and, if the past is a reliable guide, we have at hand the appropriate tools for analyzing the change.

CONCLUSION

The process of urbanization, as we have followed it through U.S. history, conveys a clear lesson. Cities are not ecological things that obey natural laws or subsocial processes. Neither are they expressions of any unique cultural urbanism independent of social classes and political struggles. Rather, as Weber saw, they are the fusion of markets, political authority, and community. They assume distinctive shapes and styles in pursuit of economic and social aims defined by the polity and community. Cities and their carrying urban systems change in the interplay of these influences. As the urbanization process moves forward, it destroys much that is in its path, including rural societies and urban forms that it previously built. Urbanization does not systematically degrade society but successively reorganizes it by realigning social classes, status groups, and ways of life in response to the political economy of each stage.

There are winners and losers in the process. The industrial city supplanted the dominant merchants of the commercial city, destroyed southern agrarian society, and displaced the family farmer. Industrialists, financiers, and a largely immigrant working class initially benefited. The corporate city abandoned the working class and ethnic minorities of central cities in favor of downtown commerce and the suburban complex of new industries and homeowners. With the emergence of the world city, entire metropolitan areas in the Northeast and Midwest, including their largely industrial labor forces, are being left behind in a movement of multinational corporations and jobs to the Sunbelt and abroad. In the new world city, Upton Sinclair's jungle of immigrant worker lodgings around the Chicago stockyards of 1900 has been transformed into the Los Angeles ethnic neighborhood of fast-food franchises and small businesses of the 1980s. Urbanization makes plain to the eye how societies change and what the consequences of those changes are in a metric of social well-being.

The advantages of the framework I have called urban political economy are that it supersedes useful approaches of the past and explains both historical and comparative differences in the experience of urbanization.

The urban structure of a given country such as the United States is revealed as a product of shifting economic and political forces. Urban communities take shape under these influences and create different worlds of experience for the slum dweller and suburbanite. Political choices figure as prominently as economic dictates in shaping these worlds. People struggle with their condition in ways that make culture and spark social conflict. Those responses change urban and rural life, even in failure.

Max Weber was right when he defined the city as a special kind of economy fused with political power and community. The physical form and social experience of urbanism varies with the intersection of these forces, starting with the fundamental conditions of economic production and political control. The hillbilly and the new Third World immigrant come to their respective cities for similar reasons: to find work denied them at home. Initially at a great disadvantage in the city, many of them survive, building new lives and distinctive communities. But the opportunities presented to those lives and communities are determined politically and by the fortunes of societies within a global political economy.

SELECTED BIBLIOGRAPHY

BELL, DANIEL, and VIRGINIA HELD. 1969. "The Community Revolution." *The Public Interest* 16:142–77.

BENSMAN, DAVID, and ROBERTA LYNCH. 1987. *Rusted Dreams: Hard Times in a Steel Community.* Berkeley: University of California Press.

BERGER, BENNETT. 1960. *Working Class Suburb: A Study of Auto Workers in Suburbia.* Berkeley: University of California Press.

CAUDILL, HARRY M. 1962. *Night Comes to the Cumberlands: A Biography of a Depressed Area.* Boston: Atlantic Monthly Press.

DAVIS, KINGSLEY. 1955. "The Origin and Growth of Urbanization in the World." *American Journal of Sociology* 60:429–37.

DEGLER, CARL N. 1959. *Out of Our Past: The Forces That Shaped Modern America.* New York: Harper & Brothers.

FIERY, WALTER. 1947. *Land Use in Central Boston.* Cambridge, MA: Harvard University Press.

FRIEDMANN, JOHN, and GOETZ WOLFF. 1982. "World City Formation: An Agenda for Research and Action." *International Journal of Urban and Regional Research* 6(3):309–43.

GANS, HERBERT J. 1962. *The Urban Villagers: Group and Class in the Life of Italian-Americans.* New York: Free Press.

———. 1962. "Urbanism and Suburbanism as Ways of Life: A Re-Evaluation of Definitions." Pp. 625–48 in *Human Behavior and Social Processes: An Interactionist Approach,* edited by Arnold M. Rose. Boston: Houghton Mifflin.

———. 1967. *The Levittowners: Ways of Life and Politics in a New Suburban Community.* New York: Vintage Books.

GITLIN, TODD, and NANCI HOLLANDER. 1970. *Uptown: Poor Whites in Chicago.* New York: Harper & Row.

GOODWYN, LAWRENCE. 1978. *The Populist Moment: A Short History of the Agrarian Revolt in America.* New York: Oxford University Press.

GORDON, DAVID M. 1978. "Capitalist Development and the History of American Cities." Pp. 25–63 in *Marxism and the Metropolis: New Perspectives in Urban Political Economy,* edited by William K. Tabb and Larry Sawers. New York: Oxford University Press.

HARVEY, DAVID. 1985. *The Urbanization of Capital.* Baltimore: Johns Hopkins University Press.

HUGHES, EVERETT C. 1928. "A Study of a Secular Institution: The Chicago Real Estate Board." Unpublished Ph.D. dissertation, University of Chicago.

LIGHT, IVAN, and EDNA BONACICH. 1988. *Immigrant Entrepreneurs: Koreans in Los Angeles, 1965–1982.* Berkeley: University of California Press.

LOGAN, JOHN R., and HARVEY L. MOLOTCH. 1987. *Urban Fortunes: The Political Economy of Place.* Berkeley: University of California Press.

MAIER, PAULINE. 1970. "Popular Uprisings and Civil Authority in Eighteenth-Century America." *The William and Mary Quarterly,* Third Series, 27:3–35.

MARX, KARL. 1867. *Capital,* Vol. 1. Reprint, translated by Ben Fowkes. New York: Vintage Books, 1977.

MOLLENKOPF, JOHN H. 1983. *The Contested City.* Princeton, NJ: Princeton University Press.

MOLOTCH, HARVEY. 1976. "The City as a Growth Machine." *American Journal of Sociology* 82:309–30.

MOORE, BARRINGTON, JR. 1966. *Social Origins of Dictatorship and Democracy: Lord and Peasant in the Making of the Modern World.* Boston: Beacon Press.

NASH, GARY B. 1979. *The Urban Crucible: Social Change, Political Consciousness, and the Origins of the American Revolution.* Cambridge, MA: Harvard University Press.

PARK, ROBERT E., ERNEST W. BURGESS, and RODERICK D. MCKENZIE. 1925. *The City.* Chicago: University of Chicago Press.

REDFIELD, ROBERT. 1954. "The Cultural Role of Cities." Reprint. Pp. 206–33 in *Classical Essays on the Culture of Cities,* edited by Richard Sennett. New York: Appleton-Century-Crofts, 1969.

SASSEN, SASKIA. 1988. *The Mobility of Labor and Capital: A Study in International Investment and Labor Flow.* Cambridge, MA: Cambridge University Press.

———. 1991. *The Global City: New York, London, Tokyo.* Princeton, NJ: Princeton University Press.

STEINBECK, JOHN. 1939. *The Grapes of Wrath.* New York: Viking Press.

SUTTLES, GERALD. 1968. *The Social Order of the Slum: Ethnicity and Territory in the Inner City.* Chicago: University of Chicago Press.

SZELENYI, IVAN. 1983. *Urban Inequalities Under State Socialism.* New York: Oxford University Press.

WARNER, SAM BASS, JR. 1972. *The Urban Wilderness: A History of the American City.* New York: Harper & Row.

WEBER, MAX. 1921. *The City.* Reprint, edited and translated by Don Martindale and Gertrude Neuwirth. New York: Free Press, 1958.

WIRTH, LOUIS. 1938. "Urbanism as a Way of Life." Reprint. Pp. 143–64 in *Classic Essays on the Culture of Cities,* edited by Richard Sennett. New York: Appleton-Century-Crofts, 1969.

ZINN, HOWARD. 1980. *A People's History of the United States.* New York: Harper & Row.

ZUKIN, SHARON. 1980. "A Decade of the New Urban Sociology." *Theory and Society* 9:575–601.

4

Social Class and Inequality

In the 1930s tenant farmers were a class in decline, displaced by corporate agriculture, mechanization, and falling commodity prices. Yet they were aware of themselves as a class and did not go down gently. (Library of Congress)

TWO NATIONS

Over the last decade, the news has rekindled a faded awareness of the meaning of class society. A representative headline from the *Los Angeles Times* reads "Millions Hit Bottom in the Streets":

> When darkness falls in downtown Atlanta, they scurry for shelter in the ruins of the "underground Atlanta" night district, lifting themselves into the deserted storefronts that still carry names like Scarlett O'Hara's and the Tutti Frutti Palace.
>
> In New York City, where their ranks are swelling rapidly, they sleep in the steam tunnels under Grand Central Station or in subway cars. One man resides above a heating vent on Park Avenue, sustained by sandwiches and scraps of *haute cuisine* fed him by doormen and high-society cooks.
>
> In Minneapolis-St. Paul, they spend the night on the basement floors of nine churches. In Boston, there is no more room at the Pine Street Inn, the city's oldest and largest shelter. On some nights there, as many as 400 persons are turned away for lack of space on the floor.
>
> These are the homeless Americans: men, women, and children forced onto the streets by physical addictions, the failure of America's mental health programs and, increasingly, by joblessness and economic despair.
>
> A survey at San Francisco's Trinity Episcopal Church, which opened its doors on Dec. 1 to provide shelter for homeless single men, produced a remarkable portrait . . . 40% of the men were between 16 and 29 years old. Another 32% were between 30 and 39. More startling still, 92% were high school graduates, 30% were college graduates and 70% possessed work skills. The church found that 72% had been employed for more than a year before they lost their jobs.
>
> "What we are seeing now is a new class of poor. . . . Our shelters are filled with them — families, good people, productive people from our own communities. These are not the inebriates, not the street people, not the hobos. They're the new homeless, the new poor. . . . The evidence is that a great many of the homeless were just a few months ago stable, responsible people who had homes and jobs but whose world collapsed around them. . . . For the "new poor" of America, those for whom a life of jobless poverty is a completely alien experience, the road to homelessness is long, and losing a place to live is only the last stage. It is preceded by loss of jobs, foreclosure or eviction from homes, expiration of unemploy-

> ment benefits, exhaustion of savings, sale of valuables and loss of credit. (*Los Angeles Times*, December 26, 1982)

At about the same time, *Architectural Digest: The International Magazine of Fine Interior Design* (December 1980) reported enviously on the furnishings of a New York City apartment overlooking Central Park. The apartment owner is a Texan and partner in a firm importing porcelain and dinnerware from Japan. His business requires that he spend half the year in the Far East and divide the rest between New York and Dallas. Although he maintains a residence in each place, "he is particularly pleased with the new apartment because the decor represents a distinctive change from his other homes and is, in his words, 'quintessentially New York. . . . I am here mostly in the winter, and naturally wanted something different from what I have in my homes in Texas and in Japan.' "

The living room is appointed in roseate sofas, chairs, and matching walls; the dining room contains oak paneling and a coffered ceiling; a recessed window in the bedroom accommodates two Louis XVI chairs and an English inlaid table. The overall mood is "comfortable, suitable, sophisticated — and not a 'trendy' " aspect.

These portraits contrast an upper and, as the journalists believe, a growing lower class in the United States of the 1980s. The existence of what we un-self-consciously call "social classes" is nothing new. The term was in common use well before Marx began writing, although his work and what followed gave it a new importance. Long before Marx's name was known to the public, the gentleman novelist and Conservative Prime Minister of England, Benjamin Disraeli, wrote in his 1845 novel *Sybil — or the Two Nations* the following:

> Two nations: between whom there is no intercourse and no sympathy; who are as ignorant of each other's habits, thoughts, and feelings, as if they were dwellers in different zones, or inhabitants of different planets; who are formed by a different breeding, are fed by different food, are ordered by different manners, and are not governed by the same laws. You speak of —
> The Rich And The Poor. ([1845] 1980, 96)

The illustration concerns social inequality and, more particularly, the divide between categories of people who differ primarily in economic means defined by measures such as income, wealth, and position. Sociologists refer to these categories as social classes, a usage that has been with us since the mid-nineteenth century.

What may be new these days is a popular style of class analysis convinced that class divisions are presently widening rather than narrowing as conventional wisdom has held for perhaps a century in the United States. In a major U.S. monthly magazine, for example, an article on "The Declining Middle" by Bob Kuttner (1983) is introduced by the statement, "with most jobs being created at the top and the bottom of the ladder, America may have difficulty remaining a middle-class society" (1983, 60). The U.S. Census Bureau's Annual Report on Poverty in America (1987), using data on after-tax income, has shown that the gap between rich and poor widened in the 1980s. The share of national income in 1986 received by the highest paid, top fifth moved upward to 44 percent, while the share going to the poorest fifth declined to 5 percent — the widest gap recorded in forty years.

The most recent evidence shows that the trend toward growing inequality that began in the 1970s continued throughout the 1980s. "From 1969 to 1989, the distribution of income was marked by an increase in overall inequality. . . . The percent [of the population] with incomes in the middle range declined from 71.2 in 1969 to 63.3 percent in 1989" (U.S. Bureau of the Census 1991, 2). The number of low-income persons, as defined for the study, increased from 17.9 to 22.1 percent, and the number of high-income persons increased from 10.9 to 14.7 percent. This reversed a trend toward greater equality that prevailed from 1964 to 1969. The change between 1969 and 1989, to be sure, is only eight percentage points, but the present trend exaggerates an already highly unequal income distribution.

Inequality, moreover, has accelerated in recent years. A startling report from the Congressional Budget Office shows that from 1977 to 1989 the after-tax income of the richest 1 percent of American families increased by 77 percent, while poor families (in the lowest 20 percent as well as the lowest 40 percent categories) actually lost (from 1 to 9 percent) real income and the middle ranks remained stationary (House Ways and Means

Committee 1991). Overwhelmed by the amount of evidence on American inequality, disheartened observers may simply say that that is the way of the world; perfect equality exists nowhere, and the United States may be better off in this respect than most countries. In fact, among the other comparable industrialized nations, the United States has by far the worst rates of poverty and inequality. A recent study by the Washington-based Joint Center for Political Studies (1991) shows that if poverty is defined as a family income less than 40 percent of the national median family income, the United States has the highest number of poor families (17.5 percent) in the developed world, twice the rate in Canada (9.3 percent) and the United Kingdom (8.6 percent) and five times the rate in Germany (3.1 percent) and Sweden (2.9 percent). Once denigrated as Marxian mischief, the idea of social class is back in fashion.

The portraits of homeless people, who sleep in the subway or laundromats, and the owner of three homes, concerned that his Manhattan apartment appear sophisticated and not trendy, are contemporary, even familiar, instances of Disraeli's two nations — between which there is no intercourse or sympathy, ignorant of each other's feelings, dwellers of different planets. The majority of individuals and families occupy a position somewhere between these two extremes, but they are constantly affected by the pressures that tend toward more or less equality. Although it is important to be reminded of these disparities, the key question about class has always been the stability of its structure and the extent of its influence on people's lives — whether distinct classes are converging in a middle mass (a diamond shape as some sociologists of the 1940s envisioned) or polarizing between a pinnacle of privilege and an impoverished mass (in the shape of a pyramid). Disraeli held guarded hope for the former, whereas contemporary journalists fear that the latter is upon us. The trouble with this question is that it yields to many answers as long as no precise meaning of social class is specified.

WHAT MAKES A SOCIAL CLASS?

The concept social class joins topical issues and modern sociological analysis. The questions that need to be answered have a practical as well as a theoretical aspect. To some it is obvious that disparities such as those between the homeless and Manhattan apartment owners exist, and the legitimate question that follows is, What does the concept social class add to one's understanding? Is this more common sense or another name for the familiar? Can it tell us where the differences came from and how the homeless and the rich fit into some structure? What is to be said of differences between these conditions that seem to have nothing to do with class, such as differences in individual ability, in life-styles, ethnicity, or culture? These are key questions because at bottom they ask, What does the notion of class itself explain? The purpose of this chapter is to answer that question.

Social class is proposed as one of the basic interpretive ideas in sociology, a concept that is also central to the classical legacy. It is a building block, even a cornerstone, in the foundation of sociological interpretation. It should help unravel the bases of social structure, the origins and trends in social inequality. But it does not explain everything, or even a great deal by itself. We do not make our homes on unimproved foundations. To the dynamics of class analysis, it is necessary to add the ideas of status, family, association, state, and other concepts — all the work of subsequent chapters. Here, we will explore the meaning of social class, asking how far it takes us toward a cogent description of social organization and explanation of social inequality.

Karl Marx on Class

The origins of the term *social class* are inconspicuous. Saint-Simon's mentors and the great English political economist Adam Smith, all writing in the late eighteenth century, divided society into orders or estates (landowners, commerce and industry, the state). Smith and Saint-Simon subdivided these orders into classes (the various working classes, propertied classes), all rather unsystematically, but also in the first explicit application of the term to economically distinct social groupings. The gradual replacement of orders with classes during the nineteenth century progressed with the crystallization of industrial society itself and with the declining fortunes of the aristocracy. Smith was a vigorous liberal critic of aristocratic privilege and its economic inefficiencies. He developed the idea that labor was the source of all value, an interpretation that Marx adopted in the most explicit analysis of classes in capitalist society.

Yet the difficulty with Marx was that he used the word *class* for many purposes: in an abstract model of the "two great classes directly facing each other: bourgeoisie and proletariat"; in a general historical description of "the owners of mere labour-power [workers], the owners of capital and the landowners [which] form the three great classes of modern society"; in concrete historical studies of the struggles among such multiple classes as rentiers, petty bourgeois shopkeepers, industrialists, workers, financiers, peasants; and in the political sense of consciously organized groups rather than mere individuals living under similar economic conditions. Moreover, Marx turned to what might have been his definitive statement on the subject in Chapter 52 of *Capital,* Vol. 3 ([1894] 1981), and posed the question, What makes a class? However, he wrote just five paragraphs and left the manuscript unfinished. Although this opened the way to a continuing debate (Ralf Dahrendorf, for example, proposed to rewrite Chapter 52 for him), it is still possible to extract Marx's central point from other works.

Marx's formulation of social class need not be obscure. He began, as we have seen in Chapter 2, with less interest in coining a new term than in deriving the consequences of the competition among capitalists for surplus value. Two important features of competitive capitalism were its tendencies to expand and to rationalize itself (or reduce costs and increase profits through greater efficiency). First, in the interests of expansion, the capitalist mode of production spread by incorporating new activities. It spread, for example, from commercial agriculture and textile manufacture by incorporating small farms, artisan trades, and merchant services — by expropriating independent producers along the way. It also spread geographically into backward and colonial areas that were transformed from subsistence and trading economies to suppliers of raw materials to the industry of the advanced areas.

Second, capitalism became more efficient or rational by creating a division of labor along factory lines. Each component task of production was subdivided, simplified, sped up, and generally made more automatic, with the result that skilled and craft workers were converted into machine tenders. Mass production was invented for and it worked best on a very large industrial scale in which the worker had an ever smaller part in the total production process. For capitalists this implied ever keener competition in which the smaller and less efficient were eliminated. For merchants, agricultural smallholders and peasants, and artisans, it also meant the progressive elimination of their occupations and ways of life. Many of these people were forced into the ranks of the working class, just as that class was seeing its own jobs degraded by mechanization and other means for exploiting unpaid labor (such as the extended day discussed in Chapter 2).

In short, old groups were being eliminated and new ones created under exploitative conditions. Marx chose to call these groups social classes, thereby implying several things in answer to his own question, What makes a social class? First, social classes are locations in the process of production, similar to but not coincident with the modern notion of occupations. Marx believed that one's location within the social relations of production was the single most important of a variety of influences on one's life. Second, classes are generated in production processes that are themselves changing, old ones waning and new ones forming, which explains why descriptions of the classes appropriately change over time. Classes are not static categories whose relative numbers can be compared from one time to another. Third, beyond objective locations, classes under capitalism are experienced social relations — relations that people feel, for example, in alienation or the estrangement from one's labor, other people, and one's own self. Fourth, the objective condition of "class in itself" may be transformed into an active "class for itself" to the extent that people in similar economic conditions become socially (or class) conscious and politically organized. Classes are potential bases of organization resulting from class struggle. Finally, for Marx, class is a value-ladened term, a concept that simultaneously analyzes and criticizes society. Just as the term is not static, neither is it purely descriptive.

Max Weber on Class Situation

Max Weber agreed with Marx about the importance of classes in industrial society but disputed the idea that they were generated exclusively in the process of production. His differences with Marx on this point stemmed from a more general difference about the underlying forces of historical change. Weber believed that the most important feature of modern society was not the growth and differentiation of classes, which is a continuous process in history, but the expansion of rational bureaucracy — the growth of large-scale, impersonal organizations in all branches of society. As we saw in Chapter 1, Weber understood the expropriation of independent producers as a general process, not limited to the economy, but the essence of the growth of the state and other social institutions (such as business enterprises).

The connection between Weber's general observation and the specific problem of social class lay in the interpretation of expanding capitalism as a process of rationalization throughout society. That is, Weber expected an expansion of bureaucratic organization; in the economy, this meant a more complex division of labor. By contrast to Marx then, he believed that the expansion of capitalism leads to a more diversified class system, including various bureaucratic or white-collar classes, rather than a polarization of two great classes (the owners and workers, or bourgeoisie and proletariat in Marxism). Moreover, some of these classes would be better off than before, accounting for the rural to industrial migration patterns he studied.

As a result, Weber introduced a new and more discerning terminology. Class situations (rather than locations in production) are determined with growing rationalization of society and diversity in the division of labor. **Class situation** is defined by the power of groups in the marketplace — the power, for example, of workers and their unions to monopolize a certain skill in the labor market, the power of manufacturers to eliminate competitors in the commodity market, or the power of financiers to dominate the credit market. Weber claims that class situation is, in this sense, ultimately market situation, thus arriving at an explicit definition of class:

> We may speak of a "class" when (1) a number of people have in common a specific causal component of their life chances, insofar as (2) this component is represented exclusively by economic interests in the possession of goods and opportunities for income, and (3) is represented under the conditions of the commodity or labor markets. ([1946] 1958, 181)

Weber accomplished a great deal in a few tightly reasoned propositions. He generalized Marx by arguing that classes are formed in the entire economy, not just in production. He distinguished sharply between the economically determined classes and status groups (which we will discuss in the next chapter) that also form a hierarchy, but one based on social honor or prestige. And he characterized the way classes arise in markets that are based on economic interests but comprise all manner of bargaining encounters. Finally, he claimed that the life chances of large aggregates of individuals are determined in these encounters to which each brings different resources — a different amount and kind of power.

People compete, for example, in a market for jobs with different skills and amounts of education and in a market for commodities with different amounts of capital and credit. In a capitalist society, their life chances are determined largely by how well they compete (or use their market power). As social categories therefore, classes are aggregations of people with similar power in markets. Weber's concept stressed that "always this is the generic connotation of the concept of class: that the kind of chance in the *market* is the decisive moment which presents a common condition for the individual's fate" ([1946] 1958, 182).

In what follows, we will speak of **social classes** as large groupings of individuals who share common experiences, life chances, and a culture by virtue of similar situations and power in the marketplace. Crucially, classes are not things but relationships within and between groupings of people. Particular individuals may occupy several market situations at once, but classes are formed by the coincidence of these situations in ways sufficient to produce a common awareness of social position. Classes in this sense are real social and historical groupings rather than categories for enumerating a population. As historian E. P. Thompson says, "We know about class because people have repeatedly acted in class ways" (1978, 147).

SOCIAL MOBILITY

In sociology, journalism, and politics, the nature of social class is always a major and hotly debated topic because it reflects judgments about social equality. Class is an academic issue and a problem of political ideology. Questions posed in one vocabulary — about social mobility, class structure, and life chances — signify in another the extent to which people enjoy economic freedom and equal opportunity.

The question of social class in the United States has always been surrounded with ambiguity. On one hand, there is a tendency to deny the existence of classes altogether or to minimize their influence. This comes from the combination of a powerful ideology of social equality and the certain fact that the United States did not begin with the same conditions of privilege and inequality (feudalism, aristocracy) found in Europe where the classical theories of class were formulated. Indeed, major European thinkers confirmed this opinion when they visited or wrote about the United States. Alexis de Tocqueville traveled the country in the 1830s and wrote his famous *Democracy in America,* which stressed the openness of U.S. society resulting from an active associational life. Marx wrote for the *New York Herald Tribune* about the undeveloped U.S. class structure. On the other hand, few sensitive observers would deny the existence of social classes in the United States, and many native-born sociologists have spoken harshly of their constricting effects. Englishman R. H. Tawney (1952) best described the dilemma:

> The word "class" is fraught with unpleasing associations, so that to linger upon it is apt to be interpreted as the symptom of a perverted and jaundiced spirit. . . . If the word "class" is one which everyone dislikes, it is also one which no one in practice can escape from using. (1952, 53)

The Mobility Controversy

Among the early North American social analysts, Thorstein Veblen was the most forthright class theorist. He viewed history as the long struggle between productive industrial and parasitic leisure classes. In the early twentieth century, he believed that the United States had experienced a marked ascension of class exploitation through the force and fraud of vested business interests and absentee ownership (his term for emergent monopoly capitalism). A leisure class had grown to bloated proportions and imposed its own standards of idleness and conspicuous consumption on other classes.

Yet Veblen was virtually alone among sociological observers in this uncompromising view. Most sociologists took a more ameliorative position consistent with progressivism. Edward A. Ross is revealing in this appraisal. Although he borrowed heavily from Veblen and was one of the more outspoken sociological critics of class privilege, he shared the moderate conclusion of his colleagues. An open class structure was developing with the equalization of opportunity, the redressing pressure of public opinion, and the social reform movement. The dominant interpretation held by Lester Ward, William C. Sumner, Franklin Giddings, and Charles H. Cooley, all with a cautionary nod in the direction of the evils of class, was that the United States was moving in the direction of a classless society and had already surmounted the hurdle of a disassociation of economic condition and social class position. The latter, they claimed, now relied more on considerations of status or prestige.

The implicit contrast everyone had in mind was between the United States and the European countries where, they felt, class theories were appropriate. This impression was fostered by an influential article in 1906 by the German economist and historian Werner Sombart. Following the suggestions of Marx and Friedrich Engels, Sombart explained the absence of a strong socialist workers' movement in the United States by the high degree of social mobility between classes. Although there was a socialist movement in the United States, the relative weakness of which resulted mainly from state repression, the popularity of Sombart's analysis testified to a strong ideological attachment to principles of egalitarianism. For example, Reinhard Bendix and Seymour Martin Lipset demonstrated that it is not so much the actual amount of social mobility, which has been significant in all industrializing countries, as "it is the American emphasis on equality as a part of the democratic credo which differentiates American society from the more status-oriented cultures of Europe" (1959, 111).

Research on social class in the United States during the 1930s affirmed all these principles. The centerpiece of this work was the "Yankee City Series" by Lloyd Warner and his associates, an elaborate study of Newburyport, Massachusetts, published in five volumes and summarized in *The Social Life of a Modern Community* (Warner and Lunt 1941). Warner referred to the classes he found in terms of levels (upper-upper, lower-upper, upper-middle, etc.) emphasizing, in a complete departure from classical definitions, that class position was determined by individual prestige — the individual's social standing in the eyes of his or her neighbors. Social mobility was common as individuals tended to move up the prestige ladder by contrast to their parents' standing. With so many moving up, the natural result was a change in the overall structure of social classes — a change in its shape from something resembling a pyramid to something else more like a diamond. The diamond-shaped image was a contribution of Charles Page's influential book *Class in American Society* (1940), which concluded that " 'social class' refers to an intracommunal group possessing a distinct status."

Three points are important here. First, the view expressed by most sociologists reflected and contributed to the dominant interpretation of U.S. society. It was not merely an academic opinion but a widely held evaluation of the open society in which the experts concurred. Second, it was a conclusion about class that departed from the classical definition of the term. Third, in this work, the very objective of class analysis had shifted from explaining structural change and social action to describing a set of fixed categories or cross-sectional profile of the classes (indeed, it was at this time that sociologists began speaking metaphorically about social "stratification" rather than about real classes). Apart from these points, the key question is, Is it true? If it is true, the concepts introduced are either wrong or inappropriate. If it is not true, the notion of social class drawn from Weber and others may help explain what is.

Some of the best evidence that U.S. sociologists had mistaken opinions about prestige for real changes in class mobility came from a new breed of social historians who began looking into the actual records on social change. The work of Stephan Thernstrom is especially important because he restudied, using original census schedules from 1850 to 1880, the same Newburyport where Warner had conducted his influential research. Thernstrom states that "Warner's interpretations rested on assumptions about the past which were demonstrably false"

(1969, 230); his error lay in the use of historical methods relying on personal notions of prestige ranking rather than on any solid information about economic condition.

Thernstrom discovered that the working class of nineteenth-century Newburyport experienced only modest upward mobility, often by the unexpected route of acquiring agricultural property while remaining in unskilled or semiskilled occupations (or by "property mobility" rather than "occupational mobility") and that this limited upward movement was seldom cumulative in the sense of carrying over from one generation to the next. Indeed, property mobility was usually accomplished at the cost of "sacrificing the children's education for an extra paycheck, and thereby restricting their opportunities for intergenerational occupational mobility" (1969, 161). Working-class efforts to achieve a better life produced mixed results, although the central fact was immobility;

> These working class families did not remain in a uniformly degraded social position. . . . On top was a small but significant elite of laboring families who gained a foothold in the lower fringes of the middle class occupational world. Below them was a large body of families who had attained property mobility while remaining in manual occupations, most often of the unskilled or semiskilled variety; these families constituted the stable, respectable, home-owning stratum of the Newburyport working class. At the very bottom of the social ladder was the impoverished, floating lower class, large in number but so transient as to be formless and powerless. . . . The contrast between the literal claims of the rags-to-riches mythology and the actual social experience of these families thus appears glaring. A few dozen farmers, small shopkeepers, and clerks, a large body of home-owning families unable to escape a grinding regimen of manual labor: this was the sum of the social mobility achieved by Newburyport's unskilled laborers by 1880. (Thernstrom 1969, 158, 163)

Newburyport, of course, is just one small town, and its working-class experience may be atypical — a consideration that did not prevent the earlier sociologists from generalizing Warner's results. Other community studies were pursued in the 1920s and 1930s, especially a series by Robert Lynd and Helen Lynd (1937) which shared equal fame with Warner's work. "Middletown" (Muncie, Indiana) did not demonstrate a great deal of social mobility. On the contrary, the Lynds concluded that mobility opportunities were shrinking noticeably.

Yet a genuine debate persists about the historical experience of social mobility. Results vary depending on the sites and the class origins or destinations chosen for study. Historian Herbert Gutman studied a rapidly expanding industrial city in the mid-nineteenth century and concluded

> that the rags-to-riches promise was not a mere myth in Paterson, New Jersey, between 1830 and 1880. So many successful manufacturers who had begun as workers walked the streets of that city then that it is not hard to believe that others less successful or just starting out on the lower rungs of the occupational mobility ladder could be convinced by personal knowledge that "hard work" resulted in spectacular material and social improvement. (1977, 232)

The differences between Gutman and Thernstrom arise from two sources. One, of course, is that they studied different towns, with different economies, at different periods. The second, however, is more artifactual. Gutman's method was to examine a select group of successful manufacturers, whereas Thernstrom's focused on the entire working-class population. To discover that among those who made it many came from the working class is not the same as saying that many from the working class make it. These kinds of differences have opened a profitable and continuing research debate among social historians and sociologists. In brief, the question that divides both disciplines is, Was upward mobility typical in the U.S. past, contributing to greater class equality, or was upward mobility matched by downward mobility producing more class inequities?

A Critical Assessment

The sociological evidence on this question is rich and controversial, but it also allows some definite conclusions. The first landmark study by Bendix and Lipset (1959) concluded that occupational mobility was common to all Western industrial nations; rates in the United States were not markedly different from those in western Europe, although the belief in mobility was stronger. The qualified interpretation of Bendix and Lipset paled in the epic study of Peter Blau and Otis Duncan. Recognizing that mobility met with occasional setbacks, Blau and Duncan were forthright about the generality of upward movement:

> The rates of upward mobility in the United States today are still high. . . . Technological advances have sometimes led to serious economic depressions, which worsened chances of upward mobility and seriously disrupted careers in general, and they have sometimes effected a routinization of formerly skilled tasks, which can only have an adverse effect on mobility. In the long run, however, *technological progress has undoubtedly improved chances of upward mobility and will do so in the future* [emphasis added]. Technical improvements in production and farming have made possible the tremendous expansion of the labor force in tertiary industries — those other than agriculture and manufacturing — and, particularly, in professional and semiprofessional services since the turn of the century. . . . This great expansion of the occupational group at the top of the hierarchy, in combination with the simultaneous contraction of the bottom strata, has been a major generator of upward mobility. (1967, 426, 428–29)

Blau and Duncan's conclusion represents the dominant view in U.S. sociology and echos in a more sophisticated form ideas advanced by earlier generations from Ross to Warner. There are, however, some important qualifications on this conclusion. First, Blau and Duncan deal exclusively with occupation rather than class in a full or proper sense. Second, they do not contend with the serious problems of inferring long-term change from cross-sectional data. Particularly troublesome is the assumption that the same occupation means the same thing (e.g., about life chances) at two widely separated times.

Social mobility is usually measured by comparing the prestige or standing of the occupations of a contemporary sample with their parents. A white-collar office worker today is assumed to have improved on the condition of his or her employed parent who was a skilled or craft worker — say, a cabinetmaker. A manager today is better off than a parent who was clerk. Generally, white-collar jobs now are better than blue-collar jobs of the past. Much of the research supporting high mobility rates rests on crude inferences about labor-force shifts from manual to nonmanual occupations. The problem is that all such inferences are dubious and, in any event, do not look into the actual conditions of the positions compared.

Illustratively, a clerk in 1900 enjoyed a relatively uncommon and rewarding position, far more so than today. In the intervening years, the entire occupational structure has shifted upward. A clerk today — say, in a grocery store — does not enjoy the same advantages of 1900; there are more of them now, and their incomes relative to other occupations are probably lower. A parent and grown child both holding a clerk's position across time very likely reflect downward mobility rather than immobility. A parent cabinetmaker may have been better off than today's office worker. Similarly, a skilled worker today may enjoy a position far superior to a parent grouped in the same category and even a better position than some of today's managers. The point is that the meaning of these positions change, casting doubt on point-to-point comparisons of presumed mobility and requiring informed evaluations of the actual experience or life chances attached to these positions across time. Any evaluation of the mobility controversy must deal with changes over time in the occupational positions of individuals as well as changes in the occupational structure as a whole.

Harry Braverman (1974) interprets the same changes that Blau and Duncan considered — upward mobility as a long-run process of labor degradation. Technological change and a once expanding economy meant that new jobs were created in a middle ground or buffer zone between top management and labor. Initially, these jobs granted a privileged market position because specialized and technical labor was "in the earlier phase of its development, at a time when the supply of such labor is only in the process of catching up with the needs of capital accumulation" (Braverman 1974, 407). Later, however, the fundamental condition of wage labor makes itself felt as a mass market of qualified workers is created, driving down relative wages, and as the new jobs themselves become mechanized, all in a cyclical process. Studies that purport to demonstrate upward occupational mobility miss this key development by making cross-sectional comparisons rather than following workers and jobs through time.

Recent analyses of social mobility are more general and perhaps even more pessimistic than Braverman's cyclical theory, which would allow for the return of upward mobility with a new technological surge. Bennett Harrison and Barry Bluestone (1988) argue that the United States is on a steady course toward deindustrialization resulting in the exhaustion of the "great American job machine." As production moves away from the old industrial towns to the urban fringe and Third World countries, "good jobs" that feature union wages and benefits are vanishing while the new "bad jobs" being created are primarily in the low-wage, part-time, nonunion service sector. This change is true not only for traditional factory jobs but also for the new forms of production in computers and electronics. As Harley Shaiken notes, "There is a growing tendency to see a 'two-tier workforce,' with a small group of creative people at the top and a large workforce of people needing relatively low job skills and being paid correspondingly low wages [at the bottom]" (1984, 30).

Which of the interpretations about mobility is correct: the technologically driven upward-mobility theory of Blau and Duncan (and Braverman under some conditions) or the two-tiered downward-mobility theory of Harrison and Bluestone and Shaiken? The answer is enormously complex and still unfolding. It involves many changes that have taken place since these writings. Moreover, labor degradation is far from a uniform process, and class determination is increasingly taking place on an international level. Nevertheless, we can draw two conclusions. First, we simply cannot know which interpretation will prove valid in the long term because the theories address historical processes that are still unfolding. Second, for perhaps the last twenty years, beginning in the mid-1970s, historically documented patterns of upward mobility stopped. They have yet to return.

Much of the argument in support of that conclusion follows in the balance of this chapter. Three factual points about the significance of the "new middle class," however, provide some initial closure on the issues presented thus far. First, as Anthony Giddens (1973) has summarized, an important segment of the newly mobile stratum is made up of low-skilled and low-paid clerical jobs, filled increasingly by women who have entered the labor force under discriminatory conditions. The buffer zone at the bottom ranks of clerical and service jobs does not represent mobility so much as a new white-collar proletariat. Second, automation and recession have reduced the ranks of middle-level supervisors and technicians. Third, all evidence on job creation in the 1980s indicates that the openings are for a few managers and a great many salesclerks, cashiers, secretaries, stockhandlers, cooks, and so forth — as the journalists have correctly reported, the trend is toward a declining middle. The class structure is changing to the disadvantage of working people in the middle.

THE EXPERIENCE OF SOCIAL CLASS

The belief that upward social mobility is the destiny of all industrious labor in advanced capitalist society implied another, even more astonishing, hypothesis: a new age of the affluent worker. The two interpretations went hand-in-hand. Occupational mobility, it was argued, brought with it a changed consciousness in the working class, a preference and aspiration for middle-class (bourgeois) values, or what was called a process of "embourgeoisement." The old working classes allegedly were becoming differentiated, richer, and more like the middle classes as a direct result. The popularity of this theory was marked but short. Beyond its ungainly name, embourgeoisement ran into factual trouble. In England, John Goldthorpe and his associates conclude

> that in the case of the workers we studied there remain important areas of common social experience which are still fairly distinctively working-class; that specifically middle-class social norms are not widely followed nor middle-class life-styles consciously emulated; and that assimilation into middle-class society is neither in process nor, in the main, a desired objective. . . . Thus, it may be claimed that ongoing trends of change in modern industry are not in fact ones which operate uniformly in the direction of reducing class differences and divisions. (1969, 157–58)

Similar conclusions poured in from other advanced industrial countries where the theory was presumed to apply. Richard Hamilton's (1972) research repudiated the idea of working-class affluence in France and the United States. Working-class families enjoyed periods of modest prosperity that depended on dual incomes, prevailed only during times of full employment, and did not result in any conversion to middle-class values or political preferences. Similar results were found in West Germany and Italy. Nowhere had the miracles of technological advance or the blandishments of middle-class life worked their presumed revolutionary effects on the structures of social class.

The Injuries of Class

As evidence refuting the generalized upward-mobility and affluent-worker hypotheses began to accumulate, a more critical interpretation gained recognition. Class differences, it appeared, were large, persistent, and consequential for people's life chances — their opportunities in education, wealth and income, occupation, health and longevity, and moral standing. It is commonplace to observe that the upper classes enjoy an abundance of wealth and worry more about how to shelter it from taxation than about its sufficiency, while the working classes live precariously at or below the margin of a decent standard, often depending on moonlighting or second and third salaries as essential sources of family income. What is less obvious is that the working classes living with chronic economic anxiety constitute almost one-half of the U.S. population — not a small minority with only itself to blame. The truly poor comprise about one-quarter of the general population. Contrary to prevailing myth, the structure of income inequality that separates the desperate from the comfortable and affluent has scarcely changed over the past fifty years. The change that has occurred is mostly cyclical; gains of lower-income groups secured in periods of national prosperity are lost in hard times.

The better one's class situation, the greater the chances of surviving infant illnesses, living in good health as an adult, and enjoying an extended life span. The working classes have bad teeth, poorer health in general, less access to medical services, and they die five to ten years sooner on average than do other classes. Their work, when they can get it, is more physically demanding and less respected. Independent of their own conduct, they are at greater risk of zealous law enforcement, criminal assault, depredations of the tax collector, and induction into the military when drafts are in effect. During the Vietnam

War, for example, the working class was greatly overrepresented among combat troops and, controlling for unequal numbers, suffered twice the middle-class death rate.

Once again, such vital statistics are generally known. Less appreciated is that people of the lower classes are taught to bear a stigma for their station in life or, at least, to deal with social blame as though circumstances beyond their control (birthplace, family background, opportunity, necessity) were matters of their own choice. Their teeth are bad, this attitude of blame supposes, not because they cannot afford the affluent dentist but because they come from a working-class or poverty culture indifferent to personal hygiene. They do manual work not because they left school to help support a family burdened with illness but because they have short time horizons and relish immediate gratification. William Ryan (1976) coined the phrase "Blaming the Victim" to designate a middle-class ideology common in schools, welfare programs, and public agencies whereby social problems are interpreted as individual deficits. Neighborhoods are run-down through no fault of the slumlords or developers but owing to the acculturation of rural migrants in the city. Ryan's phrase can be justifiably extended to the working class who are sometimes blamed for their position and, not least, often blame themselves.

Social classes, in the present usage, do not exist without class consciousness, although the coherence and contents of the latter vary with time and the class involved. **Class consciousness** is a collective awareness of beliefs and attitudes linked to the life-style of a class situation and a recognition that these are distinct from other classes. Class consciousness is many-sided, containing elements of pride, rebellion, satisfaction, and humiliation. We shall touch on all of these, but in the case of the U.S. working classes, it is important at the start to note the heavy hand of blame that pervades conscious experience.

A sensitive study by Richard Sennett and Jonathan Cobb (1972) discovered the complexity of this experience among Boston workingmen. The workers talked at length about complex and ambivalent feelings regarding their class position. They could take pride in their ability as workers, but regretted their own lack of freedom and of any conferred dignity. They had an "innate disrespect" for their class situation, and they believed that education was the most likely avenue for achieving something better. Yet they also understood the sham of class privilege and the fact that persons with an education were not superior to themselves, least of all as workers. They felt class differences intensely and recognized that those who have gotten ahead through the acquisition of knowledge did so mainly with luck and guile; yet they also blamed themselves for not having been as clever. "If only I had what it takes, things would be different," lamented a shoe salesman. Their lack of success and self-blame, however, did not mean they had given up. What is striking is that these workingmen were determined to recoup a sense of dignity by sacrificing themselves for the betterment of their children.

> The workingmen of Boston almost never voiced resignation in the face of the injuries of class. The people we encountered had a powerful, though complicated, sense of mission in their lives: they were determined that, if circumstances of class had limited their freedom in comparison to that of educated people, they would *create* freedom for themselves. By that we mean that they were resolved to shape the actions open to them so that, in their own minds, they felt as though they acted from choice rather than necessity. . . . No matter how much the world had enslaved you, in this way you kept alive your dignity. (Sennett and Cobb 1972, 120–21).

Much as that self-effacing courage may be admired, it is an unhappy bargain when it carries little real promise for the prospects of the next generation. In the end, it is a demoralizing dilemma: "Something hidden and perverse is at work in our society so that people lose a conviction of their dignity when they try to take responsibility for either an increase in or a limit on their 'freedom' as society defines that word" (p. 37). As they struggle with this paradox, workers acquire no special esteem for the middle class, nor any expectation of joining it. They covet self-esteem and autonomy on their own cultural terms but despair of achieving that save through their legacy. From this amalgam of circumstance and feeling, they fashion a consciousness of their class that is pessimistic but unresigned.

From Boston to the San Francisco Bay Area, the story is similar. Lilian Rubin reports on a California working-class family life pervaded by the exhaustion of men and women with stultifying jobs, the resentment of economic uncertainty that corrodes personal relations, conflict on the role of the working wife, sexual insecurity, alcohol abuse, and heartbreak over the likelihood that children face more of the same:

> The affluent and happy worker of whom we have heard so much in recent decades seems not to exist. . . . [The realities] call the lie to the mobility myth we cherish so deeply . . . one that avoids facing the structured reality that there's no room at the top and little room in the middle. . . . The economic realities of working-class life and the constraints they impose upon living are common ingredients from which a world of shared understandings arises, from which a consciousness and a culture grows that is distinctly working class. Whether in ways of being in the family, in childrearing patterns, in orientations toward work and leisure — common experiences create common adaptations, all responses to a particular set of life circumstances. (1976, 204, 210–11)

Working-Class Life

Yet the hidden injuries and pain of working-class life should not obscure a vital aspect of its consciousness, the wit and spiritual energy that keep people going. Studs Terkel's book *Working* allows working people to speak for themselves "about what they do all day and how they feel about what they do." Of the result, he says,

> This book, being about work, is, by its very nature, about violence — to the spirit as well as to the body. . . . It is, above all (or beneath all), about daily humiliations. To survive the day is triumph enough for the walking wounded among the great many of us. . . . It is also about a search, too, for daily meaning as well as daily bread, for recognition as well as cash, for astonishment rather than torpor; in short, for a sort of life rather than a Monday through Friday sort of dying. Perhaps immortality, too, is part of the quest. To be remembered was the wish, spoken and unspoken, of the heroes and heroines of this book. (1972, xiii)

Above all, in their work, people seek self-esteem. This is the root of varied expressions of class consciousness. What is distinctive about the working classes, however, is that the great majority is forced to accept the trivial and demeaning kinds of work on which a rationalized advanced economy is based. This is the link between working-class consciousness and social structure, the source of frustration entailed by an all-consuming stress on productivity.

The working classes, like other large groupings of people, are not by nature angry, tormented, and preoccupied by their precarious positions in the marketplace. On the contrary, they resemble the middle classes in coming hopefully to their jobs and other encounters in the marketplace. Because of the ways in which production, consumption, and the amenities of life are organized in advanced capitalist societies, however, they experience exploitation, commence to struggle, and discover themselves as classes in the process. Barbara Garson describes how this sequence of events takes place. She talked with workers across the country in settings as diverse as a tuna fish cannery and a Ping Pong factory:

> I have spent the last two years examining the way people cope with routine and monotonous work. I expected to find resentment, and I found it. I expected to find boredom, and I found it. I expected to find sabotage, and I found it in clever forms that I could never have imagined.
>
> But the most dramatic thing I found was quite the opposite of noncooperation. *People passionately want to work.* Whatever creativity goes into sabotage, a more amazing ingenuity goes into manufacturing goals and satisfactions on jobs where measurable achievement has been all but rationalized out. . . . Almost everyone wants to feel she is getting something accomplished — to see that stack of paddles, the growing pile of dark meat, or to master the job blindfolded since there's not much to master the other way. (1972, x–xi)

Ben Hamper's book *Rivethead: Tales from the Assembly Line* (1991) is the autobiography of an autoworker born, raised, and formerly employed in Flint, Michigan. Hamper is a fourth-generation "shoprat" whose great-grandfather started making cars in 1910 when the industry itself was beginning. He remembers as a child attending family night at the factory, watching his father install one windshield after another on the cars rolling down the assembly line, and wondering why one person did not build the whole car as youngsters did with model kits. During his childhood in the 1960s, Flint was still a factory town.

> Even the neighborhood we lived in was a by-product of General Motors. During the boom years of the twenties, houses had to be constructed in order to keep up with the influx of factory workers arriving from the South to find jobs. General Motors built their own little suburb on the north side of Flint. In keeping with their repetitive nature, all the houses were duplicates. Our neighborhood was strictly blue-collar and predominantly Catholic. The men lumbered back and forth to the factories while their wives raised large families, packed lunch buckets and marched kids off to the nuns [at Catholic school]. (Hamper 1991, 10)

As a child, Hamper decided that he would never become an autoworker, in part because of his father's alcoholic rebellion, which he blamed on the job, and the family instability it produced. Nevertheless, when his high school girlfriend got pregnant and, both Catholics, they got married, Hamper found himself badly wanting the GM job. At the time (1977), the automaker offered security, an attractive $12 hourly wage, and the companionship of people sharing a common working-class culture: "We belonged. There were really no other options — just tricky lies and self-soothing bullshit about 'my real talent lies in carpentry' or 'within five years I'm opening a bait shop.' We weren't going anywhere" (p. 48).

To deal with the fierce heat, noise, tedium, and mental strain of the assembly line, workers developed various strategies. By "doubling-up," a pair could take turns — each one doing two jobs for a period, leaving one free to rest or roam the shop floor while the other had to work so fast to keep up that it broke the boredom. Despite their inventiveness, "the one thing that was impossible to escape was the monotony" (p. 41). Alcohol and other drugs provided a common means of coping with the oppression: "Shoprat alcohol consumption was always a hot debate with those who just didn't understand the way things worked inside a General Motors plant. While not everyone boozed on a daily basis, alcohol was a central part of many of our lives" (p. 55). In the end, after ten years at GM, all the tricks stopped working. Hamper found himself suffering debilitating anxiety attacks that led to hospitalization and release from the job. Without self-pity, or even much conviction, he speculates that the only ones who can survive thirty years on the assembly line are those who can blot out their daily experience.

A broader irony surrounding Hamper's story is that just as he lost the job he could bear no longer, Flint was losing most of its GM assembly plants that were moving to Mexico in search of cheap labor. Hamper's friend, Michael Moore, made a documentary film about the plant closings called *Roger and Me,* and the ex-auto worker made a brief appearance as a pop-tune-singing mental patient.

Rebellion is key to understanding one important expression of U.S. working-class culture. In the past, that culture has been properly understood as a product of what William Yancey called "emergent ethnicity," or the enactment of an immigrant heritage in the new setting of U.S. urban neighborhoods and industrial workplaces. Today, it is changing from an immigrant to a more indigenous working-class culture as younger generations ironically hark back to a supposed tradition of the American West. Its symbols are the pickup truck or four-wheel drive vehicle, the cowboy hat or monogrammed baseball cap, jeans and belt buckle. It fancies the freedom of the open road and the citizens' band radio as a collective defense against police, although most of its devotees work in offices and factories. Highways and the ability to outsmart police express independence from oppressive jobs and middle-class standards. In play its tastes run to beer and country music and in worship to fundamentalism — the interdependent wages and salvation of hard living. Behind the postured urban cowboy or cowgirl is a cultural statement of a desire for respect and self-determination.

The working class expresses itself in popular culture, in the songs of artists such as Bruce Springsteen who expresses the plight of a Vietnam veteran with no prospects in civilian employment:

> Born down in a dead man's town
> The first kick I took was when I hit the ground
> You end up like a dog that's been beat too much
> Till you spend half your life just covering up
> Born in the U.S.A.
>
> (Bruce Springsteen, "Born in the U.S.A.," Columbia Records)

The Anxious Middle Class

When Americans are asked in sample surveys the social class to which they belong, up to 90 percent identify with the middle class, including many persons who by Weber's criteria would certainly be classified as upper or working class. Being middle class is the correct normative posture in the United States. There are good historical reasons for this preference. Large numbers of U.S. families have enjoyed upward social mobility, and, all things considered, life in the middle class has been good, at least until the current downturn that began in the early 1970s.

Membership in the U.S. middle class has meant intergenerational social mobility; relative job security; health, accident, unemployment, and retirement benefits guaranteed by private employers and the state; housing, often in the expanding suburbs, supported by state programs; and a college education, often at state schools. The creation of a vast middle class in the United States, particularly from the postdepression years until the 1970s, was without doubt one of the great social achievements of any nation, even after it is acknowledged that many were left out.

Today, however, the middle classes are worried, as Katherine Newman entitles her book *Falling from Grace: The Experience of Downward Mobility in the American Middle Class* (1988). Newman takes the story of one David Patterson and his family as typical of the roughly 25 to 40 percent of the population who have suffered downward mobility in the past decade. David was born into the working class, but, through effort and the discipline to avoid risky if creative career paths, he had become a highly paid executive in the computer industry. Then with little notice, his company was caught in a profit crunch and overstaffed in the new division David had just moved from California to New York to help lead. Left in an expensive new home with just four weeks' severance pay, David began looking for another job, and his wife Julia took a low-paying position. Combining Julia's salary and David's unemployment, the family was bringing in only one-quarter of their previous income. As their savings dwindled, David and Julia sold the new house but realized little cash owing to a large mortgage and the rush to sell. Soon they were living in a modest apartment and mutually embarrassed in declining contacts with old friends. The newly imposed social isolation was all the more difficult on their two teenage children who had resisted the earlier move away from their California friends and upbringing. Isolation and fear hounded the unhappy family. Despite herself, Julia began to doubt the sincerity of David's continuing inability to land a new job. Before long, David too lost confidence and became convinced that "it must be me," even though the entire computer industry was depressed during the mid-1980s.

The Patterson's story is the nightmare of the U.S. middle class, a horror brought ever closer to reality as increasing numbers of white-collar workers lost their jobs during the recessionary periods of the 1980s and 1990s. Corporate downsizing has permanently eliminated professional and executive positions in the same way deindustrialization has reduced manufacturing jobs. When jobs go, so does health care insurance and savings earmarked for the children's education. Equally frightening is the scarcity of good jobs for today's new college graduates who had come to count on middle-class membership, if not further upward mobility, as their birthright. No one was expecting what Harrison and Bluestone (1988) call the "great U-turn" in middle-class prosperity. It is often said of today's adult population that they will be the first American generation to leave their children worse off than their parents left them.

The Higher Reaches of Class

As class differences reassert themselves, their associated cultures diverge. The middle and upper classes have their own cultures and consciousness. These attitudes require less description because they are the familiar, dominant ideas behind much of the interpretation of U.S. society that has been questioned here — the ideas of mobility, the declining significance of class, and the equalization of opportunity. They pervade the mass media, schools, and government, defining appropriate ambitions and blaming other classes for their failure to fulfill those ambitions.

We understand the upper classes mostly as they like to see themselves: competent in matters of finance, generous in charity, public-spirited, rich but not gauche, tasteful (like the Manhattan apartment owner, sophisticated but not trendy), and, with the rare exception of a playboy scion, generally deserving of their position. Indeed, we see them in ghost-written biographies that celebrate the wisdom and charm of a Donald Trump, Lee Iacocca, or T. Bone Pickins — best sellers all. When Texas billionaire H. Ross Perot ran for president in 1992, his vaunted reputation as a businessman who got results was taken on faith despite his computer consulting firm's failure to reorganize General Motors with the efficiency it had once aided the Texas health care system. Such defeats were not covered in the popular biography that Perot hired novelist Ken Follett to write. The upper classes are glamorous, less resented than envied. In business and art, they are indisputably in charge, although it is arguable whether they control politics (an issue reserved for later). There is little doubt that they constitute a self-conscious, socially cohesive group.

In *The Power Elite* (1956), C. Wright Mills shows that the corporate rich attend the same prep schools as children, go to the same elite universities and law schools, join interconnected firms and social clubs, marry one another, and rotate in the same influential public posts. William Domhoff (1974) has added to this portrait showing how the likes of (former Secretary of State) Henry Kissinger and John McCone (former CIA director, head of the commission that investigated the Watts riot, and director of Standard Oil of California and International Telephone and Telegraph) mix socially in the California redwoods with the heads of most major U.S. corporations and government agencies. Because they are enveloped in glamour and envy, the legitimacy of the upper classes is not seriously challenged — or, if it is questioned, the blame is laid to mean-spirited socialists.

During the same period in which middle-class prosperity has eroded, the upper classes have enjoyed hefty increases in wealth and income. The richest 1 percent of U.S. families, whose income gains (60 percent after taxes) were discussed above, earned an annual average of $560,000 in 1989 (House Ways and Means Committee 1991). Typically, much of this is earned from property and interest. The rich, if far from idle, seldom earn their living in the form of wages. These, in turn, are people whose tax burden was steadily reduced during the 1980s on the (trickle-down) theory that they would invest in new productive enterprises that create jobs and business expansion. That did not happen. Evidently, much of the money instead went into such things as luxury consumption and real estate speculation. Seldom has a theory and experiment in social policy proved so decisively wrong. Yet it is doubtful that the policy failure will rub off on the upper classes. The rich will remain where they like to be, above public scrutiny, as labor and management blame one another, politicians blame everyone else, and the middle class blames itself.

In summary, social classes are much more than categories for describing a population. They are socially meaningful groupings that display common life-styles, a consciousness of kind, and patterned relations with other classes. They figure importantly in the life chances that people enjoy. Taken together they portray much of what we understand as a social structure.

CLASS STRUCTURE

Having described the experience of selected social classes, it is useful to develop a portrait of the overall **class structure**, the set of social groupings defining positions in the marketplace and, most important, the relations between them that reciprocally define the power and circumstances of each.

Any portrait of the class structure, as a geometric figure of a set of categories, entails difficulties. On the one hand, classes as acting and self-conscious social groupings vary in cohesiveness and salience. Factory workers and the corporate rich are clear-cut class positions replete with a well-developed consciousness and culture. Middle classes such as the executive–managerial are less internally coherent and more torn by conflicting loyalties to upper-class owners and middle-class colleagues. Combining all these in one structure means slighting, to some extent, important differences. On the other hand, joining many experiences of market power in one structure requires a common dimension of class comparison, the most convenient being occupational or economic. There is no way to escape these problems if a general portrait of class structure is desired, but it is important to remember that the structure is an approximation in meaning.

The class structure is divided into three levels on the underlying dimension of power in the marketplace, using source of income as a shorthand for designating specific positions (Table 4.1). The upper classes include the propertied rich, often old money; the corporate rich, owners of industry and large corporations, and the top financiers and owners of money. The middle classes include top- and middle-level executives, managers, and technocratic experts of the business and corporate world; owners and proprietors of small businesses that employ small numbers; independent, self-employed business and professional people (e.g., lawyers, doctors, accountants); employed professionals and highly skilled workers (e.g., teachers, engineers, medical technicians); and lower-level supervisors and managers. Other groupings, such as farm owners, could be added, although their numbers have declined drastically. The

TABLE 4.1 Contemporary Class Structure in the United States

Class	*Percentage of Population**
Upper Classes	*2*
Property owners	
Corporate and business owners	
Large employers	
Financiers	
Middle Classes	*52*
Executives, managers, technocrats (top and middle levels)	12
Small-business owners and employers	6
Self-employed business owners and professionals	7
Employed professionals, skilled craftspeople	10
Supervisors	17
Working Classes	*46*
Industrial workers	16
Trade and construction workers	12
Clerks, attendants, office and service workers	14
Manual and agricultural labor workers	4

*Estimated percentages

SOURCE: Erik Olin Wright et al. 1982. "The American Class Structure," *American Sociological Review* 47:709–26; U.S. Department of Commerce, Bureau of the Census, 1980. *United States Census of Population*. Washington, D.C.: Government Printing Office.

working classes include industrial and factory workers; trades and construction workers; office workers, attendants and service workers; and manual and agricultural labors (of which there are still many now employed by corporate and industrialized agriculture). In addition, perhaps "beneath" these is a sizable underclass or *lumpen proletariat,* in Marx's words, including the permanently unemployed, migrant and casual laborers, and some career criminals.

The major markets in which these classes compete and develop their varying degrees of cohesiveness are the labor market for jobs and earned income, the credit market, and the commodity market for goods and services. The most dramatic changes in the labor market over the last few generations have been the increase in wage earners by contrast to the declining numbers of independent professionals, tradespeople, farm owners, and small-business owners. At the same time, however, industrial employment, the traditional blue-collar job, has declined with automation and the transfer of production abroad. The industrial working class has been the principal victim, while the technical and skilled middle class and the corporate rich have been the beneficiaries. The market power that industrial workers once had by virtue of their skills and their unions has been severely eroded. The industrial working class of the midwestern rust bowl are losing this struggle, particularly in steel and automobiles, with large reductions in the labor force and in wages and benefits.

In the credit market, the financial upper class enhanced its power during the period of high-interest rates in the 1980s, while most of the working and middle classes either paid that interest or were forced out of the market altogether, as in home buying. Interest rates came down sharply in the late 1980s and early 1990s, but by then a recession dampened enthusiasm for borrowing and new home purchases. It was the case, however, that many average-income families refinanced their existing home at this time, taking advantage of a rare moment in which the credit market turns to their advantage. Lost power in the credit market has lately been the ruin of many small businesses and of the few remaining farm owners. In the 1980s, the number of bankruptcies, traceable in important part to interest rates, reached its highest level since the 1930s depression. Financiers, however, have also been playing a risky game in high-interest loans, particularly to Third World countries, coming more than once to the brink of international monetary collapse. Domestically, the Federal Deposit Insurance Corporation, which insures commercial banks, reported that the number of banks in serious financial trouble had risen sharply (up from 385 in 1976 to 597 in 1983). Actual bank failures quadrupled in the early 1980s (from 10 in 1981 to 42 the following year). By 1990 the great speculative boom in the real estate industry collapsed and with it hundreds of savings and loan institutions. Although a few of the perpetrators of this fraud have been caught and fined, most of the money is gone and the taxpayer is reimbursing savings and loan depositors to the tune of $200 billion — so far.

Commodity markets have followed a pattern typified by the growing difficulty in buying a new home. Unemployment, reduced wages, and inflation have cut severely the purchasing power of the middle and working classes. In this sense, commodities such as the education of children are increasingly out of reach, compounding disadvantages in other markets. Meanwhile, with corporate taxes declining as a proportion of national income, middle-class taxpayers have been squeezed to the point of tax revolt. The victims of resultant public fiscal crises have been other middle-class civil servants (e.g., teachers), working-class service and clerical people, and lower-class beneficiaries of public programs for medical care, Social Security, and nutrition, among others.

CLASS STRUGGLES

These examples show how class situations are determined in markets, and they hint strongly at the forms of exploitation that help produce open class conflicts. The class situations described illustrate the bases of conflict that nevertheless are beneath the surface — the continuing sources of disadvantage and inward struggle that contrast with episodic class fights. The essential point here is that classes take their most decisive form when they come to struggle for changes in these conditions. Historian E. P. Thompson notes that

> classes do not exist as separate entities, look around, find an enemy class, and then start to struggle. On the contrary, people find themselves in a society structured in determined ways (crucially, but not exclusively, in production relations), they experience exploitation (or the need to maintain power over those whom they exploit), they identify points of antagonistic interest, they commence to struggle around these issues and in the process of struggling they discover themselves as classes, they come to know this discovery as class consciousness. Class and class consciousness are always the last, not the first, stage in the real historical process. (1978, 149)

The classical, and once most common, form of open conflict is the struggle between labor and capital, between corporate and industrial owners and the working classes. Other forms are increasingly common today, including struggles between certain upper and middle classes.

A celebrated instance of the first type was the 1972 strike at the General Motors plant in Lordstown, Ohio. The notoriety of Lordstown stemmed from the fact that a new generation of autoworkers confronted the corporate upper class with protests about the quality of their jobs in the face of measures for automating production. The Lordstown factory opened in 1966, billed as a model plant and designed to turn out 60 standard Chevrolets each hour. The first workers considered themselves privileged to have these jobs and, at an average age of thirty-five, represented the traditional industrial working class. By 1970 GM was feeling the competition of foreign imports and converted the Lordstown facility for accelerated production of the Vega economy car. Physically demanding work and high labor turnover by then had altered the work force to an average age of twenty-five, including many Vietnam veterans, hillbillies, hippies, and minorities — the Woodstock of the working class, some said. With the help of automated welders and paint dipping, the line was now to produce 101 cars, rather than 60, per hour. In many instances, this could be accomplished only through more intensive exploitation of labor. Stanley Aronowitz describes the arduous working conditions:

> For example, within a minute on the line, a worker in the trim department had to walk about 20 feet to a conveyor belt transporting parts to the line, pick up a front seat weighing 30 pounds, carry

> it back to his work station, place the seat on the chassis, and put in four bolts to fasten it down first by hand-starting the bolts and then using an air gun to tighten them according to standard. It was steady work when the line moved at 60 cars an hour. When it increased to more than 100 cars an hour, the number of operations on this job were not reduced and the pace became almost maddening. In 36 seconds the worker had to perform at least eight different operations including walking, lifting, hauling, lifting the carpet, bending to fasten the bolts by hand, fastening them by air gun, replacing the carpet, and putting a sticker on the hood. Sometimes the bolts fail to fit into the holes; the gun refuses to function at the required torque; the seats are defective or the threads are bare on the bolt. But the line does not stop. . . . "You really have to run like hell to catch up, if you're gonna do the whole job right." (1973, 22–23)

The highly publicized technology of the reorganized factory was not the basic source of its productivity. Economist Emma Rothchild (1972) notes that Lordstown technology is the speed-up, as developed by Henry Ford, and methods that follow from the earliest techniques of mass production. General Motors tried to gain a competitive edge over Japanese manufacturers by extracting more work from the people on the assembly line, by reducing what was thought to be their "idle time." But the new breed of worker rebelled, as one of Terkel's conversations revealed:

> So what happened? A guy faced up to the facts. . . . If he were young and married, he had to do one thing: protect his pace. He had to have some time. The best way is to slow down the pace. He might want to open up a book, he might want to smoke a cigarette, or he might want to walk two or three steps away to get a drink of water. He might want to talk to the guy next to him. So he started fighting like hell to get the work off him. He thought he wasn't obligated to do more than his normal share. All of a sudden it mattered to him what was fair. (1972, 259)

The result was a bitter strike. The local United Auto Workers union voted 97 percent in favor of a strike and was out for twenty-two days before reaching an agreement that restored back pay to workers previously disciplined for violating work standards. UAW locals in ten other GM plants around the country followed suit with strikes related to speed-ups. In the long run, however, little was done to improve working conditions. Grudging concessions were at first made, but as recession and unemployment mounted in the 1970s, the screws of productivity were retightened, accompanied now by actual wage and labor-force reductions that the union was forced to accept rather than face plant closings. The long-term result of the Lordstown struggle, and others based on similar issues, was working-class defeat and lost power in the market.

In the twenty years since the Lordstown strike, strikes and other forms of labor protest have declined markedly. During the 1980s, the Reagan administration took a vigorously antilabor position, which included breaking the (PATCO) aircraft controllers' strike. In this climate, the courts supported Eastern and Continental airlines that declared bankruptcy to nullify their labor contracts, only to open the next day under "reorganization" and a free hand in labor disputes. The meatpacker's union lost a protracted strike at Hormel. Some unions were, in effect, locked out by companies seeking to provoke a strike as the first step toward winning wage and benefit give-backs from labor.

With changes in the economy and increasing production overseas by U.S. multinational corporations, the size of the industrial labor force is declining steadily. The fastest growing working class is devoted to services and clerical work, much of it in the public sector. Accordingly, since the mid-1960s, a common form of class struggle has been the fight for wages and working conditions by public service unions. James Green explains:

> During the fifties, strikes by government employees were rare, but in the following decade 119 walkouts occurred in the public sector, involving 70,000 workers. The strikers ranged from professionals, especially teachers and nurses, to blue-collar workers such as garbage collectors and fire fighters. Violating laws that prohibit public employees from striking, the American Federation of Teachers adopted militant tactics and gained thousands of new members. The American Federation of States, County, and Municipal Employees (AFSCME) also used the strike effectively, increasing its membership from 182,000 in 1960 to 350,000 in 1967. . . . City workers faced increasing job pressures during the sixties, whether they were bus drivers wrestling with poor equipment, teachers contending with growing classroom violence, or mental health

> workers dealing with the large number of people who could not cope with the urban "rat race." While public employees absorbed the human and social effects of urban disintegration, they also faced increasing demands for productivity and efficiency. As public-sector bureaucrats adopted strategies for control used in the private sector, government workers rebelled. Public-school teachers, who had enjoyed a good deal of control on their jobs, fought to remain in charge of their classrooms and to reduce the power of administrators. By the end of the sixties, public employee unions had become powerful. (1980, 234–35 passim)

One of the rare examples of a vital working-class organization is New York City's Local 1199 of the Drug, Hospital and Health Care Employees Union. Of 100,000 members, a great many are women, and approximately 80 percent are black or Hispanic. In June 1992, when most unions were in retreat, Local 1199 was celebrating a new contract with the city's League of Voluntary Hospitals, which covered 40,000 of their members in the service and maintenance divisions of the hospitals. Without a strike, the union and hospitals agreed to a new wage and benefit package that included a 12 percent pay increase, pension fund improvements, employer-funded child care, limitations on the use of temporary workers, and more worker participation in decisions affecting the workplace. At the same time, negotiations were underway for Local 1199 to merge with the health care units of the American Federation of State, County, and Municipal Employees (AFSCME) that would create a 600,000-member union active in thirty-nine states (Roberts 1992, 14).

What explains the difference between the successful struggles of health care workers and the quiescence of the traditional labor movement in manufacturing? We will return to this question at the beginning of Chapter 5, but three factors deserve mention here. First, the health care industry is profitable (overly so, critics would say) and expanding by contrast to the declining manufacturing sector. Hospitals need qualified workers and have been expanding their staffs. Second, the union has a predominantly minority membership who express a high degree of solidarity and participation. Their union is an occupational and a civil rights group. Third, Local 1199 is a political force in New York City; it organizes to prevent budget cuts that would affect health care, registers voters, supports political candidates, and maintains a close alliance with Cardinal O'Connor and the Roman Catholic Church. According to union President Dennis Rivera, 1199's ability to register voters "impressed both politicians and employers with our political clout" (*Local 1199 News*, June 1992, 5). Indeed, the hospitals and city government view Local 1199 as an ally in efforts to secure greater state funding for health care, according to Rivera.

The foregoing examples of class struggle touch only the most prominent forms and portray a two-party contest in the labor market. The picture is always more complex. First, the struggles involve additional parties. In the dispute between autoworkers and the corporate owners, the other upper classes side with industrial management. So, too, do many middle-class managers, skilled workers, and technicians — all finding themselves in an ambiguous (in the words of Eric Wright) or conflicting class situation. The struggles of public service workers pit working- and middle-class unionists against other middle-class bureaucrats and taxpayers. Corporate and propertied upper classes have benefited indirectly from middle class–led tax revolts — just as the working and lower classes dependent on public services have suffered.

Second, class struggles take place in other, interpenetrating markets. In the commodity market, price increases by dominant producers is met with consumer protest or boycott. General Motors met with consumers turning to cheaper imported cars in the commodity market and acted on this problem in the labor market in the specific ways that Lordstown demonstrates.

A topical and instructive example of protest in the commodity market is the utility-rate rebellion that spread across the United States in the 1980s. Energy is provided chiefly by publicly regulated monopoly corporations. Nevertheless, these are owned and managed by the corporate rich and financial classes who hold the largest blocks of stock. With rising petroleum prices, inflation, and costly investments in new, often unworkable, nuclear plants, the major utilities began raising their rates, sometimes doubling or tripling the monthly cost to ratepayers. Protests spread and took many forms. The *Wall Street Journal*, for example, perceived the connection between the struggle and finance capital:

> Rebels are rising in the Pacific Northwest. Sporting nasty buttons, they threaten to undermine banks, to unseat politicians and, for the prime target they call "Whoops," to deal the harshest blow — refusal to pay their bills.
>
> The action is part of the continuing drama centered on the Washington Public Power Supply System. Until now the drama has been set on Wall Street, where investment bankers struggle to patch together financing for the giant public agency, and at two construction sites in Washington state, where mismanagement, snafus and cost overruns killed two of the system's five partially built nuclear power plants.
>
> Now an open rebellion against skyrocketing electric rates is gathering momentum in Washington, Oregon and Idaho communities large and small. Jolted by the doubling and tripling of electric bills and faced with the prospects of similar stiff hikes this year, customers are organizing, marching, packing legislative hearings and withholding payment of rates. (March 19, 1982)

There is little doubt that the poor, the elderly, and people on fixed incomes have suffered most from rising utility rates. The *Los Angeles Times* reported "seniors living in bathrooms with kerosene heaters. They don't even turn on their furnace, they don't use lights at night" (December 26, 1982). Nevertheless, the speeches and tactics of the ratepayer revolt suggest that it is largely a middle-class movement. Organized petition drives, bill burnings, sieges on utility commission hearings are the methods of a nonalienated, nonapathetic, politically socialized middle class using the conventional weapons at hand to confront the regulators of monopoly corporations. In many instances, the movement succeeded at winning rate rollbacks or freezes. Although these small victories may not last, they contrast sharply with the recently bleak record of working-class struggles in the labor market.

SUMMARIZING TRENDS

From all these illustrations of class conflict, it appears that both the structure of social classes and the nature of their struggles are changing. In order to conclude this discussion with a sense of what the near future holds, the conception of social class that has served this far need only be applied to current trends. If we can discern the changing shape of markets, and thereby of class relations, we should be able to infer their practical consequences. A good place to begin is in the labor market.

Over the last generation, the labor forces of the advanced industrial countries have changed fundamentally with the rise of a service economy. In the United States, Germany, France, and England, the number of industrial workers has declined. Between 1940 and 1990, the U.S. labor force (including middle- and working-class persons) engaged in manufacturing dropped from 34 to 17 percent, while the proportion in commerce and services rose from 46 to 73 percent. More revealing is the fact that 90 percent of all the new (public and private sector) jobs created during the last two decades were in commerce and services — including in this omnibus category such things as wholesale and retail trade, banking, insurance, government, repair services, and many more.

Rothchild (1981) has captured this change with an impressive contrast. Speaking only of the private economy, from 1973 to 1980, 70 percent of all new jobs were in services and retail trade, and of these 40 percent occurred in just three areas: eating and drinking places, health services, and business services. Considering only the first of these, "the increase in employment in eating and drinking places since 1973 is greater than total employment in the automobile and steel industries combined" (p. 13). By contrast to the old blue-collar jobs, which are unionized and tended to win favorable wages and benefits, the outstanding features of the new jobs include an overrepresentation of women, low pay, short hours or part-time character, "deadend" prospects, and the relative absence of protections and benefits: "The United States, in sum, is moving toward a structure of employment ever more dominated by jobs that are badly paid, unchanging, and unproductive" (Rothchild 1981, 13). Recent evidence suggests that the trend continues (Harrison and Bluestone 1988).

Two more trends explain and compound the picture. Obviously, the United States continues to be an industrial society and to produce an enormous quantity of manufactured goods. Much of the actual production, however, has moved out of the country in search of cheap labor markets accessible to U.S. multinational corporations. Nearly all of the television sets and radios, and increasing proportions of the clothes, shoes, toys, computers, appliances, steel, and even automobiles are all being manufactured in low-wage countries (Taiwan, Mexico, the Philippines, Malaysia, Hong Kong, etc.) and reimported into the U.S. market under preferential tariff arrangements. Since the 1970s when this practice became common for the largest U.S. manufacturers, literally millions of domestic jobs have been eliminated, and, as the trend continues, the market power and the share of national income going to labor have plummeted. Increasingly, class determination lies in the international division of labor (see Chapter 7).

Meanwhile, in the altered labor market at home, the second compounding trend is toward further mechanization now in the service economy. Business services are computerized and offices mechanized with word processors and photocopiers. Banks are becoming fully automated. Franchise food operations build standardized outlets (one hesitates to say restaurants) that mechanize assembly-line food. Medical services depend increasingly on machines for treatment and laboratory analysis. Transportation introduces computerized trains and containerized cargo handling attached to mechanized warehouses. Repair services, notably for automobiles, are dispensed by national chains using factory methods to restore transmissions, tires, brakes, and mufflers. Supermarkets and gas stations are automated to the extent that attendants and checkers can now be replaced by plastic "debit cards" that effect purchases through electronic communications between banks.

Sociologists such as Daniel Bell (1973) have spoken enthusiastically about this process, hailing it as a "Post-Industrial Society" in which new and exciting jobs will exist in the information industry. If the computer industry is prototypical, however, new jobs are not produced in numbers anywhere equal to those lost. To stay in the competitive race, companies go abroad to create their jobs (as in the celebrated case of Atari), and their production and programming tasks are further mechanized. Ernest Mandel has captured better the significance of these changes: "Far from representing a 'post-industrial society,' late capitalism thus constitutes *generalized universal industrialization* for the first time in history" (1972, 387).

The broad results of these trends for the structure of classes are now readily observed. As new service jobs replace old industrial ones, work is degraded as a net result. This is not a uniform "deskilling" process. Some displaced industrial workers and, particularly, new entrants to the labor force get decent jobs in the service or "information" industries. But the typical transition is from skilled industrial work to attendant or clerk in retail services — illustratively, from steelworker to liquor-store clerk. Because fewer of the new jobs are unionized, organized labor as a proportion of the total work force declines (down from 33 to 18 percent in the last thirty years) and with it the pay and benefit levels of the working class erode. Job security is declining. Real wages dropped over the course of the 1980s. In 1983, for example, the average U.S. worker earned $281 per week; by mid-1990 that figure was down to $267 per week in constant (inflation-adjusted) dollars (Friedman 1990). In summary, for the majority of the population — all of the working classes and middle-class civil servants, small-business and some corporate employees — power in the labor market has declined. With that, of course, their class situations and life chances have worsened.

Class relations have become more polarized, notably during recessionary years. Open class conflict is more likely from now until the end of the century than it was during the boom decades after World War II. But the locus of that conflict is also shifting. The struggles of labor and capital in industry are less frequent because the threat of more automation and of offshore production have chastened labor into fighting a rearguard action to protect the remaining jobs even at the cost of wage reductions. Public service unions are on the defensive as a result of fiscal strains in government. These facts do not mean that conflict has been suppressed as the utility ratepayers' movement shows. Rather, the conflicts take different forms and are displaced to other market situations.

Recent struggles in the labor market have focused on plant closings — to prevent them, to provide for retraining and community reparations when they occur, or to promote employee buy-outs that would keep firms operating. In the publicized case of the Chicago and Northwestern Railway, venturesome employees successfully bought out and restored the company to prosperity. Few buy-outs are so successful, but the idea is attractive to communities threatened by plant closings. All of this suggests that, as good jobs become scarce, future class struggles in the labor market will center on the provision and rights of employment. Migrant labor, exported production, plant closings, and automation are specific targets for new organizing strategies.

Finally, class struggle appears in new forms beyond the labor market. It focuses on commodities, broadly understood, such as housing, education, clean air and water, health care, and protection of the poor and elderly. Economic austerity and the suppression of struggles in the labor market have had the ironic effect of spreading the potential for new coalitional struggles. Whole communities suffer from inflation, unemployment, and plant closings. The ratepayers' movement, with its middle-class leadership, is also a civic rebellion aimed at broad local welfare goals. Its largest national confederation, the Citizen–Labor Energy Coalition, includes 200 elderly, community, and labor union organizations. Like the classes themselves, class struggle persists and adapts to the changing structure of the economy.

Social class is a basic interpretive tool in the sociological tradition that provides lucid summations of what is going on in the world. In the process of social mobility, the changing occupational world, or the personal and cultural experience of distinct social groupings, social classes distinguish the foundations of social structure and explain the sources of inequality. An understanding of classes helps penetrate durable myths about universal mobility and the equalization of opportunity, whether those are presented historically as a steady progression toward an egalitarian society or contemporarily as a placid postindustrial order. The idea of social class uncovers a different and more complex human reality. Classes explain a good deal about why the homeless live alongside the wealthy. Classes explain power, humiliation, self-satisfaction, hard-won mobility, and the manner in which people struggle for a livelihood with some dignity.

The metaphor of class stratification or census accounts of occupational distribution tells us something about contemporary society, but they also risk obscuring a vivid social life. The world of social class is people making their own culture by acting in class ways. Any metaphor transforms its subject, but we would gain some understanding by replacing class categories or strata with an image of classes as poker players. The players, each with different styles and resources, engage in a high-stakes game governed by rules that favor the house and the highest bidder; but a game is always susceptible to intrepid plays and collective disputes over the rules.

SELECTED BIBLIOGRAPHY

ARONOWITZ, STANLEY. 1973. *False Promises: The Shaping of American Working Class Consciousness.* New York: McGraw-Hill.

BELL, DANIEL. 1973. *The Coming of Post-Industrial Society: A Venture in Social Forecasting.* New York: Basic Books.

BENDIX, REINHARD, and SEYMOUR MARTIN LIPSET. 1959. *Social Mobility in Industrial Society.* Berkeley: University of California Press.

BLAU, PETER M., and OTIS DUDLEY DUNCAN. 1967. *The American Occupational Structure.* New York: Wiley.

BRAVERMAN, HARRY. 1974. *Labor and Monopoly Capital: The Degradation of Work in the Twentieth Century.* New York: Monthly Review Press.

DISRAELI, BENJAMIN. 1845. *Sybil — or the Two Nations.* Reprint. Penguin English Library Edition, 1980. London: Penguin Books.

DOMHOFF, G. WILLIAM. 1974. *The Bohemian Grove and Other Retreats: A Study of Ruling-Class Cohesiveness.* New York: Harper & Row.

FRIEDMAN, BENJAMIN M. 1990. "Reagan Lives!" *New York Review of Books* (December 20):29–33.

GARSON, BARBARA. 1972. *All the Live Long Day: The Meaning and Demeaning of Routine Work.* London: Penguin Books.

GIDDENS, ANTHONY. 1973. *The Class Structures of the Advanced Societies.* New York: Barnes and Noble.

GOLDTHORPE, JOHN, DAVID LOCKWOOD, FRANK BECHHOFER, and JENNIFER PLATT. 1969. *The Affluent Worker and the Class Structure.* Cambridge, MA: Cambridge University Press.

GREEN, JAMES R. 1980. *The World of the Worker: Labor in Twentieth-Century America.* New York: Hill & Wang.

GUTMAN, HERBERT G. 1977. *Work, Culture, and Society in Industrializing America: Essays in American Working-Class and Social History.* New York: Vintage Books.

HAMILTON, RICHARD F. 1972. *Class and Politics in the United States.* New York: Wiley.

HAMPER, BEN. 1991. *Rivethead: Tales from the Assembly Line.* New York: Warner Books.

HARRISON, BENNETT, and BARRY BLUESTONE. 1988. *The Great U-Turn.* New York: Basic Books.

HOUSE WAYS AND MEANS COMMITTEE. 1991. "Background Material on Family Income and Benefit Changes." Print 102-30 (December 19).

JOINT CENTER FOR POLITICAL STUDIES. 1991. *Poverty, Inequality, and the Crisis of Social Policy.* Washington, D.C.: The Joint Center for Political Studies.

KUTTNER, BOB. 1983. "The Declining Middle." *The Atlantic* 252(1):60–72.

LOCAL 1199 NEWS. June 1992. "New York: Drug, Hospital, and Health Care Employees Union."

LYND, ROBERT S., and HELEN M. LYND. 1937. *Middletown in Transition: A Study in Cultural Conflicts.* New York: Harcourt Brace.

MANDEL, ERNEST. 1972. *Late Capitalism.* London: New Left Books.

MARX, KARL. 1852. *The Eighteenth Brumaire of Louis Bonaparte.* Reprint. New York: International Publishers, 1963.

———. 1894. *Capital,* Vol. 3. Reprint, translated by Ben Fowkes. New York: Vintage Books, 1981.

MILLS, C. WRIGHT. 1951. *White Collar: The American Middle Classes.* New York: Oxford University Press.

MILLS, C. WRIGHT. 1956. *The Power Elite.* New York: Oxford University Press.

NEWMAN, KATHERINE S. 1988. *Falling from Grace: The Experience of Downward Mobility in the American Middle Class.* New York: Free Press.

ROBERTS, SAM. May 10, 1992. "A New Face for American Labor." *New York Times Magazine.*

ROTHCHILD, EMMA. 1972. "GM in More Trouble." *New York Review of Books* 19 (March 23).

———. February 5, 1981. "Reagan and the Real Economy." *New York Review of Books* 28.

RUBIN, LILIAN BRESLOW. 1976. *Worlds of Pain: Life in the Working-Class Family.* New York: Basic Books.

RYAN, WILLIAM. 1976. *Blaming the Victim.* New York: Random House.

SENNETT, RICHARD, and JONATHAN COBB. 1972. *The Hidden Injuries of Class.* New York: Random House.

SHAIKEN, HARLEY. 1984. *Work Transformed: Automation and Labor in the Computer Age.* New York: Holt, Rinehart and Winston.

TERKEL, STUDS. 1972. *Working.* New York: Avon.

THERNSTROM, STEPHAN. 1969. *Poverty and Progress: Social Mobility in a Nineteenth Century City.* Cambridge, MA: Cambridge University Press.

THOMPSON, E. P. 1978. "Eighteenth-Century English Society: Class Struggle Without Class?" *Social History* 3(2):133–65.

U.S. BUREAU OF THE CENSUS, Department of Commerce. 1991. *Trends in Relative Income: 1964 to 1989.* Washington, D.C.: Government Printing Office.

WARNER, W. LLOYD, and PAUL S. LUNT. 1941. *The Social Life of a Modern Community.* New Haven, CT: Yale University Press.

WEBER, MAX. 1946. "Class, Status, Party." Reprint. Pp. 180–95 in *From Max Weber: Essays in Sociology,* translated and edited by Hans Gerth and C. Wright Mills. New York: Oxford University Press, 1958.

WRIGHT, ERIK OLIN, CYNTHIA COSTELLO, DAVID HACHEN, and JOEY SPRAGUE. 1982. "The American Class Structure." *American Sociological Review* 47:709–26.

5

Social Status and the Struggle for Equality

On August 26, 1977, modern feminists marched in Washington, D.C., on behalf of the ERA. Dressed in white, they were resuming a 1913 women's suffrage parade that had been halted by violence. (UPI/Bettmann)

BEYOND CLASS

Thirty years ago, no one in the United States, or in the quietly industrious town of Delano, California, would have hesitated to describe the Mexican–American farm workers there as working class according to their effort and lower class by their returns. Yet, no one who saw them singularly as workers or expected them to act only as a class would have understood what they were about to do.

During the spring of 1965, Filipino grape pickers in the southernmost parts of the state had walked out of the fields under the leadership of the AFL–CIO Agricultural Workers Organizing Committee (AWOC). Their simple grievance was that domestic workers paid at piece rates were earning an average hourly wage of $1.20, while Mexican guest workers (*braceros*) were guaranteed by law $1.40 an hour. The AWOC walkout won equal wages, but as farm workers moved northward to the Great Central Valley with the advancing harvest season, they encountered in Delano the same differential wage scale.

Delano was different in other ways. It was home to a tough-minded community of local growers, many of them second-generation Yugoslavians who, with guts and determination, had wrested their farms from the semidesert. It was also home to a young Mexican–American community organizer named Cesar Chavez who had come to the Central Valley on behalf of the Community Services Organization — a populist group animated by the radical methods of Saul Alinsky. The growers were tight-fisted, parochial, and clannish; only two of thirty-eight commercial farms were absentee-owned by large corporations. The Mexican–American community was bone poor, seasonally employed, and unorganized. Chavez had established the National Farm Worker's Association (NFWA) in Delano, which, despite its auspicious name, was a fragile volunteer agency devoted to organizing through provision of essential services such as legal aid, a credit union, and a co-op gas station and garage. In the fall of 1965, Chavez estimated that any militant class action on the part of the NFWA was at least three years away.

The plan Chavez had laid was disrupted when the Filipino-led AWOC struck Delano growers on September 7, 1965, over the differential wage issue. Paternalistic with their workers and contemptuous of labor organization, the growers brought in grape pickers from surrounding towns, many of them also Mexican Americans. As the chief AWOC organizer explained, "That's when I went to see Cesar and asked him to help me." For Chavez, according to John Gregory Dunne, "the strike began three years ahead of schedule":

> Chavez was in a quandary. Relations between AWOC and the NFWA had never been strong. Chavez had twice refused offers to join AWOC, and at the rank-and-file level, there was an undertow of racial antagonism between the predominantly Filipino AWOC and the predominantly Mexican NFWA. More importantly, Chavez was reluctant to incur the risk of failure by committing his ill-prepared forces to a situation he did not control. "That morning of September 8, a strike was the furthest thing from my mind. . . . "
>
> Within the NFWA, however, there was a feeling that if Chavez did not call a strike now, ready or not, he would be forever loath to take the chance. Backed into a corner, his hand forced, Chavez finally decided that it would be far worse to ignore the strike than to join it. On the night of September 16, he called a strike vote. Hundreds of people were packed into the Filipino Hall on the West side of Delano. Chavez stood before the crowd, dressed in work pants and an old sport shirt. "You are here to discuss a matter which is of extreme importance to yourselves, your families, and your community," he said. "So let's get to the subject at hand. A hundred and fifty-five years ago, in the state of Guanajoto [*sic*] in Mexico, a padre proclaimed the struggle for liberty. He was killed, but ten years later Mexico won its independence. We Mexicans here in the United States, as well as all other farm workers, are engaged in another struggle for freedom and dignity which poverty denies us. But it must not be a violent strike, even if violence is used against us. Violence can only hurt us and our cause. The law is for us as well as the ranchers. The strike was begun by the Filipinos, but it is not exclusively for them. Tonight we must decide if we are to join our fellow workers." (1967, 79–80)

So began another chapter in the civil rights struggle and, after a protracted campaign with many reversals, the first successful organization of farm workers in U.S. history. But those enormous achievements are not what makes the story of special interest here. Farm workers showed the unusual group solidarity and political energy of a movement organized on a common ethnic culture and experience.

The symbols and composition of the Delano strike distinguish it from conventional class struggles. In his successful call for a discussion and strike vote, Chavez, an Arizona-born U.S. citizen, recalled Father Miguel Hidalgo and Mexico's struggle for independence from Spain. As the strike unfolded, its participants knew it by the Spanish name *La Huelga.* Marches and picket lines displayed the flags of Mexico and the United States but usually as backdrop to the banner of the Virgin of Guadalupe, patron saint of Mexico and symbol of her poor. The Delano strike became a national crusade for the rights of Mexican Americans and farm workers, merging at times with the national civil rights movement and drawing support from Gringo liberals. Its profound aspiration was dignity, the word repeated most often in connection with what the struggle sought.

Once it became clear that the strike would succeed, that farm workers would at last be organized, the Teamsters Union tried, with some success, to steal the victory by signing agreements with growers. One unwittingly perceptive Teamster official complained of the UFWA, "They're not even a union. They're a civil rights organization" (Dunne 1967, 158). A local grower, more consciously insightful, said of Chavez, "He doesn't want a union, Jack, he wants a social revolution" (Dunne 1967, 126). They were saying, in other words, that Delano was something more than a class conflict.

Mexican Americans and farm workers are not unique in this ability to mobilize class struggles on the basis of cultural symbols and solidarity. The Drug, Hospital, and Health Care Employees Union discussed in Chapter 4 owes its present success to a long tradition of interethnic cooperation. The union was founded in 1929 by Leon Davis, a Russian–Jewish immigrant trained as a pharmacist and labor organizer. Racial and anti-Semitic restrictions on medical school admissions in the early part of this century led to a high concentration of professional blacks and Jews in pharmacy. Davis's original Local 1199 of the Retail Drug Employees Union was a rare example of an interethnic movement, initiated during the 1930s depression and expanded to black and Hispanic, principally women, hospital workers in the late 1950s.

Hospitals, like Chavez's fields, were considered impossible terrain for labor organization. Typically, they were operated by voluntary church and charity groups with a staff of dedicated and well-paid professionals. Yet this image obscured the much larger number of maintenance, housekeeping, kitchen, orderly, and nursing employees who worked long and irregular hours for near-poverty wages. By virtue of their class position, hospital workers were ripe for organization, but it was their ethnic and gender status that enabled their mobilization. Local 1199 played a key role in the civil rights movement. It provided the first national demonstration of "union power, soul power" when hospital workers in Charleston, South Carolina, sought 1199 support in a successful and highly publicized clash between the state's white political leadership and national guard arrayed against 1199 and the Southern Christian Leadership Council (Fink and Greenberg 1989).

Farm workers and health care workers provide an instructive comparison. They are different in many ways: one rural and the other urban; one principally male and Mexican American, the other female, black, and Puerto Rican; one facing the threat of harvest mechanization, the other expanding with each technological advance. Indeed, in recent years Chavez's American Farm Workers (AFW) have turned inward in the face of declining contracts and membership, while Dennis Rivera's health care workers have become a potent force in city and state politics. All these differences, however, are circumstantial by contrast to the key similarity: They were successful labor movements because they were also social movements based on ethnic solidarity and a common culture.

Reflection on these events from the standpoint of social classes and class struggles produces some analytic discomfort — a suspicion that economic interest and power in the marketplace (the defining features of class in Chapter 4) fail to capture the essence of the farm workers' and health care workers' movements or, for that matter, of the civil rights struggle as a whole. On one hand, although many Mexican Americans, Puerto Ricans, and blacks are certainly poor, ill-housed, and underemployed, the inequalities they suffer extend beyond the marketplace. In important part, their plight results precisely from the fact that they are minorities — that is, members of a social category whose history of incorporation into the United States and present social status carry disadvantages. Those disadvantages (of prejudice, second-class citizenship, and discrimination) did not originate in the marketplace, although they were later exploited, no doubt amplified, in the labor market. On the other hand, the unifying bases of action by these minorities are uniquely compounded of particular histories, religions, and ethnic cultures that are certainly brought to the marketplace, as in the Delano and Charleston strikes, but not born there. The roots of and responses to social inequality overlap with class issues, but they include much more.

In summary, this illustration suggests that group organization and social action are products of the interplay of class and status. The purpose of this chapter is to describe status groups and to show how they interact with social classes.

THE MEANING OF STATUS GROUPS

Having devoted Chapter 4 to an explanation of inequality based on social class, we now confront a problem. A good deal of the inequality experienced in modern societies stems from sources other than social class. Moreover, when struggles are mounted to eliminate social inequality, they draw their energy and methods as much from the traditions of religion, ethnicity, nationalism, and community, as from the episodic solidarity of social class. We need a broader vocabulary to explain these things.

The classical tradition provides a solution. Karl Marx knew that there were a great many group loyalties that competed with, even obscured, class awareness, but they played no important role in his theory of change. Max Weber ([1946] 1958) went a step further by suggesting that there were other bases of group formation every bit as important as class interest. Weber drew a sharp distinction between **class situation** and **status situation** as conditions that may give rise to organized **social classes** or to **status groups**, respectively. Social classes, as we saw in Chapter 4, are based on economic interests and formed by people sharing common life chances in the marketplace. Class is a powerful but highly specific basis for group formation. Classes, accordingly, are not communities with diffuse ties of social solidarity for Weber. Status situation and status group are very different, not so much opposed to class as broader unifying conditions resting on more diverse interests. Weber explains the difference:

> In contrast to the purely economically determined "class situation" we wish to designate as "status situation" every typical component of the life fate of men that is determined by a specific, positive or negative, social estimation of *honor*. This honor may be connected with any quality shared by a plurality, and, of course, it can be knit to class situation: class distinctions are linked in the most varied ways with status distinctions. . . . In contrast to classes, *status groups* are normally communities. ([1946] 1958, 186–87)

The key phrases in this passage are *social honor* and *any quality shared by a plurality*. Social honor is roughly equivalent to prestige: high status or low status in ordinary language. Qualities shared by a plurality are potentially great in number, but those typically singled out for conferral of positive or negative status include race, ethnicity, gender, age, religion, nationality, and the like. Societies differ in the bases on which they award status. Language, tribe, or regional origin are as important in some as race or gender in others. Among exemplary status groups in history, Weber listed preindustrial aristocracies and craft guilds, slaves, pariah people like Jews, and, in the fullest development, the closed castes of India.

Modern status groups are based on a distinctive set of qualities. Race, ethnicity, and national origin are everywhere important. Religion and language define prominent status groups in some European countries (Belgium, Spain) and Canada. Increasingly, gender and age have shifted from potential bases of group formation (status situations) to well-defined status groups. Contemporary pariah peoples in the United States include Native Americans, African Americans, Mexican Americans, and many of the "new ethnics" (Cubans, Vietnamese). Equally important, of course, are the status groups that enjoy positive estimations of social honor: society people, the jet set, debutantes, the "beautiful people" chronicled by *People* magazine, or, as C. Wright Mills called them, "the celebrities . . . all those who succeed in America — no matter what their circle of origin or sphere of action . . . [members of] the American forum of public honor" (1956, 71).

Status and Class

The defining features of status groups distinguish the concept from class. First, the qualities to which (positive or negative) social honor attaches are diverse — from the seemingly objective signs of race to the subtleties of success. Second, status groups vary in prominence and coherence. Homosexuals, ex-convicts, and priests are bona fide status groups, although they lack the perceived importance of other groups united by ethnicity or gender. Third, as Weber stressed, "In content, status honor is normally expressed by the fact that above all else a specific *style of life* can be expected from all those who wish to belong to the circle. Linked with this expectation are restrictions on social intercourse" ([1946] 1958, 187). Status groups are identified by the conventions they observe, by distinctive behaviors and customs. In contrast to these status group characteristics, classes are defined by a more specific quality (power in the marketplace), limited number, and varied life-styles.

Finally, status groups differ from classes by not obeying the economic logic of the marketplace. Indeed, they may hinder free development of the market. Status represents a monopoly of privilege or disadvantage that is determined by prestige rather than profit and loss. For example, low status of blacks and women in the occupational world means that they are not given the opportunity to compete freely with other ethnic groups and males as an unimpaired labor market would dictate. Similarly, celebrities go to the head of the line and command special attention when they sell products or politicians, while ex-convicts form at the rear. People who enjoy high prestige are expected to be above crass material considerations. Weber captured the point nicely when he said, "The notion of honor peculiar to status absolutely abhors that which is essential to the market: higgling" ([1946] 1958, 193).

Despite these differences, status groups are less the opposite of classes than alternative bases of stratification. Sometimes these contrasting principles of group formation conflict with one another, as in the case of status impeding the free play of the market — hindering, for example, a free labor market in which Mexican Americans or women have equal access to equal jobs based on their economic value as workers. At other times, the two principles reinforce each other. Status, to some extent, is a commodity that can be bought and sold in the market where it is embodied in automobiles, homes, clothes, baubles, and, generally, a purchasable life-style. Nevertheless, disdain for status climbers in some higher circles indicates that honor cannot be bought at will, suggesting again the autonomy of status groups.

Conversely, as Weber noted, "An 'occupational group' is also a status group" ([1946] 1958, 193). Independent professionals (e.g., doctors and lawyers) and factory workers are distinct classes in the marketplace that also carry different prestige and constitute, for other purposes, distinct status groups. Within limits, class and status position are complementary. In the case of a successful black businesswoman whose class position carries some prestige, class and status are noncomplementary. Together they form an incongruity. More tellingly, however, because this woman is apt to enjoy prestige and privilege owing to her success in business, the incongruity is reduced by the greater importance of social class in this instance.

Status groups are arrayed, from exalted celebrities to disdained criminals, on the unique dimension of social honor. This chapter focuses on the struggle for status and the contribution that status makes to explaining social inequality (i.e., what it explains beyond class). For that reason, the focus is also on the large and disadvantaged status groups — on race, gender, and ethnicity. Inferences can be drawn from this material about high-status groups, but their specific situations are treated in later chapters.

Social class and status group, in sum, are distinct, but interactive dimensions of group formation. They combine in different ways depending on circumstance — the host society, the character of status groups, the fluidity of class structure, and the dynamics of politics. The relative importance of class and status also varies in time and with economic conditions. Weber hypothesized that "stratification by status" dominates in periods of social and economic stability, whereas "naked class situation" comes to the fore during "periods of technological and economic transformation" ([1946] 1958, 193–94). At the end of this chapter, we will have the opportunity to evaluate this important proposition.

The plain distinction between class and status allows us to supersede barren debates over whether classes exist or class analyses are refuted by the shifting impact of status. Instead, we can assess their relative and shifting importance as joint determinants of social inequality. This avoids the error of confusing prestige with class and mistakenly concluding that stratification by classes is no longer important. If class and status are employed as analytic principles in tandem, determining the actual importance of each is possible. Weber's proposition, moreover, provides a foothold on the question of why modern society has witnessed such a dramatic resurgence of status or civil rights struggles — why the politics of equality dominated the 1960s yet gave way to the politics of austerity in the 1980s and 1990s.

STATUS STRATIFICATION IN WORK AND COMMUNITY

Despite the absence of aristocracy and the eventual abolition of slavery, the United States has always had status groups and practiced status stratification. Black Americans have suffered most, but the same process has victimized women, ethnic minorities, the aged, religious and linguistic communities, and a host of others. It is useful to indicate some of the broad outlines of status stratification here, although a more rigorous examination of recent changes and the current situation fits better in a subsequent section.

The Changing Labor Force

Perhaps the boldest statistics in U.S. records are those showing that working blacks and women earn less than whites and men — disparities that persist even within the same occupations and educational levels. In 1950 the average black family or individual earned just over half the income of whites. Women in general did somewhat better, receiving slightly more than 60 percent of the income of men. The postwar years saw a number of important changes, which are demonstrated in a few simple tables. From 1955 to 1991, certain income differences were reduced, while others remained. As Table 5.1 shows, black–white income differences, especially among women, narrowed. Black family income has fluctuated between 58 and 66 percent of white family income during the period without showing any tendency toward sustained improvement. Black men started to gain and then stalled in comparison to white men. Black women made the greater gains nearly reaching income equality with white women in 1979 (57 to 95 percent) before losing substantial ground in the early 1990s. Women workers as a group made few gains relative to men until the 1980s, when (from 1979 to 1991) they closed the gap more rapidly (to 74 percent of male earnings). In 1991, for the first time on record, women earned almost three-quarters as much as men, although that is not equality and almost forty years have been required to narrow the gap by 10 percentage points.

These results involve more than a simple redistribution or convergence of incomes. The class structure and status order were changing. Table 5.2 shows some of the dramatic changes in the labor force that lie behind the outcroppings of status stratification. As the total column indicates, more people are full-time workers than ever before. From 1954 to 1989, the

TABLE 5.1 Black and Female Income for Full-Time Workers as a Percentage of White and Male Income: 1955–1991

	1955	*1969*	*1979*	*1986*	*1991*
Black men (percentage of white men)	55	64	73	72	73
Black women (percentage of white women)	57	82	95	91	86
Black families (percentage of white families)	58	65	63	57	66
All women (percentage of men)	64	61	63	69	74

SOURCES: U.S. Department of Commerce, Bureau of the Census. 1981. *Current Population Reports,* Series P-60, No. 129, Tables 11 and 67. Washington, D.C.: Government Printing Office; U.S. Department of Labor, Bureau of Labor Statistics. October 1980. *Perspectives on Working Women: A Databook,* Bulletin 2080, Table 68. Washington, D.C.: Government Printing Office; U.S. Department of Commerce, Bureau of Labor Statistics. June 1992. *Monthly Labor Review.* Washington, D.C.: Government Printing Office.

TABLE 5.2 Civilian Labor Force Participation Rates by Gender and Race, in Percentage (Age 16 Years and Over): 1954–1989

January of	*All Men*	*All Women*	*White Men*	*Black Men*	*White Women*	*Black Women*	*Total*
1954	86	34	86	85	33	47	59
1960	84	37	83	83	36	47	59
1970	80	43	80	78	42	50	60
1978	78	49	79	72	49	53	63
1989	76	57	77	71	57	59	67

SOURCES: U.S. Department of Commerce, Bureau of Labor Statistics. September 1982. *Labor Force Statistics Derived from the Current Population Survey: A Databook,* Vol. II, Table D-2. Washington, D.C.: Government Printing Office; U.S. Department of Commerce, Bureau of Labor Statistics. November 1989. *Monthly Labor Review.* Washington, D.C.: Government Printing Office.

percentage of the population in the labor force has risen from 59 to 67 percent, representing a much larger increase in absolute numbers given the population growth (in 1992 there were 117 million employed people in the U.S. labor force and another 10 to 12 million unemployed). The modest overall increase, however, is compounded of an actual decrease in the proportion of working men (down from 86 to 76 percent) and therefore an enormous compensating increase in the number of working women (up from 34 to 57 percent).

For an understanding of the women's movement, it is important to recognize that the greater part of this change took place in the 1970s. In the preceding twenty years, the percentage of working women inched up from 34 to 43 percent in 1970, but in the next decade it moved much faster, bringing us to the present circumstance in which over half of all women are employed full-time. From near equality of participation, the reduced number of black working men (down to 71 percent) is greater than that of white men (down to 77 percent). Black women have changed least in this respect; 47 percent were working in 1954 and 59 percent in 1989. The proportion of white women working now is nearly equal to the historical levels of black women. In general, the U.S. labor force has become more feminized, closer to parity in sex composition, and more white — owing to big gains among white women and significant losses for black men while their counterparts were changing more slowly.

Returning to the question of income differences, many of the new women entrants to the labor force have been assigned low-paid, low-skilled "women's work" that contributes to the disparity of female income as a percentage of male. For example, in the last decade women comprised 99 percent of all secretaries-typists, 97 percent of all registered nurses, 92 percent of all bank tellers, 87 percent of all cashiers, and 71 percent of all elementary and high school teachers but only 11 percent of all physicians and osteopaths, 9 percent of all lawyers and judges, and 3 percent of all engineers. Black women, on average, have longer experience in the job market and have gained access to many of the middle-class female occupations (e.g., nursing and teaching). The rapid increase of white female participation in the labor force means that some women workers have less experience on the job, which is a second factor beyond discrimination explaining income differentials. Indeed, the gains women have made in the 1980s are probably explained by the growing experience and promotion of those in the labor force as well as the educational qualifications equal to men of the new entrants.

As a result, women have closed the income gap more than black men, and black women have made the greatest gains. But, there is a disturbing irony here. Although black women may have more job experience, they are still paid slightly less than white women. Moreover, the gains of black women really mean that they are approaching women's lower standard by contrast to white men. Claims by the women's movement about the absence of comparable worth have statistical support.

Table 5.3 completes this picture, indicating the occupational positions of the labor force by race and gender. Although the number of comparisons that can be made within this table is very large, the major ones elaborate the foregoing interpretation of how occupational (and class) stratification interacts with status inequality.

Looking first at the contrasts between white men and women, their proportions are about the same in the professional–technical ranks, but there are more men in managerial and craft occupations, along with many more women in clerical and service jobs — and, of course, the last two are lower-paying. These differences are pronounced in more detailed occupa-

TABLE 5.3 Occupational Distribution by Gender and Race, in Percentage: 1959–1991

	White Men		*White Women*		*Black Men*		*Black Women*	
	1959	*1991*	*1959*	*1991*	*1959*	*1991*	*1959*	*1991*
Professional–technical	11	12	13	15	4	7	6	11
Managerial–administrative	15	15	6	12	3	6	2	7
Sales	6	14	9	17	1	9	1	13
Clerical	7	5	33	28	5	9	8	25
Craft	20	20	1	2	10	14	1	2
Operatives	19	13	16	6	24	21	14	11
Nonfarm laborers	6	6	0	2	25	9	1	2
Service	5	9	19	17	15	19	59	28
Farm	10	6	4	1	15	5	10	0
Total number (in thousands)	39,493 (100%)	58,805 (100%)	18,512 (100%)	45,670 (100%)	3972 (100%)	6080 (100%)	2652 (100%)	6111 (100%)

SOURCES: U.S. Department of Commerce, Bureau of Labor Statistics. October 1980. *Perspectives on Working Women: A Databook*, Bulletin 2080, Tables 74 and 62. Washington, D.C.: Government Printing Office; U.S. Department of Commerce, Bureau of Labor Statistics. August 1991. *Employment and Earnings*. Washington, D.C.: Government Printing Office.

tional breakdowns, adding force to the claim of discrimination against women in the job market. As suggested previously, the occupational distribution for black women resembles that of white women, although blacks are somewhat less represented in the professional (11 vs. 15 percent) and clerical (25 vs. 28 percent) categories and more numerous in services (28 vs. 17 percent) than white women. Much sharper differences appear between black men and women. Black men are spread mainly and fairly evenly across the categories of craft, operative, labor, and service, categories that are not typical of black women, except for services.

On an impressionistic basis we can conclude from Table 5.3 what more rigorous statistical analyses discover — namely, that race and gender interact in affecting occupation. That is, similarities exist among the distributions for women irrespective of race and origin, just as there are similarities among blacks by contrast to whites. The patterns are too varied to be captured in any single (or simple) summary precisely because *both* gender and race influence occupation — just as occupations discriminate among workers in varied ways. To illustrate that point with a final irony, all women receive a fraction of male income even within the same occupational category. White women are slightly better paid than black women. However, white women receive only about 70 percent of the income of white men, while black women earn 93 percent of the income of black men (as of 1986, not shown in the table) — a fact explained by both the lower wages of minority men and the greater similarity of male–female occupational distributions for minorities.

Focusing just on income inequality, status and class (as measured very roughly by occupational category here) have separate *and* combined effects. Moreover, the combined effects vary depending on the combination. Illustratively, being female is a handicap in the labor market. So is being black or a laborer. But, being a black woman is less of a handicap by contrast to all female workers than is being a black man by contrast to all male workers. At least, it is a different kind of handicap, recognizing that black women earn less than black men.

The effects of status stratification are best known through their appearance in the labor market because those inequalities are more tangible, widely felt, readily documented, and open to redress by civil rights groups and political reformers. Yet social honor affects most aspects of people's lives often in ways that are more direct because they are unmediated by class and more damaging because they invade personal realms where self-esteem is fashioned. Where and with whom we live, home and community, are as important as the job. Social class has a decisive effect on residential choice, but status groups figure here too.

Minorities in the Community

In the early twentieth century, most U.S. cities embraced a variety of ethnic neighborhoods (shaped by the narrow choices of class and kinship), heterogeneous (though overwhelmingly white) middle-class communities, and the enclaves of luxury that were both white and Christian. Although the privileges of class helped perpetuate this pattern, it was ensured by restrictive covenants — agreements among home owners that they would not sell to undesirables such as blacks and, particularly, Jews who were the main targets of the covenants because some of them could afford to integrate the neighborhood. Although tightly knit ethnic neighborhoods and restrictive covenants have largely passed from the urban landscape, U.S. cities are more segregated today than ever. Status groups are at the heart of this change.

Residential segregation is a product of two broad trends: the overall urbanization of the United States, made possible by transportation technology and economic centralization, and the migration of southern blacks to northern cities, occasioned by the commercialization of agriculture. In 1900 just 40 percent of the U.S. population lived in urban places. Today the figure is 75 percent. In 1910 80 percent of the black population lived in the South, and 70 percent lived in rural settings. Today 50 percent of U.S. blacks live outside the South, and 80 percent live in cities. In short, while the nation became urbanized, the black population became more so, a great part of it moving to northern cities.

In an earlier benchmark study of the changing urban population, Reynolds Farley (1970) took the largest U.S. metropolitan areas and compared the relative size of the central-city black population in 1950 and 1970. In just twenty years, it doubled from 13 to 26 percent and, updating his figures, was almost 30 percent in 1980. If the Hispanic and black central-city populations are combined, they comprise 39 percent of the largest U.S. cities. Illustratively, comparing the black population in 1950 with the combined black and Hispanic populations of 1980, some of the central-city changes are as follows: New York, 10 to 45 percent; Chicago, 14 to 54 percent; Detroit, 16 to 66 percent; Washington, D.C., 35 to 73 percent; Baltimore, 23 to 56 percent; Miami, 16 to 81 percent. Eight of the top ten central cities have a combined black and Hispanic populations in excess of 40 percent.

The important point in all this is not simply the faster rate of urbanization for minority groups but their increasing segregation. The U.S. Census Bureau measures the urban population in several ways (Chapter 3), including the central cities just mentioned and the urbanized area comprising the central city and suburban ring. In a related study Farley (1977) showed the curious pattern of growth and decline that ends up producing greater segregation. The central cities are losing their white and upper-income groups and gaining blacks, Hispanics, elderly, and people with lower incomes. Some cities have experienced a *net absolute* population loss and therefore a very large *percentage* increase in minorities. Meanwhile, the suburbs have grown in *both* black and white populations. But, black suburban population growth has been mainly in either all-black developments or old suburbs that have shifted to industrial and commercial use, leading to white abandonment. Naturally, some middle-income blacks have moved into integrated suburbs, but their numbers are small compared with the all-black new developments and old suburbs. The net result is that the central cities are becoming sharply segregated minority enclaves, while, in Farley's words, "residential segregation patterns of central cities are reappearing within the suburbs" (1977, 527).

Detailed analyses of ethnic segregation based on the 1990 census are not yet available. Nevertheless, Douglas Massey and Nancy Denton developed several projections in a 1987 study showing, first, that the segregation of blacks was much worse than earlier studies had demonstrated and, second, that although Hispanic and Asian segregation is far less extensive, enclaves separating these groups are beginning to appear. By any measure, the residential segregation of minorities is getting worse.

The mechanisms through which this pattern of segregation is realized tell us more about status stratification. The sheer cost of housing and availability of jobs account for a good deal strictly on a class basis. Where that does not reach, status discrimination comes into play.

At one time, status discrimination was practiced widely and openly in restrictive covenants and racial cruelty — from refusals to show residential property to minorities in white-only areas to Ku Klux Klan–style cross burnings in the front yards of blacks who did manage to integrate neighborhoods. With growing legal curbs on such practices, new ones came along in more sophisticated forms of *institutional racism* — that is, routine practices in law, government, or the private economy that have the effect of racial discrimination, whether or not they are explicitly so intended. Minorities in the housing market found themselves the object of "steering"; they were shown homes for rent or sale only in segregated neighborhoods or ones in transition due to white flight. The pervasive practice of redlining got its name from the red lines that lending institutions drew on city maps to indicate areas in which they would make no mortgage loans. Their rationale was that because those areas were in transition, property values were likely to drop once the minorities took over, leaving the lenders with less valuable collateral. The self-fulfilling effect of redlining was that neither blacks nor whites could purchase homes in central-city areas with conventional financing. Once-viable neighborhoods began to deteriorate as older homes were converted to multiple-occupancy rental units.

The financial institutions helped create new residential segregation by ignoring the minority home buyer, disinvesting in the troubled city, and reinvesting their funds (often generated from inner-city savings and loan companies) in the suburbs where blacks were restricted by costs or steering. As the effects of lending practices took hold, other institutions added their weight, particularly industry and business that began moving to suburban quarters, thereby depriving the cities of jobs. Under the press of black urban migration and institutional neglect, the ghettos enlarged.

The critical point in this example is that class stratification working alone would have produced a different result. Whites and minorities would have been disproportionately mixed in cities and suburbs, according to their incomes and market capacities. When status group considerations compound the process, even high-income blacks have difficulty finding integrated suburban housing. Low-income whites cannot find mortgages or jobs in the cities, even when they want to stay there and maintain their ethnic communities. Status stratification exacerbates and distorts the pattern. It creates anomalies like low-income black suburbs, the exclusion of many middle-income blacks from the average suburb, and once-fashionable central cities that have been abandoned by higher-income whites except as daytime (and nocturnally dangerous) centers of commerce and government. This is what Weber had in mind when he said that class and status group sometimes conflict.

We have focused here only on the gross effects of status stratification. Volumes are written on the inequalities suffered by status groups. Jews are still restricted from certain elite clubs. Pacific Coast Japanese Americans were interned and stripped of their land in 1942, not, as we now learn, because they were a wartime security threat but because they were highly successful agriculturalists. Small fishing towns on the Atlantic and Gulf Coasts have tried to banish the competition of Vietnamese immigrants. Today blacks and Hispanics hold less than 2 percent of elective posts, although they comprise nearly 20 percent of the population and a majority in many cities and electoral districts. The Equal Rights Amendment failed to win approval of the necessary number of state legislatures. Bills aimed at outlawing employment discrimination against homosexuals have been defeated by conservative groups led by fundamentalist churches. Ex-convicts, disabled persons, the elderly, and Moslems (especially when they are black) all live with the occupational and civil consequences of disrespect.

The beneficiaries of all this are the dominant status groups: whites, men, Protestants, the middle-aged, and so forth. These are the people who routinely enjoy status honor in the sense of full civil liberties, political representation, job and mobility opportunities. Indeed, one of the peculiarities of U.S. society is that the presumably "average American" portrayed in the mass media (e.g., in the central roles of television shows) is usually a relatively affluent male professional. Several years ago, a popular television show was modeled after the real-life Japanese–American coroner of Los Angeles County. On television the coroner was a lovable middle-aged white male doctor with a dutiful Japanese assistant. This is the society's model of respect or status honor, although white, male, Protestant professionals comprise only about 10 to 15 percent of the population.

The general pattern of invidious status distinctions is not new or peculiar to the United States and western Europe. What is new, and peculiar to the more developed countries, is that discrimination is now recognized as an abridgment of rights. Claims for justice made by status groups have legitimacy in the sense that they can be made, are usually heard, and, if valid, can persuade many that they should be redressed. The moral order has changed in the course of this century to the point that equality of status honor is politically a rightful aspiration. Certainly, that does not mean that status stratification is being eliminated wherever it is challenged. It does mean that denial of equal rights is no longer taken for granted and readily justified on the basis of some natural law or economic expedient. Opponents are on the defensive.

The rights of minorities and other status groups to equal treatment (and, slowly, to equal outcomes) are no longer debated. The debate has shifted to whether real equality exists. In something of an extreme example of public tolerance, a bill guaranteeing employment rights for homosexuals was recently vetoed by the governor of California after passing both houses of the state legislature. Although this was a defeat of social rights, the veto was defended, not on the basis of principle (that homosexuals should or should not enjoy those rights), but on the allegedly factual claim that sufficient evidence of discrimination did not exist. This change in the cultural understanding of social rights is the result of a political struggle and equality revolution that swept over the United States and Europe in the 1960s.

THE EQUALITY REVOLUTION

I borrow, and recommend, the phrase *equality revolution* from an essay by the sociologist Herbert Gans. Although he was writing in the midst of the controversy, Gans saw a broader meaning:

> Someday, when future historians write the history of the 1960s, they may describe it as the decade when America rediscovered poverty still in its midst and when social protest, ranging from demonstrations to violent uprisings, reappeared on the American scene. But these historians may also note a curious fact, that the social protest had very little to do with poverty. . . . The social protest that began in the 1960s had to do with *inequality*. So far the demand for greater equality has come largely from the young, from the black, and from women, but other groups have asked for more autonomy or control over their own lives, for more liberty and democracy. . . . In the years to come, I believe America will face more such demands from many other people, which will be widespread enough that they might be described as an "equality revolution." ([1968] 1974, 7–8)

From a longer historical vantage, in 1950 the British sociologist T. H. Marshall observed that citizenship has three elements: the civil liberty composed of rights necessary for individual freedom, the political right to participate in the exercise of power, and the "social element [meaning] the whole range from the right to a modicum of economic welfare and security to the right to share to the full in the social heritage and to live the life of a civilized being according to the standards prevailing in the society" ([1950] 1964, 71–72). Marshall thought that civil rights were won in the eighteenth century, political rights in the nineteenth century, and social rights are the essence of twentieth-century movements. Gans and Marshall agree that the struggles over inequality and social rights represent a historical watershed in the philosophical principles of collective action. Although this provides a valuable historical perspective on status struggles, it is also true that the battle for social rights did not begin in the twentieth century or in the 1960s. These movements have a long history. What changed in the recent period was that social rights came to be accepted as a legitimate political end.

Black Civil Rights

The centerpiece of the equality revolution was the civil rights movement. The uniqueness of this period, however, was that other movements grew up to supplement and reinforce one another, notably the women's and the peace movements. Each of these had a long pedigree. Prior to the Civil War, slave rebellions rocked the South (Nat Turner's revolt of 1831, John Brown's Harper's Ferry raid of 1859, and many more), and after reconstruction civil rights organizations became active in the North (the National Association for the Advancement of Colored People [NAACP] was founded in 1910). Antiwar movements grew from protest of U.S. involvement in the Philippines in 1898 to demonstrations and refusals to serve in the "imperialists'" World War I. The first Woman's Rights Convention was held in 1848 and grew into the women's suffrage movement which reached its zenith in 1920 with the passage of the Nineteenth Amendment and women's right to vote.

In the 1960s, these movements reappeared with new vigor and began joining forces to an unprecedented degree. Martin Luther King, Jr., pledged his forces to ending the Vietnam War, which was a moral and a black issue. Women formed interracial alliances. People spoke simply of "the movement." The times encouraged other insurgencies: the American Indian Movement, Chicano Power, Gray Panthers, and Gay Liberation. This was the equality revolution.

These years of turmoil changed the face of U.S. society. Yet, if they constituted a revolution of sorts, they did not end in *the* revolution. On the contrary, their peculiar brand of success lay in a new accommodation with conventional society. How is this explained? To be more precise about the question itself, it should be divided into three parts: Why did the equality revolution occur in the 1960s? What did it accomplish? Why was it transformed and in some ways reversed in the 1980s? The interaction of class and status contributes much to an explanation.

The immediate background to the civil rights movement included changes, particularly from the 1940s onward, that affected the U.S. population distribution, economy, and a growing moral issue — what Gunnar Myrdal's classic book of 1944 called *An American Dilemma: The Negro Problem and Modern Democracy*. A world war had just been fought to eliminate religious persecution and the doctrine of a superior race, a war in which U.S. blacks played a central role. How could the country continue to deny basic liberties to its minority populations and pretend to call itself the land of the free? Myrdal understood this as a corrosive conflict of values, a cultural conflict between manifest racism and the values of social equality. "The American Negro problem is a problem in the heart of the American" (1944, xvii).

It was also a dilemma at the heart of the U.S. economy. The modernization of southern agriculture had precipitated much of the black urban and northern migration. Blacks were not only crowding the cities, but also were now contributing a large slice of the urban labor force — men as industrial labor and women in services, as the earlier tables show.

The black movement had been slowly gaining political strength since before the turn of the century. From the 1890s until 1920, Booker T. Washington urged black education for equal employment opportunities and a National Negro Business League. From 1917, with the formation of the Universal Negro Improvement Association, through the 1920s, Marcus Garvey brought a new militance to the cause by urging "integral nationalism" and a network of self-sufficient black enterprise. Sociologist W. E. B. Du Bois left academic life to lead the Niagara Movement and the NAACP, which included white liberals, all working for equal rights. During World War II, a foundation was laid for the modern movement. The Congress of Racial Equality (CORE) was formed in 1942 and joined A. Phillip Randolf's segregated yet powerful Brotherhood of Sleeping Car Porters in demanding employment in wartime industry. Randolf threatened a march on Washington and pressured President Franklin Roosevelt to create the first Fair Employment Practices Commission.

Black discontent grew apace of urbanization and wartime industrialization. A rash of race riots in 1919 was repeated in Harlem in 1935 and during the war. Growing black rage in the northern ghettos was expressed not only in the sensational novel *Native Son* by Richard Wright (1940) and the urban sociology that Robert Park inspired. In 1945 St. Clair Drake and Horace Cayton published *Black Metropolis* profiling migration to Chicago and the disenfranchisement of blacks in the northern economy. Indeed, as perhaps the first prominent black novelist, Wright not only shocked white America by portraying the anger of a black youth who killed two whites, but he also noted that he came to understand the black plight by reading the Chicago urban sociologists. In the introduction to *Black Metropolis,* Wright says that he came to Chicago to tell his story:

> But I did not know what my story was, and it was not until I stumbled upon science that I discovered some of the meanings of the environment that battered and taunted me. I encountered the work of men who were studying the Negro community, amassing facts about urban Negro life, and I found that sincere art and honest science were not far apart, that each could enrich the other. The huge mountains of fact piled up by the Department of Sociology at the University of Chicago gave me my first concrete vision of the forces that molded the Negro's body and soul. (Cited in Drake and Horace Cayton 1945, xvii–xviii)

Although it is important to see these links between sociology and the public understanding fostered through art, neither the growing black militance, a moral dilemma, nor the maturing political organization precipitated a civil rights movement at this time. Instead, it was the changing conditions of national politics that lay behind the breakthroughs of the 1950s. The race issue had divided the Democratic party in 1948 when the southern "Dixiecrats" split from the party over civil rights issues. As Frances Piven and Richard Cloward observe, "The Republicans saw opportunities in the difficulties being experienced by the Democratic Party over the race question" ([1977] 1979, 213). Republicans were not sure whether they had more to gain by appealing to dissident southern Democrats or northern blacks. Pushed by President Eisenhower's commitment to civil rights, "the Republicans finally cast their lot with the potential for gain among northern blacks, for the congressional Republicans were all from northern states where blacks were concentrating" ([1977] 1979, 214). Although the NAACP had promoted legal challenges to discrimination for many years, it was the 1954 U.S. Supreme Court decision outlawing school segregation that symbolically began the modern civil rights struggle.

None of this should minimize the prolonged struggle of black people. That history was a precondition of the equality revolution. But as long as the white power structure could ignore the struggle with no risk to its own supports, the revolution could be held off. Economic and demographic shifts culminating in a postwar urban and industrial boom now meant that the race question could no longer be dodged by politicians who hoped to hold their jobs. It was the interplay of the continuing civil rights movement and the changing bases of political power that brought success in the late 1950s and 1960s.

Yet, the changes were far from preordained. The civil rights struggle had to escalate the pressure as it did with the Montgomery, Alabama, bus boycott in 1955. Suddenly the black struggle was mobilizing thousands of people drawn from its own middle class and Baptist church leadership, personified in Dr. Martin Luther King, Jr. Montgomery led to the Little Rock, Arkansas, fight for school integration (1957), public accommodations sit-ins (1960), freedom rides including northern and white liberal supporters (1961), and the decisive campaign in Birmingham, Alabama, which brought open clashes between civil rights marchers and southern law enforcement agencies, forcing the Kennedy administration in 1963 to act by proposing a comprehensive civil rights bill.

The Civil Rights Act of 1964 did not quell the black uprising. On the contrary, the summers of 1965–1968 witnessed the most extensive urban riots in history as the oppression of a century since the abolition of slavery took its violent revenge. But in 1964, both President Johnson's compensatory War on Poverty and the ghetto revolt burned into the political conscience a conviction that blacks and organized minorities had to be accommodated — in part to preserve the social order and its economic base and in part to resolve the moral dilemma. Political and industrial leaders, when challenged and moved to assure their own positions, were among the first to understand this. Blacks won political recognition of their rights.

Mexican Americans

The struggle of Mexican Americans bears some similarities to the civil rights movement. One key difference, however, is that instead of arriving in the United States as slaves, Mexicans and their homeland in the Southwest were acquired by conquest in the Mexican–American War of 1848. An enormous territory in then northern Mexico was annexed to the United States (including present-day Texas, New Mexico, Arizona, Utah, Nevada, California, and parts of Colorado and Wyoming — one-half of Mexican territory in 1847). Many of its 75,000 residents were more Spanish than Mexican because by 1848 Mexico had enjoyed only twenty-seven years of independence and the northern settlements dated back to Spanish land grants.

Similarities between the black and Hispanic populations began to appear with the twentieth-century trends in migration and urbanization. In 1920 there were about 1 million Mexican Americans in the United States. By 1960 the number had grown to 4 million and today it is 12 million — a figure that includes only citizens and permanent residents. If we add to this other Spanish-speaking people (Cubans, Central and South Americans, Puerto Ricans) and the estimated number of undocumented migrant Mexican workers, the size of the Hispanic population doubles and comes close to the number of blacks. As with blacks, 80 percent of Hispanics, in general, and of Mexican Americans now reside in urban areas. As blacks moved from the rural South to urban areas and northern central cities, Mexican Americans left the rural Southwest for its cities and for the northern climes of Chicago, Denver, and the like. Equally important, rapid urbanization in the Southwest spread to incorporate many of the tattered "Mextowns" that languished on the outskirts of cities. In addition, Hispanics closely resemble blacks in occupations and earnings.

Mexican Americans pursued uncelebrated struggles for economic and social rights beginning with mutual benefit societies in the 1910s, a League of Latin-American Citizens (LULAC) in Texas and a militant Confederation of Mexican Workers' Unions in southern California in the 1920s. World War II provided the same impetus for political organization as it had among blacks. The American GI Forum founded in Texas and the Community Services Organization in California devoted themselves to voter registration, community organization, and discrimination in the late 1940s. The GI Forum was established when a Texas funeral home refused to bury a Mexican–American veteran. Yet it was not until the 1960s that the renamed Chicano movement came to prominence with broad mobilization by the Mexican American Political Association (MAPA), Viva Kennedy clubs, and the Political Association of Spanish-speaking Organizations (PASO). Mexican Americans moved squarely into the civil rights movement with the stimulus of black pride and strong affinities for the youthful and Catholic symbols of the Kennedy administration.

Mexican Americans differed from blacks in the scope and divisions within their movement. An established middle class pushed successfully for electoral representation through the older groups. Chavez and the farm workers focused their efforts on unionization, allying more with church, labor, and Anglo liberal groups than with Chicanos. In New Mexico, the intrepid *Alianza Federal de Mercedes* led by Reies Lopez Tijerina captured a county courthouse to protest the expropriation of Mexican lands and grazing rights. The Raza Unida Party organized in south Texas, taking control of Crystal City through local elections and aiming for repeat performances elsewhere. A youth movement spawned Brown Berets in the urban barrios and the student-based MECHA (*Movimiento Estudiantil Chicano de Aztlan*).

Despite their differences, the gains by Chicanos matched those of blacks because, in each case, they accrued as much from the changing political environment as from the strategies of the movements themselves. As the Democratic administrations of the 1960s sought new urban constituencies and loyalists for their War on Poverty, Mexican Americans became valuable allies — particularly as swing voters in large states like Texas and California. In a revealing analysis, Craig Jenkins and Charles Perrow conclude about the farm workers something that was true for Chicanos as a whole. The 1960s did not differ greatly from earlier periods in the level of minority protest: "What changed was the political environment — the liberal community was willing to provide sustained, massive support for insurgency. . . . The dramatic turnabout in the political environment originated in economic trends and political realignments that took place quite independent of any 'push' from insurgents" (1977, 263, 266). Once more, this should not detract from the courageous stand of Mexican Americans but emphasize the necessary interplay of movement demands and divisions in the political elite that require reformed coalitions.

Native Americans

Unlike blacks and Hispanics, Native Americans predated colonialism and the modern state, governing vast areas of North America through diverse tribes and confederations prior to European settlement in the late 1500s. From the original colonies in present-day Virginia and New Mexico to the last Indian war at Wounded Knee, South Dakota, in 1890, the only ambition European settlers had for Native Americans was elimination. The first 300 years of Native American policy followed a sequence repeated in each new region of settlement: armed conquest, political suppression, exploitation of native labor (as slaves, in fur trading, and later in agriculture), and eventually removal of natives from one area to another farther west or to some unwanted tract of land designated by treaty as a reservation. In hundreds of treaties, European colonists and later the U.S. government dealt with Native Americans as separate nations. The treaties, of course, were unceremoniously broken as additional white settlers sought room for expansion and the U.S. government sold off parcels of land seized from Native Americans as a principal source of state funding. In the process, nevertheless, the U.S. government also fortified tribal identities through the system of treaties and reservations.

The period of conquest was followed by an attempt at detribalization through the 1887 General Allotment (or Dawes) Act that allotted parcels of land to individual owners rather than tribal groups. The government intended to extinguish tribal identities and treaty claims in this fashion. But the tribal-based social organization persisted because Native Americans lived in isolation among members of a common culture and defended their treaty rights to land and independence from U.S. laws. The first signs of supra-tribal organizations since the Native American confederations that were defeated in nineteenth-century wars appeared in political organizations founded by Native Americans who had been educated at boarding schools off the reservations. In 1911 the Society of American Indians (SAI) was founded in ideas of the Progressive Era, including belief in progress through education and integration rather than tribal separation. The Indian Defense League founded in 1926 sought to preserve treaty rights that allowed New York tribes to pass freely back-and-forth to Canada.

The largest political mobilization of Native Americans during this period came in 1934 when the New Deal reform administration passed the revolutionary Indian Reorganization Act (IRA). In a direct reversal of detribalization and integration, the IRA advanced tribal self-government, collective ownership of land and resources, and education in native history and culture — all protected by the oversight of the Bureau of Indian Affairs (BIA). Native Americans divided on the desirability of these changes, some in the American Indian Foundation (AIF) founded in 1934 opposing the New Deal and others in the National Congress of American Indians (NCAI) founded in 1944 stressing the preservation of tribal rights independent of the BIA. Despite this opposition to the IRA, Native Americans made gains during this period, which began to be threatened in the 1950s. Annoyed by continuing Native American claims based on treaty provisions, Congress now proposed an ominous scheme for "termination," which meant federal withdrawal of any role in Native American communities, including cessation of health services, education, grants, and sundry programs of the BIA. When termination was actually carried out with the Menominees in Wisconsin, the result was disastrous. Native Americans around the country began to mobilize against this new threat in meetings, such as the 1961 American Indian Chicago Conference (Cornell 1988).

The present Native American movement began with this new urban initiative. Over the past fifty years, Native Americans have become urbanized along with the rest of the population — from less than 10 percent urban in 1940 to nearly 60 percent urban (and off-reservation) in 1990. With termination threatening and the civil rights movement well underway, the American Indian Movement (AIM) was founded in Minneapolis in 1968. Subsequently, AIM became famous for militant occupations of Alcatraz Island in San Francisco Bay (1968), BIA headquarters in Washington, D.C. (1972), and BIA facilities at Pine Ridge, South Dakota (1973). Since 1976, however, AIM has been on the defensive after a shootout with FBI agents at Pine Ridge and lack of support among some elements of the traditional Native American communities. Yet AIM was only the dramatic expression of a political protest movement that began in the 1960s, with fish-in demonstrations for treaty rights, and continues in the 1990s, with tax protests, land and water claims, environmental lawsuits, and a remarkable resurgence of Native American culture. It would be difficult to evaluate how much the new Native American movement has accomplished since AIM was founded and how much was lost at the same time to reservation poverty, emigration, tribal-council corruption, and environmental pollution of Native American land (Matthiessen 1992). By way of gains, however, when the 1990 census discovered that the size of the Native American population had jumped dramatically to over 2 million (from 1.36 million in 1980), they realized that a good deal of the increase was explained by people who were proud to identify themselves as Native Americans for the first time.

Feminists

The women's movement, as we have seen, began in the 1830s and continued to the present with periods of quiescence during the abolitionist struggle of the 1850s and after suffrage in 1920. The modern movement coincided with the equality revolution and is conveniently dated by the heralded book of Betty Friedan, *The Feminine Mystique,* in 1963. Friedan argues that women, credited with some general competence and independence through the 1940s when they played key roles in wartime and defense industries, were denigrated in the next fifteen years. The feminine mystique comprised a series of ideas that added up to a patronizing celebration of femininity in the exclusive role of housewife and mother. Much of the responsibility for this change in women's roles Friedan assigns to the mass media and

> experts [who] told them how to catch a man and keep him, how to breastfeed children and handle their toilet training, how to cope with sibling rivalry and adolescent rebellion; how to buy a dishwasher, bake bread, cook gourmet snails, and build a swimming pool with their own hands; how to dress, look, and act more feminine and make marriage more exciting; how to keep their husbands from dying young and their sons from growing into delinquents. They were taught to pity the neurotic, unfeminine

> women who wanted to be poets or physicists or presidents. They learned that truly feminine women do not want careers, higher education, political rights — the independence and opportunities that the old-fashioned feminists fought for. . . . All they had to do was devote their lives from earliest girlhood to finding a husband and bearing children. (1963, 15–16)

Beyond capturing the mood of a generation — the "problem with no name" that haunted and rang true for many women and some men — Friedan locates the problem in time and hints at its causes. A change had occurred around 1945, and another one was in the offing. "*In the fifteen years after World War II,* this mystique of feminine fulfillment became the cherished and self-perpetuating core of contemporary American culture" (1963, 18, emphasis added). Yet by placing so much emphasis on experts and media stereotypes, Friedan does not satisfactorily explain the timing of the changes. Why did the experts and media begin promoting the feminine mystique after World War II?

Without pretending to have a complete answer to that question, it seems plausible that the changing economy and social structure explains a good deal. Large numbers of servicemen returned looking for jobs, some of which women had been holding or might compete for. Women were not needed in the labor force, at least not in their wartime numbers. Consequently, many women were removed or demoted from these good-paying jobs and occupational gender-typing was restored to some extent.

Meanwhile, U.S. industry had to convert from a wartime economy to peacetime production. This required creation of a massive new domestic market in products for the home and family (automobiles, appliances, television sets, etc.). An emphasis on domestic production for home and family consumption suggested the importance of homemakers to whom the media could appeal with its products and stereotypes. A key segment of the postwar economy was precisely the construction of suburban housing, assisted by government financing (Chapter 3). Women were now needed in a social role that provided the demand and symbol of the suburban consumer family. Moreover, a baby boom followed World War II, providing a good many more young mothers with larger families to whom the media and manufacturers could appeal with the suggestion that they should consume. The economy required, or at least benefited handsomely, from the newly conceived feminine role.

The rebirth of the feminist movement in the 1960s was also assisted by the equality revolution. A conventional explanation is the presumed explosion of women in the labor force. This, no doubt, is part of the picture and one reason that the feminine mystique was grating to so many women. But, as the previous tables show (especially Table 5.2 under the "White Women" column), women in the labor force during the years prior to the new movement edged up only gradually and did not take off until the 1970s. Changes in the number of working women alone do not explain the resurgent feminist movement.

What caused the *Rebirth of Feminism* (1971), as Judith Hole and Ellen Levine entitle their informative book? Consistent with the previous observations about social movements in the 1960s, the new feminism was occasioned by the interplay of women's advocacy and political groups that moved, however reluctantly, to incorporate new constituents. The advocacy drew on the facts that the new feminists were mainly educated, white, middle-class, and, typically, employed women. To the extent that these women did participate in the labor force, they experienced inequality more acutely than working-class women. As can be inferred from Table 5.3, most employed women were assigned to the lower rungs of the occupational ladder or generally to working-class positions. Working-class, black, and Hispanic women may have accepted that as normal, whereas the typical recruit for the new feminism understandably perceived it an injustice given their reference group of middle-class working men. Consequently, it was not so much the numbers of women who were now working, but, as Gans says, the experienced inequalities of their circumstance. Juliet Mitchell (1971) develops this interpretation by adding that the inequality went further to include a contradiction between the idealized femininity and the productive role women were expected to fulfill in the workplace and the home. Even homemakers would find it difficult to look feminine and build a swimming pool with their own hands.

In the changing political environment, many women had been active in the civil rights movement (and encountered discrimination even there), acquiring organizational skills and a clear sense that the rights of women were an integral, if neglected, part of that movement. Just as Democratic administrations had sought the support of blacks and Hispanics in a new coalition, women were courted, particularly in 1961 when Kennedy created a President's Commission on the Status of Women. As Hole and Levine point out,

> Kennedy had appointed Esther Peterson, a member of his campaign staff, and a long-time labor lobbyist, to head the Women's Bureau. It was she who originated the idea of a President's Commission, and it was she who convinced Kennedy to set it up. Some of the suggested "political" reasons for establishing such a Commission are: it would allow Kennedy to discharge political obligations to women who had worked hard for his election in 1960 (none of whom, with the exception of Peterson, had been appointed to policy-making positions in his administration); it would result in the creation of a strong political block (women) to campaign for Kennedy in 1964; it could be used to get his administration "off the hook" on the Equal Rights Amendment which, as in the past, had been included in the Democratic Party platform . . . since Kennedy was beholden to labor for electoral support and labor was dead set against the Amendment. (1971, 19–20)

This commission, of course, did not initiate or fully represent the new feminist movement. But it did signal encouragement for women to join the growing liberal coalition, and it advanced measures (such as the Equal Pay Act of 1963 and Title VII of the 1964 Civil Rights Act that prohibited race and sex discrimination in employment) that came to define many aims of the movement. The National Organization for Women (NOW), formed in 1966 with Betty Friedan as its first president, drew many of its initial supporters from the commission and developed a Bill of Rights focused largely on employment issues, along with advocacy of the Equal Rights Amendment and women's control of their reproductive lives. The new liberal political coalition did not dictate the aims of the feminist movement, but it had a great deal to do with explaining when and why that movement was reborn.

Explaining the Revolution

We may now return to the first of three questions posed earlier, namely, why did the equality revolution occur in the 1960s? Obviously, there is no simple explanation for so complex an event. Several sets of causes operated at once and converged. First, changes in population and social structure formed a backdrop to the decade. The most dramatic of these was migration and rapid urbanization, prompted on one hand by the commercialization of agriculture and reflected on the other by growing differentiation between central cities and suburbs. After World War II, educational levels advanced rapidly as returning veterans with the GI Bill and others began attending college in record numbers. The birth rate and family size increased for a time. The ratio of men to women in the labor force shifted slowly until the 1970s when the shift became dramatic. And the economy boomed, despite cyclical recessions, on the basis of postwar consumer demands and international trade. It was, in John Kenneth Galbraith's phrase, an "affluent society." But prosperity was unequally distributed on class lines as long as the major status groups continued to suffer economic discrimination.

A second set of forces was political in nature. The social structural changes disrupted the traditional and New Deal bases of political power. Rural and southern political districts lost population and, with reapportionment, their political leverage. Voter registration drives enfranchised minorities, many for the first time. All disadvantaged status groups saw their opportunity to organize from below in this fluid situation. The Democratic party, given its reliance on the South and working-class districts in the cities, was most shaken by these changes and moved more decisively than did the Republicans to reorganize from above. But Republicans saw the opportunity to profit from Democratic losses and began promoting civil rights. In a relatively short time, the political scene became a competitive struggle among divided elites for the allegiance of the newly emboldened status groups. There was a democratic opening — short-lived, perhaps — that was not an opening at the top, but a huge change in the numbers of people who were allowed into the mainstream and whose right to be there was established.

The last set of forces behind the equality revolution was sociological. It involved changing ways in which people experienced their circumstances and defined their rights. As Gans ([1968] 1974) said, the social protest of the 1960s had to do with inequality. Status groups perceived their lack of equality in relation to others — whites, Anglos, men. And they argued that such inequalities should not exist on the basis of cultural values, including justice, freedom, dignity, and liberation. When Myrdal (1944) characterized the American dilemma as a conflict of cultural values, he saw ahead two decades to the wellsprings of the equality revolution. Blacks and Mexican Americans served in the war to end the moral blight of racism. The Supreme Court of the United States said that they should receive equal treatment. Women, too, served in that war and in industrial jobs on the home front. The president seemed to say that they deserved equal employment opportunities. As these status groups were activated by cultural values, each could draw on a rich subcultural history of political organization and struggles for justice. And, paradoxically, as the political system continued to maintain many barriers to equality, its own opportunism also encouraged in other ways the minority insurgency.

LEGACY OF THE EQUALITY REVOLUTION

Returning to the second of three questions posed earlier, what did the equality revolution accomplish? In one sense, I have answered this in the claim that a democratic opening occurred and a new conception of social rights was established. What about the more concrete effects in the areas of income, education, and employment that were the specific targets of so much movement activity?

Contrary to some critics, minorities and, particularly, blacks made important material gains in the 1960s that were not reversed in the recessionary 1970s and early 1980s. That statement is controversial and viewed in some quarters as an apology for the racism and sexism that undeniably persist. Consequently, much care is required in assessing the changing pattern of inequality. In fact, the picture is complex. Gains in some areas have been offset by losses in others. Some clear gains have not reached parity, and improvement often means slightly less inequality. Most important, some gains of the 1970s have been lost, refuting the optimistic notion that equality is the long-term trend, only a matter of time. At bottom, large inequalities among status groups still exist — racism and sexism (and ageism) are still routine practices and lively moral issues, but the pattern has changed. Unless we examine carefully the pattern, which includes important improvements, we will fail to understand the contemporary nature of inequality.

As Table 5.1 shows, and evidence from the 1980s confirms, black women have made gains in income, while ironically the median black income lost ground from its position in the early 1970s. Black men closed ground on whites more slowly, but there are differences here by education. Young, black male, high school graduates earned 74 percent of their white counterparts in 1979, while college graduates were at 84 percent. Among blacks the most disturbing statistic is a return in median family income to the level of 1955 relative to whites. This seemingly inconsistent fact is a result of the higher rate of single-parent families among blacks and therefore fewer two-income families, which has been the decided trend for whites.

Against these mixed accomplishments in income, a pattern of unalloyed gains appears in other important areas. In education, from 1968 to 1987, the proportion of black high school graduates increased faster than for whites; blacks went from 56 to 83 percent and whites from 76 to 87 percent. The number of blacks completing college rose from 5 to 9 percent between 1968 and 1987. Whites increased from 17 to 25 percent as of 1982. This is typical of what gains mean in this context. In the mid-1980s, the percentage of black college graduates was only about half that of whites, but it had increased from one-third of that figure in 1970. Worrisome, however, is a decline in the number of black high school graduates entering college by 1987 (down from 34 to 26 percent since 1976). The black–white gap in occupational prestige has narrowed, especially for women. Black registered voters have increased over the past decade and nearly equaled white proportions, and black elected officials have increased sharply from 1970 to 1987, including a rise from 48 to 303 in the number of mayors — big-city mayors at that, from Los Angeles to Chicago, Atlanta, and Washington, D.C.

In a valuable article, Farley (1977) asks whether the gains of the 1960s disappeared in the following decade. The significance of the question lies in the fact that the 1960s witnessed an economic boom when minorities would be expected to gain, if only by trickle down, while the 1970s, 1980s, and early 1990s saw a series of recessions in which, some sociologists claim, minorities are bound to lose (on the principle of last hired, first fired). Farley's challenging conclusion is that this did not happen. His book, *Catching Up: Recent Changes in the Social and Economic Status of Blacks* (1983), shows that gains continued, along with some losses, particularly in black male employment.

Finally, some status inequalities either did not improve with the equality revolution or they got worse. Gaps in the income and occupational levels of women and men have not narrowed significantly. Black males in the labor force have declined, and this is due in large part to the increasing rate of unemployment among black youth. In 1954 the unemployment figures for black and white youth (sixteen to twenty-four years) were 15.8 and 9.9 percent, respectively — a difference of 5.9 percentage points. In 1992 the comparable figures were 33.9 percent and 13.2 percent, respectively, or a much larger difference of 20.7 points (U.S. Department of Labor 1992, 36). The unemployment gap for black youth increased nearly fourfold. In the area of residential segregation, John Logan and Mark Schneider (1984) reanalyzed for 1970–1980 the previously mentioned absence of black gains with suburbanization. Once again, suburban blacks were found mainly in older and declining communities, while white suburbs had gained few black residents: "The typical pattern appears to have been consolidation of black population in segregated communities" (1984, 875).

Generally, however, in the early 1980s sociologists were mildly optimistic as a result of research like Farley's that suggested some cumulative gains. In a few years, that assessment changed with a closer look at what was going on within the black population. William Julius Wilson (1987) argues that black society is increasingly divided into three parts: a middle class moving upward in education and income and outward from the inner city to suburbs, a working class struggling to hold its gains in the face of deindustrialization, and a large lower class worse off than ever because it is abandoned in the jobless ghettos. The newly isolated black lower class concentrates problems of high unemployment (up to 40 percent), female-headed households, teenage pregnancies, and children born out of wedlock — many of these resistant to the public policies that worked only for those who have since moved up.

From this mixed bag of results, can we arrive at any unambiguous conclusions about the accomplishments of the equality revolution? The answer is yes, and it comes in three parts. First, on a number of measures, the circumstances of disadvantaged status groups improved (e.g., women's income and participation in the labor force, black education). Second, on other measures, the situation remained the same (e.g., residential segregation) or got worse (e.g., black family income). Third, the overall change defies simple additive description because the core explanation for the shifts is the changing interaction of status group characteristics, class, and the measures of inequality. We encountered this circumstance previously in the description of status stratification, and it applies as much to how the pattern changes.

The dominant pattern can be illustrated rather simply with some of the discoveries by Erik Olin Wright and Luca Perrone (1977), who have compared the effects of class and status stratification. The familiar facts are that race (black), gender (female), and low education all mean low income. There are relatively few blacks and women in such higher-class positions as management (managers and executives are considered a distinct grouping within the middle class in Chapter 4). However, comparing just those people who have become managers, black and white males have similar incomes that are significantly greater than the incomes of female managers. "Sex differences among managers are considerably greater than the race differences" (Wright and Perrone 1977, 53). The fact that this may not be true for a different (nonmanagerial) occupation constitutes another kind of interaction. More generally, Wright and Perrone conclude that

> the class differences between workers and employers are considerably greater than the differences between men and women or between blacks and whites within the working class. . . . Racial discrimination operates more in sorting people into class positions in

> the first place than in giving them lower incomes for lower levels of education and skills once they are in a class position. (1977, 53–54)

Wilson has made the same point in a previous book whose controversial title, *The Declining Significance of Race: Blacks and Changing American Institutions* (1978), has misled some critics. Although Wilson has been charged with minimizing the harmful effects of race, which he does not do, his central point is the new interactive pattern. The changing significance of race for Wilson means that by contrast to 1940, when race was close to the sole determinant of black life chances, today class influences combine with race and, in some circumstances, override them.

The key point in my argument, based on the ensemble of evidence, is that *status and class operate jointly* on the outward consequences of inequality (income, wealth, residence). The effect of the equality revolution has been to *reduce certain status group disadvantages* that nevertheless have been *replaced* in many instances by *class inequalities.* Status group inequality has lessened in some respects, but class inequality has reasserted itself, sometimes canceling the categorical gains of working-class minorities.

THE COST OF PROGRESS

The last of our three questions about the equality revolution asks why it was transformed and in some ways reversed in the 1980s and early 1990s. The answer should begin with a qualification. The movements discussed here were never united with one another or even within their own ranks. As early as 1869, the women's movement divided in two camps: an assertive wing headed by Susan B. Anthony and Elizabeth Cady Stanton, who founded the National Woman Suffrage Association, and the more "respectable" branch gathered around Lucy Stone and the American Woman Suffrage Association. In 1920 blacks were split between the Booker T. Washington moderates emphasizing education for integration and Marcus Garvey's black nationalism. The same tendencies characterized Mexican–American civic associations such as LULAC by contrast to the labor movement activism of the Confederation of Mexican Workers.

The divergence appears again in the 1960s. Civil rights groups agreeably divided in tone and strategy between the mainline NAACP–Urban League–Southern Christian Leadership Conference (SCLC) and the militant Student Nonviolent Coordinating Committee (SNCC)–Black Panthers. Similarly, the Mexican American Political Association (MAPA) and the Political Association of Spanish-speaking Organizations (PASO) took the moderate side by contrast to the Brown Berets. The women's movement was different, less publicly divided than organizationally segregated with mainliners in the National Organization for Women (NOW) and radical feminists gathered in a variety of decentralized support groups.

The equality revolution was special because of the relative unity within and between status groups. For a decade or so, the movements were in step. There were overarching interests, notably the peace movement that unified liberal and civil rights groups. Black and white women joined hands. Differences within groups were set aside as the NAACP and SNCC participated in the same freedom marches. But the unity was short-lived. Splits that appeared in the late 1960s began to separate groups in the next decade. Actually, two events happened in the 1980s — a general demobilization of the movement and a fragmentation of surviving loyalists.

The most important division across status groups was between those who responded to concessions, deciding to work within or for the system, and those who endeavored to push the movements beyond a victory of political compromise. I want to avoid any implication that the partisans of these divergent movement strategies can be easily classified as sell-outs or true believers. As the gains of the equality revolution were registered and the movements looked toward an agenda for the 1980s, good arguments were made for each divergent strategy. Movement participants had no way of knowing that reversals were in store, and, had they seen this coming, it is not at all clear that one or another strategy could have avoided that fate. If hindsight tells us anything, it is that the revised course of the late

1970s and 1980s was set by conditions in the broader society rather than by the movements themselves.

The civil rights movement reached a turning point precisely because it succeeded in its legislative and legal aims. The 1964 Civil Rights Act was an enormous achievement, not least because it came through a recognition by the political parties that their own futures depended on a new guarantee of social rights. Blacks followed the victory with massive voter registration drives and political pressure for inclusion in President Johnson's Great Society programs. As Piven and Cloward perceptively remark, "The disintegration of the movement was inevitable given the integrative impact of the very concessions won" ([1977] 1979, 253).

Ironically, demands for black power were mounted after this stage of the struggle for civil rights. Stokley Carmichael and Charles Hamilton explained: "The concept of Black Power rests on a fundamental premise: Before a group can enter the open society, it must first close ranks. . . . Solidarity is necessary before a group can operate effectively from a bargaining position of strength in a pluralistic society" (1967, 44). Piven and Cloward conclude that black power actually contributed to the integrative thrust of the movement: "Defined in this way the concept was especially suited to the ideological needs of a black leadership stratum seeking to exploit the new possibilities for electoral and bureaucratic influence" ([1977] 1979, 253). One study of the demise of CORE by August Meier and Elliott Rudwick (1975), for example, indicates that the new antipoverty programs of the Johnson administration began hiring civil rights activists and leaders who felt that their positions ruled out continued participation in protest organizations.

The incorporation of many blacks into ameliorative programs, in combination with the documented gains, understandably led to a conservative appraisal of the changes by some middle-class blacks. Thomas Sowell, in his widely heralded book *Ethnic America: A History* ([1978] 1983), argues that black gains were actually generated by the operation of a competitive economy in which discrimination had proved inefficient and government programs beneficial to only a few. Christopher Jencks (1983), among others, has demonstrated the many fallacies in this argument and supported the proposition that blacks did benefit generally from Great Society programs. What is significant about the conservative argument for present purposes is that during the late 1970s and 1980s the U.S. economy and social structure changed in ways that created an audience receptive to Sowell's viewpoint — a socially mobile black middle class which had, indeed, worked hard for its gains and a group of conservative politicians who wanted to give credit for that achievement to the free-enterprise system.

These broad changes in social structure and political climate help explain subsequent events such as the infamous confrontation of Supreme Court nominee Judge Clarence Thomas, law professor Anita Hill, and the U.S. Senate in 1991. One aspect of the Senate hearings on Thomas's confirmation was to consider the merits of Hill's charges of sexual harassment ten years earlier when she worked for Thomas, then head of the federal Equal Employment Opportunity Commission. Both Hill and Thomas were beneficiaries of the equality revolution, blacks from humble origins who had achieved distinguished positions in government and education. Now these two essentially conservative blacks were divided on a gender issue — whether Thomas had made sexual advances to Hill that were improper under the hazy definition of harassment or whether Hill (for reasons of unspecified connections to pro-choice groups opposed to the nomination) was trying to sabotage Thomas's confirmation to the Court. After three days of testimony that transfixed a national television audience, neither side could produce corroborating evidence. The all-male Senate committee discounted the charges and in a split vote endorsed the nomination, despite other prior arguments that Thomas was poorly qualified for the job. We still do not have evidence of who was telling the truth and who was lying (Jane Mayer of the *Wall Street Journal* is writing a book on the question), but national polls at the time provide revealing evidence about who people thought was telling the truth. Blacks tended to believe Judge Thomas, perhaps because they supported a black on the Supreme Court and were more attuned to ethnic than gender issues. Whites more so than blacks believed Hill, perhaps because they identified more closely with discrimination against women. No one had much sympathy for the senators. The incident suggests new and diverse divisions of status group politics.

Mexican Americans made gains in the 1960s and found themselves, like blacks, incorporated into reform programs. Indeed, when the militant spark went out in the civil rights movement, all status groups faced a new political prospect. Just as the change was setting in, John Womack noted that

> Chicanismo has left many in the movement cold, like those . . . who gladly call themselves Chicanos but reckon that a new alienation would ruin la raza. And it has scandalized the LULACs, GI Forumites, MAPAs, and PASOs, who now have their investments in "Mexican Americanism." Since their first encounter with him the nationalists have repelled Chavez, who is committed far beyond Atzlan to building solidarity among farmworkers of every complexion. (1972, 15)

In some respects, the problems of Mexican Americans were different from those of blacks. They had not suffered slavery or constitutional disenfranchisement. They enjoyed a strong cultural heritage in their southwestern communities, the base on which their movement was founded — and one that was lost to some through migration and urbanization in what Octavio Paz called a *Labyrinth of Solitude* (1961). By the 1970s, Mexican Americans had their own U.S. senators and representatives. This is not to say that they suffered less from discrimination, only that they responded to it in ways that necessarily differed from blacks. Beyond electoral gains, many Mexican Americans found positions in education, particularly in programs for bilingual education mandated by the federal government. By the mid-1970s, Chavez and the United Farmworkers of America (UFW; previously, NFWA) had won their struggle for legally recognized unions in California. In Texas the Farah Slacks workers gained national recognition and a boycott-supported victory for their union — with the support of the Catholic church and Amalgamated Clothing Workers. While programs devoted to Chicano studies wrestled with undeniably vexing questions of cultural identity, these working-class movements were helping create a new one.

In the 1980s, fragmentation in the Chicano movement was less a product of the conservative reaction of a Sowell than it was a statement about the effects of integration. The artful autobiography of Richard Rodriguez ([1982] 1983) is the story of a boy raised in Sacramento, California, in the 1950s by hard-working middle-class parents and parochial school teachers. The confusion Rodriguez faced stemmed from the fact that the private Spanish language of his home was discouraged in the public setting of his education. He was torn several ways. Although he had gained an excellent parochial education, he enjoyed the benefits of concessions to disadvantaged minorities by being given a scholarship.

Rodriguez represents a minority within the Mexican–American minority. The barrios of Sacramento include many truly disadvantaged youth and, coincidentally, fifty years earlier had produced Ernesto Galarza who earned a Ph.D. at Columbia University in the 1930s, directed a branch of the Pan American Union, returned to his native state in 1946 to work for the unionization of farm workers, and wrote the autobiography *Barrio Boy* (1971). The differences between Galarza and Rodriguez show the changing conditions of Mexican–American generations and something of the present divisions within their communities. Galarza chides the psychologists who have spread the rumor that Mexican Americans have lost their self-image: "I, for one Mexican, never had any doubts on this score. I can't remember a time when I didn't know who I was" (1971, 2). Rodriguez complained of an experience that came much later when he was treated as something he was not, simply because he had a Spanish surname. His identity problem stemmed from well-meaning, compensatory affirmative action programs aimed at the inequalities of poverty and social class but awarded on the basis of ethnicity, thereby including studious middle-class Chicanos. His experience is not typical, but it does help to explain the source of contemporary divisions among Mexican Americans.

Divisions among Native Americans were perhaps more severe than they were for other minorities owing to the special governing arrangement in which factions favorable to the Bureau of Indian Affairs and its policies were able to control tribal councils despite the opposition of "traditionals." It was against the corruption of these "BIA Indians" that the American Indian Movement rallied, although in the end AIM itself proved divisive for its tactics and the official repression it generated. Most of these battles, however, were fought during the mid-1970s, and twenty years later Native Americans have successfully rid themselves of corrupt leaders, launched new economic development plans, and healed some of the divisions.

Finally, the most publicized fragmentation has been within the women's movement — a split between "radical" feminists and the career and political women that continue to support NOW. Ironically, although it was Friedan's book *The Feminine Mystique* (1963) that did so much to advertise and articulate the aims of the women's movement, her 1981 sequel *The Second Stage* attacks the *feminist* mystique as the new barrier to liberation. The radical feminists, according to Friedan, go too far with their sexual politics and repudiation of male and female needs for intimacy in the family. Feminists like Judith Stacey (1983) answer that their aim is real equality for women and an egalitarian family. The difference is that Friedan and her colleagues argue for the dual-career family in which men and women play mutually supportive roles. Men and women should take advantage of flexible economic arrangements to meet their mutual career needs and rely more on self-help than on government programs. Unreconstructed feminists argue that this plan retreats from the original and unmet aims of women's equality in the face of political unpopularity and would reinforce gender as a social-tracking system.

Susan Faludi (1991) argues that a good deal of the alleged division within the women's movement in fact derives from the criticisms of its opponents rather than the mood of the majority of women. Opponents of feminism, she argues, claim that the movement has been harmful to women, but those claims rest on bogus research and media sensationalism. Faludi identifies four major "myths of the backlash": a man shortage endangering women's opportunities for marriage, a devastating plunge in women's economic status following divorce under no-fault laws, an infertility epidemic affecting women who postpone childbearing until their thirties, and growing emotional depression or burnout among professional women caught in the "equality trap." Although each claim has been reported in social and demographic research, the initial evidence was slim, and none of the claims has held up under closer examination. Indeed, some are the direct opposite of results produced in more rigorous studies — men, to a greater extent than women, are suffering these days from depression and job stress, for example (Faludi 1991). Yet in each case, the antifeminist reports were widely publicized, while the more reliable studies were ignored.

For example, on February 18, 1982, the *New England Journal of Medicine* reported a French study claiming that a woman's chances of conceiving a child dropped abruptly after the age of thirty. Women between the ages of thirty-one and thirty-five had a 40 percent chance of being infertile. The *New York Times* featured the story on its front page the same day, thus initiating the now commonplace concerns about a woman's "biological clock." What the *Times* called an "unusually large and rigorous study" turned out to be an inappropriate piece of research for the conclusions drawn from it. A sample of 2000 Frenchwomen was selected from couples in which the man was sterile because the object of the study was to determine which methods of artificial insemination might produce pregnancy. Three years later, the U.S. National Center for Health Statistics published its study showing that women between the ages thirty and thirty-four had only a 13.6 (not 40) percent chance of being infertile and this figure was only 3 percent higher than for women in their twenties. The news release received no coverage in the newspapers.

Like the supposed infertility epidemic, efforts by the women's movement to win greater equality in the workplace are alleged to have produced stress disorders, role confusion, loneliness, family dysfunction, and many such maladies. The feminist backlash would have us believe that "women have 'made it.' . . . Women have so many opportunities now that we don't really need equal opportunity policies . . . [but] behind the celebration of American women's victory another message flashed. You may be free and equal now, it says to women, but you have never been more miserable" (Faludi 1991, ix). Behind the myth, women want what they have always wanted — to be taken seriously as equals. That message, however, is not getting through a divided women's movement and those interests that resist further gains in employment and reproductive rights.

If these brief descriptions serve to persuade that the current thrust of status group politics is a reversal from an equality revolution in the 1960s and 1970s to fragmentation and compromise in the 1980s and 1990s, the question returns: Why? The foregoing explanation for the rise of the equality revolution relied heavily on political changes in the broader society. The same applies here. The fragmentation of status group politics resulted precisely from the gains it won — new employment opportunities, broader political representation, marginal to significant improvements in economic circumstance, a piece of the action. Recall, once more, Gans's ([1968] 1974) observation that the equality revolution "had very little to do with poverty" as opposed to the inequality experienced by status groups that enjoyed fewer advantages than their reference groups. Once the political system, for its own reasons, began responding to these claims in ways that made a difference for some, it won converts. Blacks who saw their income increase, women who began getting jobs, Mexican Americans who unionized or entered the educational establishment — many of these the leaders of the equality revolution — knew that they had gained something worth preserving.

It is also true that the benefits of these status-based movements are not broadly distributed. Particularly, they touch only indirectly the bases of class privilege. Rodriguez is correct when he notes that

> remarkably, affirmative action passed as a program of the Left. In fact, its supporters ignored the most fundamental assumptions of the classical left by disregarding the importance of class and by assuming that the disadvantages of the lower class would necessarily be ameliorated by the creation of an elite society. The movement that began so nobly in the South and the North came to parody social reform. Those least disadvantaged were helped first, advanced because many others of their race were more disadvantaged. ([1982] 1983, 151)

But, it is equally true that this was not the fault of the status groups that fought the equality revolution with the weapons they had at hand — namely, the ethnic, gender, and cultural resources that enabled them to unite and act against great odds. We have to bear in mind two conclusions despite the moral tension between them: The equality revolution made a difference, winning converts to the system that embraced it and splitting its original foundations, but it did not begin to wash away the footings of class inequality.

This lesson was made acute by recurring worldwide recessions in the last two decades that have reaffirmed class divisions and further divided status groups as those who had made their own gains fought to hold on to them. Unemployment climbed to 11 percent in 1982, declined, and then returned to almost 8 percent in mid-1992, and real income dropped over the decade. Those who had advanced during the previous decade came to realize that they now occupied a new class position with its attendant interests. Status groups made the equality revolution, and social classes captured it.

The 1980s and 1990s, accordingly, witnessed another experience. Class inequalities returned with a vengeance, but status anxieties did not disappear. On the contrary, a far-flung backlash to the equality revolution appeared. In defense of a status hierarchy that had been under siege, traditional values were reasserted: religious fundamentalism, patriotism, family, sexual probity (with antiabortion linked to several of these), free enterprise, an ethical defense of capitalism. Liberalism as a political philosophy was denied by Republicans and Democrats alike. The shift was facilitated by more than a "pendulum effect" or a vaguely cyclical notion of backlash. On one hand, dominant status groups reorganized themselves: men's rights, reverse discrimination, racial quotas, abortion, busing versus neighborhood schools — all became passionate concerns. On the other hand, many of the very beneficiaries of the equality revolution joined the establishment in position and in principle. Many Hispanics became Republicans; some blacks and women who attained better positions took on the values attached to the new status rather than to the movement that made it all possible. The balance of political power shifted, at least for a time.

CLASS AND STATUS

Chapter 4 argued that a resurgent pattern of class stratification characterizes the United States in the 1980s and early 1990s.

More evidence of the sort reviewed in this chapter supports the same conclusion. Writing in the *New York Times* for March 14, 1984, the economist Lester Thurow indicates that if the U.S. middle class is defined by households with an income of $15,000–25,000 (a very rough definition for sociological purposes, but an indicator), then the size of that stratum has shrunk. Between 1967 and 1982, it was reduced from 28.2 to 23.7 percent of the population. During the same period, households with incomes in both the lower- and the upper-income ranges increased. That is, household income tended to become more polarized between rich and poor. The same conclusion was reached in a later study by the U.S. Census Bureau cited in Chapter 4, showing that from 1969 to 1989 the proportion of the population with incomes in a broad middle range declined from 71 to 63 percent (U.S. Department of Commerce 1991).

Now, if we add to these indirect measures of class the facts reviewed here about status group gains that have not been reversed during the same period, the resultant picture is one of a reorganized class and status structure. The current hierarchy is more balanced by race and gender and more rigidly stratified by material rewards and life chances. Class differences are more often the bases of sharp inequalities than are status differences.

The new and large question we are left with after this analysis is, Why do these shifts occur in the relative importance of class and status determinants of inequality? As discussed earlier in this chapter, Weber hypothesized that systems of social inequality are produced by changes from economic stability (and stratification by status) to "technological and economic transformations" (and class stratification). Weber qualified this by admitting that "only very little can be said" from our present knowledge of why such changes occur. Yet Weber's proposition is generally supported in the postwar United States — the economic boom and technological advance that prevailed in this period did help break down a pattern of status stratification and "push the class situation into the foreground." As Wilson (1978, 1987) has argued, blacks made gains, with many achieving middle-class conditions and breaking down stratification by homogeneous status groups. But, for the period following the equality revolution, Marx's expectation that classes would become more polarized with advancing capitalist development is equally plausible on the evidence.

The problem with both of these grand generalizations is that they miss the key discovery in this material — namely, the *constant interaction of class and status* in a pattern of reciprocal influence. The economic and technological transformations that Weber stressed have recently brought class back to the forefront of stratification. Yet a similar transformation in the early part of this century that modernized southern agriculture had the effect of reinforcing race status as a feature of the urban and industrial economy. Moreover, at any given time, it is the combination of class and status characteristics that explains the life chances of particular groups — explains, for example, why black women made more progress than black men or women in general during the equality revolution. Stated more simply, it is the interaction of class and status that reveals the nature of any system of social stratification, that tells us how racist, sexist, or class-based that system is.

When we ask, then, what determines the blend of class and status, the clear answer from this material is politics — the political struggles of groups that find it necessary or effective in a given situation to organize themselves on the basis of class or status themes. Political struggles based on class or status take different forms, drawing on economic groups or on religious and cultural influences. And they seek different ends — material benefits or social rights. These differences, indeed the ultimate success of the struggles, depend on the structure of political power in society, including what appears in this analysis as the moral dilemmas and political strategies of ruling groups.

In summary, the status group is an essential parallel to social class in any sociological analysis of group formation, political action, and social inequality. The relative importance of class versus status changes over time, requiring an explanation of its own and reminding us of the rule that adequate sociological explanations must also be historical. The main result of this analysis, however, is that class and status cannot be unraveled and each weighed for its individual importance. This is because the most discerning explanations of social action or inequality derive from the interaction of class and status, an interaction that is causally specific to certain situations and changing in ways that mark off distinctive eras. Failure to understand this principle has led to many unnecessary controversies between social reformers (militants for working-class solidarity vs. pragmatists for ethnic–religion coalitions) and social analysts (Marxists vs. culturalists). Once the interaction principle is understood, however, better explanations are generated, and new questions come forward.

As the studies in this chapter show, during the last thirty years an equality revolution was fought and, in some senses won, for the twin reasons that the status groups fighting it had the cultural energy of justice on their side and the elites who accommodated it discovered their own interests in a new class order. The state was changing and with it the social structure.

SELECTED BIBLIOGRAPHY

BANKS, OLIVE. 1981. *Faces of Feminism: A Study of Feminism as a Social Movement.* Oxford, England: Martin Robertson.

CARMICHAEL, STOKELY, and CHARLES HAMILTON. 1967. *Black Power: The Politics of Liberation in America.* New York: Random House.

CORNELL, STEPHEN. 1988. *The Return of the Native: American Indian Political Resurgence.* New York: Oxford University Press.

DRAKE, ST. CLAIR, and HORACE R. CAYTON. 1945. *Black Metropolis: A Study of Negro Life in a Northern City.* New York: Harcourt Brace.

DUNNE, JOHN GREGORY. 1967. *Delano: The Story of the California Grape Strike.* New York: Farrar, Straus & Giroux.

FALUDI, SUSAN. 1991. *Backlash: The Undeclared War Against American Women.* New York: Crown.

FARLEY, REYNOLDS. 1970. "The Changing Distribution of Negroes Within Metropolitan Areas: The Emergence of Black Suburbs." *American Journal of Sociology* 75(4):512–29.

———. 1977. "Trends in Racial Inequalities: Have the Gains of the 1960s Disappeared in the 1970s?" *American Sociological Review* 42(2):189–208.

———. 1983. *Catching Up: Recent Changes in the Social and Economic Status of Blacks.* Cambridge, MA: Harvard University Press.

FINK, LEON, and BRIAN GREENBERG. 1989. *Upheaval in the Quiet Zone: A History of Hospital Workers' Union Local 1199.* Urbana: University of Illinois Press.

FRIEDAN, BETTY. 1963. *The Feminine Mystique.* New York: Norton.

———. 1981. *The Second Stage.* New York: Summit Books.

GANS, HERBERT J. 1968. "The Equality Revolution." Reprint. Pp. 7–35 in *More Equality,* edited by Herbert J. Gans. New York: Vintage Books, 1974.

GREBLER, LEO, JOAN W. MOORE, and RALPH C. GUZMAN. 1970. *The Mexican–American People: The Nation's Second Largest Minority.* New York: Free Press.

HOLE, JUDITH, and ELLEN LEVINE. 1971. *The Rebirth of Feminism.* New York: Quadrangle Books.

JENCKS, CHRISTOPHER. 1983. "Discrimination and Thomas Sowell." *New York Review of Books* 30(3):33–38.

JENKINS, J. CRAIG, and CHARLES PERROW. 1977. "Insurgency of the Powerless: Farm Worker Movements (1946–1972)." *American Sociological Review* 42(2):249–68.

LOGAN, JOHN R., and MARK SCHNEIDER. 1984. "Racial Segregation and Racial Change in American Suburbs, 1970–1980." *American Journal of Sociology* 89(4):874–88.

MARSHALL, T. H. 1950. *Class, Citizenship, and Social Development.* Reprint. Garden City, NY: Doubleday, 1964.

MASSEY, DOUGLAS S., and NANCY A. DENTON. 1987. "Trends in the Residential Segregation of Blacks, Hispanics, and Asians: 1979–1980." *American Sociological Review* 52(4):802–25.

MATTHIESSEN, PETER. 1992. *Indian Country.* New York: Penguin Books.

MEIER, AUGUST, and ELLIOTT RUDWICK. 1975. *CORE: A Study in the Civil Rights Movement, 1942–1968.* Urbana: University of Illinois Press.

MILLS, C. WRIGHT. 1956. *The Power Elite.* New York: Oxford University Press.

MITCHELL, JULIET. 1971. *Woman's Estate.* London: Harmondsworth.

MYRDAL, GUNNAR. 1944. *An American Dilemma: The Negro Problem and Modern Democracy.* New York: Harper & Brothers.

PIVEN, FRANCES FOX, and RICHARD A. CLOWARD. 1977. *Poor People's Movements: Why They Succeed, How They Fail.* Reprint. New York: Vintage Books, 1979.

RODRIGUEZ, RICHARD. 1982. *Hunger of Memory: The Education of Richard Rodriguez.* Reprint. New York: Bantam Books, 1983.

SOWELL, THOMAS. 1978. *Ethnic America: A History.* Reprint. New York: Basic Books, 1983.

STACEY, JUDITH. 1983. "The New Conservative Feminism." *Feminist Studies* 9(3):559–83.

WEBER, MAX. 1946. "Class, Status, Party." Reprint. Pp. 180–95 in *From Max Weber: Essays in Sociology,* translated and edited by Hans Gerth and C. Wright Mills. New York: Oxford University Press, 1958.

WILSON, WILLIAM JULIUS. 1978. *The Declining Significance of Race: Blacks and Changing American Institutions.* Chicago: University of Chicago Press.

———. 1987. *The Truly Disadvantaged: The Inner City, the Underclass, and Public Policy.* Chicago: University of Chicago Press.

WOMACK, JOHN, JR. 1972. "The Chicanos." *New York Review of Books* 19(3):12–18.

WRIGHT, ERIK OLIN, and LUCA PERRONE. 1977. "Marxist Class Categories and Income Inequality." *American Sociological Review* 42(1):32–55.

U.S. BUREAU OF COMMERCE, BUREAU OF THE CENSUS. July 1983. *America's Black Population: 1970–1982.* Washington, D.C.: Government Printing Office.

U.S. DEPARTMENT OF COMMERCE, BUREAU OF THE CENSUS. December 1991. *Trends in Relative Income.* Washington, D.C.: Government Printing Office.

U.S. DEPARTMENT OF LABOR, BUREAU OF LABOR STATISTICS. October 1980. *Perspectives on Working Women: A Databook.* Washington, D.C.: Government Printing Office.

U.S. DEPARTMENT OF LABOR, BUREAU OF LABOR STATISTICS. June 1992. *Employment and Earnings.* Washington, D.C.: Government Printing Office.

PART III

INTEGRATION

6

Social Organization and the State

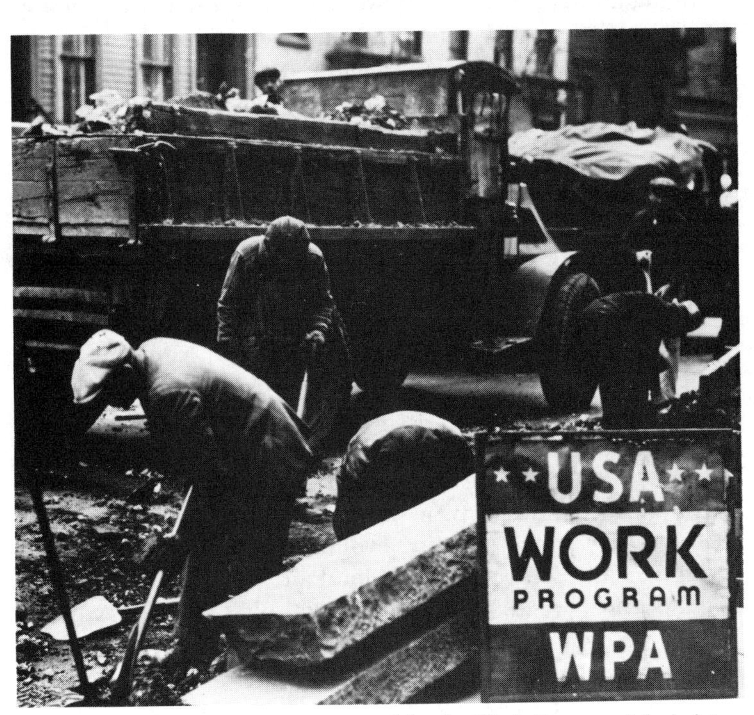

The public works programs sponsored by the Work Progress Administration produced the nation's highway system and urban renovation on a scale unmatched since the depression. (The Bettmann Archive)

PRIVATE LIFE

On Long Island, New York, in the fall of 1983 an infant was born severely disabled, yet destined for notoriety. Baby Jane Doe, as she became known to the world, was delivered with an open spinal cord, fluid on the brain, an abnormally small head, and an underdeveloped brain. With routine care, she would live perhaps two years. Surgical intervention to close the hole in her back and drain the excess brain fluid might extend her lifetime. She would never be "human." Her parents anguished over what to do and discussed it with their doctors, friends, and clergy. In the end, they decided against surgery.

Baby Jane Doe's condition was told to a lawyer and antiabortion activist by an anonymous informant. The lawyer, a specialist in litigation on behalf of newborns, filed suit in the New York State Supreme Court claiming that the infant's right to life was being violated, that the court had a responsibility to take the decision out of the hands of the parents and to require the doctors to perform surgery. The state court rejected this argument noting that the lawyer was an outsider with no standing in the case or right to sue. That did not end the public drama. The lawyer went to federal court demanding an injunction that would force the hospital to fight any parent's choice to reject life-prolonging surgery.

More important than the lawyer's tenacity, however, was a decision by the federal government to enter the now-publicized case on behalf of the people of the United States. The government brought suit against the hospital, where Baby Jane Doe lay quietly through all this, seeking access to her medical records. The government's case was based on the Rehabilitation Act of 1973 that includes a provision (and here an unanticipated consequence of the equality revolution) stating that it is illegal for any institution receiving federal aid to discriminate on the basis of race, creed, color, religion, ethnic origin, or *handicap*. The government requested access to medical records to determine whether the infant was a victim of discrimination by virtue of being denied medical treatment solely because of her disability. Baby Jane Doe's mute life became a civil rights issue because the administration decided to contest the judgment of attending physicians and the family at the behest of politically organized pro-life groups. The family found itself face-to-face with the state in a tormented contest over who bore ultimate responsibility for the unfortunate baby's life.

Cases like Baby Jane Doe's are commonplace with large hospitals and obstetricians. Traditionally, they have been handled in a manner determined by the doctors and family. There is no question that grotesquely malformed and hopelessly retarded infants are allowed to die, just as it is true that modern medicine and extraordinary efforts are now devoted to abnormal newborns that were once written off. Down syndrome babies, once discounted at birth or permanently institutionalized, are now raised in families. Although pro-life groups claim that there are 300 victims of infanticide each year, no consensus exists on what to do with babies like Jane Doe — who decides and who bears the costs for the prolongation of life.

The federal government has no position on what should become of the life prolonged by nondiscrimination. Indeed, it is still a mystery how discrimination could be established. Because the treatment involved would never be administered to an infant without the disability, it is difficult to say that it is denied solely because of the disability. Without answers to such thorny issues and with the officials steadfastly pursuing public jurisdiction over these decisions, the only suggestion agreeable to physicians and government is the establishment of quasi-official bio-ethical committees composed of clergy, medical, legal, and ethical experts to judge individual cases. However the issue is resolved, it represents a significant extension of state authority into the realm of personal choice.

In this illustration, I have referred knowingly to "the state" despite the elusiveness of the term. In ordinary usage, it ranges from a vague synonym for government to a critical composite such as the welfare state or capitalist state, and, in the extreme, a futuristic nightmare like the vision of George Orwell's novel *1984* — an oppressive, big brother state based on the complete usurpation of individual will. Orwell's allegory, written in 1948, transposed two digits and projected a trend toward enforced conformity into an imaginary totalitarian state that made decisions for individuals, provided them diversion, and brooked no independent thinking or unofficial recollection of history. Orwell provided a warning rather than a prediction. When 1984 came, his warning was as important as ever, although events appeared differently. The scope of the state, as Orwell saw, had expanded enormously — it had even marched into family and medical choices — and so, as Orwell hoped, had the conflict over what the state should and should not do.

THE STATE

The state is both a set of organizations possessing legitimate authority and an institution. **Organizations**, public bureaucracies and agencies, possess legitimate authority when, in the eyes of the citizens, they have the right to command (as do the police or Internal Revenue Service). **Institutions** exist at a higher level as established ways of doing things. Representative government, for example, is a way of exercising public authority, and marriage a way of organizing companionship and reproduction. Organizations are concrete representations of institutions, but institutions are foremost ideas about how society should be organized. This standard sociological usage differs from ordinary-language phrases, such as mental institution, which confuse an important distinction. By distinguishing between concrete organizations and institutional ideas, it is possible to investigate how one influences the other and how both change under a variety of influences.

The **state organization** includes government in all its branches and levels. It is the congress and the executive departments, as well as state, county, and municipal governments. But the state is more than representative governing institutions. It is also the military, police, public financial institutions (the Treasury and Federal Reserve Board), the courts and officers of the court (lawyers), prisons and welfare agencies, regulatory commissions, public utilities, and a variety of independent agencies whose authority stems from government (the Federal Deposit Insurance Corporation, Environmental Protection Agency, National Science Foundation, Federal Election Commission, Commission on Fine Arts). When nongovernmental bodies act to ensure that public mandates are carried out, as bio-ethical committees might do, they act for the state.

As an institution linking these diverse organizations and agencies, the state is the expression of an established and legitimate way of conducting the public's business. As Alan Wolfe puts it in a valuable article on the theory of the state, the state is embodied in public bureaucracies, "yet that thing called government can only do what it is supposed if behind it is an apparatus responsible for the reproduction of the social system within which the government operates. That other thing, which cannot in fact be directly touched or seen, is the state" (1974, 149). More simply, the state is civic authority, law and administration, as it is practiced by a great variety of organizations.

This chapter develops three general themes about the state and society: (1) The state is a crucial determinant of social organization, particularly in modern societies — although conventional sociology has neglected this fact, leaving an important part of its own responsibility to disciplines such as political science; (2) the state has vastly expanded, particularly in the twentieth century, in tandem with the maturation of industrial capitalism; (3) the state is presently changing in fundamental ways owing to a crisis of the welfare state as an institution.

Social organization, defined in the standard text by Leonard Broom and Phillip Selznick, "is the pattern of individual and group relations" in a society; relations that ensure cooperation as well as those that "isolate people and groups and that foster disharmony" (1968, 14). These relations exist in interpersonal behavior (in social roles), groups (families, peer and work groups), and broader settings (communities, organizations, bureaucracies). Social organization is a pattern of relationships with many loci, a pattern that is constrained by basic human needs (nourishment, reproduction, etc.) and the requisites of social order (production, law, etc.). Accordingly, it is a pattern negotiated from the bottom up and the top down. Social organization includes those patterned relationships we learn in families, schools, churches, communities, careers, and political life.

The connection we want to pursue is that between the state and social organization. Social life is led mostly in organized settings — in work, for example, to which we bring social relationships developed in family, school, and community. Historically, family and community were interwoven and, together, pretty much exhausted social organization of the preindustrial "world we have lost" (to borrow a phrase from the English historian Peter Laslett). In modern society, by contrast, we are less engaged by family and community, more so by school, work, career, and organizational life. The state, in turn, has steadily taken responsibility for regulating all these institutions. Stated differently, contemporary society presents us with a world that is organized differently owing to the intrusion of the state into so many other institutional spheres. Social organization is changing in consequential ways that we can only get at by reconceiving the institutional order, seeing it as a new set of relationships in which the state is critical in its own right and a potent influence on the character of other institutions.

Social thinkers have struggled with this issue, usually by posing grand typologies. Émile Durkheim, as we saw (Chapters 1 and 2), viewed industrial society as moving toward state centralism by eliminating the "intermediate powers" that promoted group cohesion and associational life (and also curbed "anomie"). His contemporary Ferdinand Tonnies noted the same shift from personal bonds and community life to a more formal and distant society — or, in his own German terms, from *gemeinschaft* to *gesellschaft.* Max Weber spoke of the transition from traditional to rational–legal authority. More recently, North American sociologists, including C. Wright Mills, suggested a shift from local to mass society. With certain variations, all these polar types claim that the old order is disappearing mainly as a result of the decline of the institutions of community and association, or the buffers, between family and state.

These formulations have the virtue of capturing the general sense that something important is afoot, that the world is changing rapidly around us in ways that we do not fully understand. They also have a few drawbacks. They tend to ignore the pole that Durkheim identified as the state, concentrating on what is supposedly disintegrating below. They disagree on the implications of the change, some anticipating more involuntary (Orwellian), integration and others less. And, according to a good deal of research, they are factually in error. All those intermediate groups are not disappearing so much as changing in function. The expected mass society did not come, although the society organized around class and state that did come seems no more agreeable.

THE STATE IN SOCIAL LIFE

If the foregoing terms are necessarily abstract, it will be useful to indicate some of the countless and often unacknowledged ways in which the state affects our daily lives. In U.S. society, there is a peculiar schizophrenia about the state. On one hand, we may think about it very little or think it a nuisance, a source of bureaucratic meddling in private affairs if not an Orwellian beast. On the other hand, we want a strong state to protect us from foreign and domestic enemies and to guarantee a far-flung collection of rights. We are ambivalent about many things that the state does, except when it ensures our own interest. The sociological implication of that somewhat obvious statement is that from an individualistic standpoint we tend not to understand the causes of state expansion because our interests appear limited and stable. Ralph Miliband's important book *The State in Capitalist Society* opens with the following observation:

> More than ever before men now live in the shadow of the state. What they want to achieve, individually or in groups, now mainly depends on the state's sanction and support. But since that sanction and support are not bestowed indiscriminately, they must, ever more directly, seek to influence and shape the state's power and purpose, or try to appropriate it altogether. It is for the state's attention, or for its control, that men compete; and it is against the state that beat the waves of social conflict. This is why, as social beings, they are also political beings, whether they know it or not. It is possible not to be interested in what the state does; but it is not possible to be unaffected by it. (1969, 1)

Consider some of the ways in which the state routinely affects our lives.

- *Life.* As the Baby Jane Doe case indicates, the state has recently entered the field of supervising the care of newborns to ensure the prolongation of any life, or at least of those lives that an arm of the state (such as a bio-ethical committee) may decide have a right to prolongation. This initiative furthers standards that the state has already set limiting legal abortions by barring their performance under federal medical assistance programs. In some instances therefore, the state has assumed the responsibility for deciding who lives.

• *Death.* As the Baby Jane Doe case was argued in the New York courts, a young woman in California was legally restrained from allowing her own life to expire. Elizabeth Bouvia, a cerebral palsy victim and quadriplegic from birth, who had nevertheless completed college and married, decided that she no longer wished to live totally dependent on others. She entered a hospital asking to be kept comfortable while she starved herself to death. Having already admitted her but refusing this request, the hospital was in the uncomfortable position of being able neither to put her out on the street nor to force-feed her without legal authority. In a ruling sought from the California Superior Court, the judge said that Ms. Bouvia "does have a right to terminate her existence, but not . . . with the assistance of society." The hospital won state support to force-feed her as long as she remained (later she left the hospital, but changed her mind about wanting to die). This is not a unique case. Contrary to popular belief, neither the suffering individual nor the family can choose to terminate an unconscious or painful life without medical sanction. We cannot simply pull the plug where doctors and hospitals are concerned. They must make or approve that choice, and their approval is constrained by state and court-established precedents on medical malpractice. In the end, the state still watches over us, regulating where burials are permitted or ashes spread.

• *Marriage and Divorce.* The state, of course, grants licenses for marriage and officially confers divorce, in both instances under terms that bind the partners to legal responsibilities (monogamy in marriage or child support in divorce). People may enter into marriages without the consent of the state, as common-law and homosexual marriages demonstrate, yet those unconventional unions are not free of state influence. In the eyes of the state, a common-law marriage becomes legal after seven years, meaning that spouses may call on the state to intervene in the relationship, say, to require alimony after separation or legal inheritance with the death of one partner. Conversely, because the state does not recognize homosexual marriage, those couples are not entitled to the forms of intervention just mentioned or to certain marital privileges such as joint ownership of property. People can marry with or without the state, but the state's attitude toward the union has personal consequences in both cases.

• *Child Rearing.* The state cannot dictate how families must raise their children, but under a variety of circumstances it may usurp parental responsibility (in cases of family destitution, unfit parents, child abuse). Conversely, parents may relinquish the duties of child care to the state by convincing the appropriate authorities that a son or daughter is incorrigible. Beyond the extremes of family disruption, parents cannot keep minor children out of school or put them to work without the risk of state intervention. The government tries to require conventional forms of medical care for children. In one case of a Seventh Day Adventist family that chose prayer over chemotherapy for a child with cancer, the government lost a court contest to overrule the family because that was held a violation of religious freedom. Presumably, a family choice based on another standard with fewer constitutional safeguards could be reversed.

• *Sex.* Studies of sexual behavior dating from the Kinsey reports in the 1950s show that people routinely engage in practices that are illegal in certain jurisdictions (e.g., oral copulation, homosexuality). In general, the state, particularly the police, confines attention to prostitution when it regulates sexual conduct. Yet the state may use its authority to intervene under other circumstances, such as questions of public health. Concern about the transmission of AIDS has led to a federal information campaign aimed at changing sexual practices; in some local jurisdictions, such as San Francisco, legislation outlaws sex in public places. In 1986 the Supreme Court in a five to four decision upheld the right of the state of Georgia to outlaw private homosexual acts by consenting adults.

• *Education.* The state is now responsible for most primary and secondary schools, as well as much of the advanced education in colleges, universities, and graduate and professional schools. People are trained for specific jobs in the military, in special manpower programs, and with the help of government loans (once titled and justified as national defense loans). Governments at various levels are also involved in credentialing schools or determining that they meet certain standards of content, coverage, and competence in their programs. Among other things, this involves assuring that schools are offering courses of instruction mandated by the state. Even private schools and universities are critically dependent on the federal government for financial support (through program and research grants).

• *Occupational Life.* By itself and in collaboration with professional associations, the state sets standards for a variety of occupations and participates in the process through which people are permitted to practice. Physicians, real estate agents, accountants, attorneys, building contractors, civil servants, and many more are required to pass state-supervised examinations before legally pursuing their occupations. Public school teachers and counselors must obtain a state credential based on specific training. Psychologists and beauticians are licensed by the state. State supervision affects not only who may enter a vast number of occupations but also their conduct and the circumstances under which the right to practice a profession may be withdrawn.

• *Leisure Life.* Even our recreation is shaped by state control. Camping permits, wilderness passes, hunting and fishing licenses all regulate use of the outdoors. Some hobbies such as operating a ham radio require licenses. The state regulates when and where people can drink and dance. Where leisure life may include travel to foreign countries, the state issues passports and keeps tabs on exit and reentry, including the personal property one may travel with. Until recently, U.S. citizens were barred from legal travel to certain countries, and even today there are restrictions.

• *Income and Wealth.* For most people who enjoy an income or net worth above a certain minimum, taxes are the most forceful reminder of state authority. We are taxed not only by the many levels of government and charged for all those licenses, we are also taxed in ways that vary with our life-styles. All wage earners are taxed according to a theoretically progressive schedule — the greater the income, the greater the tax bite. That principle, however, is compromised by another — the greater the income, the more available are opportunities for tax exemption and "sheltering." Low-income people who cannot afford to buy a home must rent with no tax advantage.

Higher-income people can purchase homes and deduct mortgage interest payments from their income taxes.

• *Retirement.* As in work, the state influences how people spend their retirement. All those who pay into the Social Security fund are eligible for retirement income. The state determines the level of that income as well as the addition (and, more recently, the subtraction) of associated benefits such as medical care. Social Security sets limits on any additional earned income. Retired people who have difficulty living on their pensions and want to supplement that income must forego part of their retirement pay or break the law by not reporting extra earnings, even when those just meet essential living expenses (property taxes or food and medical bills that have risen faster than retirement incomes). The state's generosity determines how well or poorly 36 million Social Security recipients live.

• *Times of Need.* In 1983 the Census Bureau published a report centered on the impressive datum that 30 percent of the U.S. population currently receives direct benefits from the federal government. Social Security and Medicare top the list of these benefits, but they are followed closely by (means-tested) programs for the needy such as Food Stamps and Medicaid. Studies of cyclical poverty, such as those carried out at the University of Michigan, show that in the United States during any given eight-year span, one family in ten is poor over *all of those years,* but more than one-third of all families are poor *for at least one* of the eight years. Accordingly, a substantial number of people at one time or another depend on state benefits for unemployment, disability, aid to dependent children, or food stamps. Once receiving those payments, the individual's life is subject to close official scrutiny and stigma. Welfare recipients are required to follow budgets drawn up by social workers, and mothers of children receiving aid cannot live with whom they please.

• *Consumption.* State agencies provide consumer protection against unsafe products and fraud. Presumably, products sold to the public must meet standards set by the government, such as those for automobile safety. We are taxed differentially on consumer goods, including a judgmental luxury tax, while the state subsidizes certain (e.g., agricultural) producers to control prices. Various rates and fees, such as for shipping or utilities, are set by the state.

• *Information.* Modern societies are increasingly dependent on the electronic media for information. The airwaves, in turn, are considered public domain under state supervision. Broadcasting requires a license and adherence to a state-enforced code. In the United States, three large corporations dominate radio and television broadcasting, and a fourth public system is the only general source of variation in coverage or editorial slant. New networks require large amounts of capital and state certification, neither of which are easy to get as the battle over cable television franchises shows. In the last few years, congressional critics of administration policy in Central America have claimed, on the basis of clear evidence, that the state (or, perhaps, an outlaw group within it) has resorted to a public disinformation campaign.

• *Religion.* The First Amendment of the U.S. Constitution begins "Congress shall make no law respecting an establishment of religion, or prohibiting the free exercise thereof. . . ." What constitutes a religion or its free exercise, however, is sometimes a matter of dispute. Recently, the state has denied the normal tax-exempt status of churches to new and prosperous fundamentalist groups that use their facilities for profit-making activities. The Heavyweight Champion of the World was once stripped of his title (the state regulates sports too) for claiming military draft exemption as a conscientious objector. Many conscientious objectors have served jail terms for failing to persuade the state of their religious convictions. Conversely, courts have set aside convictions of immigrant Jamaicans who use drugs in their religious cult ceremonies.

• *Environment.* State responsibility has extended to the air we breathe. Health hazards of pollution and the depletion of natural resources have brought the state into the business of regulating use of the natural environment. Because the potential sources of environmental damage are vast, the state must create special agencies to deal with air pollution, toxic and atomic waste, contamination of waterways and aquifers, use of chemicals in industry and agriculture, depletion of the ionosphere, and an orbiting junkyard of spent space equipment. The manner in which the state deals with environmental abuse affects the individual's health, enjoyment, and pocketbook.

The foregoing list is weighty, fitting a burdensome subject. The point is not simply to give examples, but to show their scope by elaborating them over a variety of areas. The list names fifteen different ways in which the average person is affected by the state, intimating both the pervasiveness of and the mixed feelings we have about state regulation. Once that fact is established, it will also be clear that the world was not always organized in this fashion; the vast expansion of the state is historically recent, explaining, perhaps, why social science has yet to absorb the changes into its analytic schemes.

There was a time when the functions of the state were minimal by today's standards. In the United States, there were no public schools until the 1820s, when they were started in New York exclusively for children of the indigent classes. The first operating state university began in 1795 with the University of North Carolina, but it was not until the 1860s and the Public Land-Grant College (Morrill) Act that the major state systems of higher education took shape. The first state mental hospital opened (or closed) its doors in 1835 in Worchester, Massachusetts. Social welfare was a function of families, churches, and mutual societies until the late nineteenth century. Roads, steam and electrical power, worker housing, and, generally, all of the industrial infrastructure considered today as essential public services were once provided by factory owners themselves.

Personal income tax, the source of almost half of U.S. government revenues, did not exist until 1913; another thirty years passed before it was withheld from wages. What is called in some quarters the welfare state did not take its definitive form until the unemployment crisis and economic depression of the 1930s. During previous periods of recession and high unemployment in 1893, 1914, and 1921, Congress failed to approve bills providing federal aid to the unemployed. Frances Piven and Richard Cloward note that "during the severe depression of 1914–1915, for example, it was left to the New York Association for Improving the Condition of the Poor to initiate work projects for the unemployed in New York City" (1971, 48).

Much of the specific character of the modern United States as a state and a society derives from our historically recent experience of state building. The changes began with an elaborate system of pensions for Civil War veterans and took institutional form in the Progressive Era (1870–1920). It reached a zenith with the New Deal Administration of Franklin D. Roosevelt beginning in 1932. Particularly, the modern welfare state took shape with the Social Security Act of 1935 that created federal responsibility for unemployment insurance, old age pensions, and aid to the unemployable (the blind, orphaned, aged not eligible for pensions). The New Deal had precedents, but it was a qualitative break with the past in its redefinition of state responsibility. Once the state accepted the principle that it was responsible (at least as a regulator) for the public's economic and social welfare, its accelerating growth and entry into issues as far-flung as consumer and environmental protection were ordained.

Two summary points can be drawn from all this: First, the state universally influences individual and group life; second, this is new in the sense of a rapid expansion of its scope in the twentieth century. The latter implies that the new role of the state had identifiable causes, some of which are hinted at with the observation that state expansion followed economic crisis. If this much is granted, the three themes that introduced this chapter now appear as three questions that must be answered: What explains the growth of the state? What are the implications of that growth for social organization? How are state–society relations changing — where are we headed? These questions are treated in the next three sections.

Answers to the first question are easier once the ideological underbrush is cleared away. An intuitive sense of its importance has led conservative, liberal, and radical social critics to cast the state in the image of their respective demons. Although its origins are kinder, the term *welfare state* typically occurs in conservative criticism of big government, misguided initiatives that undermine free enterprise, and costly expenditures that should be replaced by workfare. A new variation on this theme is Christopher Lasch's (1977) criticism of a "therapeutic state," the extensive social service agencies staffed by liberal professionals that are allegedly destroying the family. Liberals like

John Kenneth Galbraith and Gunnar Myrdal point even-temperedly to the "new industrial state" or the "post-welfare state" as the unavoidable creature of advanced capitalism that requires a broadly planned economy. Ralph Miliband, Gabriel Kolko, and other radicals understand the "capitalist state" as a marriage of big business and government, a means of rationalizing the economy and stabilizing politics in a manner that safeguards the elite interests of both in a new "political capitalism."

There is truth in each of these views, rather more in the last I believe. But there is also much simplification. Group perceptions can be mistaken, but it is instructive to find that sharply opposed political camps separately view the state as a tool of their antagonists. The problem is partly perceptual and partly a result of complexity. The state serves many contradictory interests and sometimes contradicts itself. Moreover, caricatured liberal and conservative positions contain their own contradictions.

This is apparent if we return for a moment to the case of Baby Jane Doe. Conservatives generally stand for less government intervention and more free choice; liberals favor state responsibility for the general welfare. Yet the conservative pro-life groups that took up the baby's case want the state to ensure the rights of infants and to outlaw abortion. The liberals who stood up for the Doe parents insist on individual choice free of state coercion. The pro-life debate is not unique in this respect. Conservatives and liberals differ uncharacteristically on questions like farm subsidies, defense spending, and loans to large corporations.

The lesson to be drawn from these examples is that the state cannot be properly understood as merely the instrument of particular, and sometimes contradictory, interests. Karl Marx was certainly wrong in his polemical writings when he said, "The executive of the modern state is but a committee for managing the common affairs of the whole bourgeoisie." The state is more than that for several reasons. First, its own legitimacy depends on the fact or appearance of serving "the people." Second, concrete actions that appeal to one constituency, such as the pro-life conservatives, are bound to antagonize liberals and other conservatives (e.g., some physicians, hospital and health insurance managers). By attempting to find the least hazardous path through this thicket, the state is bound to act in ways that do not represent anyone. For example, in the Doe case, the government did not intervene to save the baby but to establish its own right to review medical records. Third, in the near-impossible task of trying or claiming to represent all groups, the state develops its own distinct interests in legitimacy and political support.

These arguments become clear when we examine actual state action in the economy and society. That is, instead of trying to explain state formation with a theory about something else (such as capitalism), we need a state-centered approach — one that traces how the state expands in specific areas and then asks, Why?

THE STATE AND ECONOMY

During the first year of its operation in 1789, the U.S. government collected $5.72 million in revenues and spent $5.78 million. By 1991 the budget was $1.3 *trillion,* which exceeded federal income by $270 billion. Times had changed with 200 years of inflation and the expansion of the union from thirteen to fifty states, but neither of those changes match the budget that increased 232,000 times over. What changed was the number of obligations the state had assumed. Moreover, much of the expansion occurred in recent years. Leaving aside periods of civil and world war when expenditures skyrocket, the greatest increases came in the 1930s and 1960s. Federal expenditures increased 1000 percent between 1955 ($64 billion) and 1981 ($660 billion) — when inflation rose about 300 percent — and it doubled again between 1981 and 1991.

A simple picture of federal finances is provided in Table 6.1. Of the $1.05 trillion in 1991 revenues, most came from the combined sources of individual income and Social Security (insurance) taxes ($864 billion when combined). Corporation income taxes amounted to 21 percent of what individuals paid and 9 percent of total revenues — figures down from the 26 percent and 13 percent, respectively, in 1980. By far the largest of $1.3 trillion in federal expenses are Social Security and

TABLE 6.1 U.S.Government Receipts by Source and Outlays, by Function, Fiscal Years 1980 and 1991 (in Billions of Dollars)

	1980	*Percentage*	*1991*	*Percentage*	*Percentage Difference*
Receipts					
Individual income taxes	244.1	47.2	467.8	44.5	-2.7
Corporation income taxes	64.6	12.5	98.1	9.3	-3.2
Social insurance taxes and contributions	157.8	30.5	396.0	37.6	+7.1
Excise taxes	24.3	4.7	42.4	4.0	- .7
Estate and gift taxes	6.4	1.2	11.1	1.1	- .1
Customs duties	7.2	1.4	15.9	1.5	+ .1
Miscellaneous receipts	12.7	2.5	22.4	2.1	- .4
Total	517.1	100%	1053.8	100%	—
Outlays					
National defense	134.0	22.7	272.5	21.0	-1.7
International affairs	12.7	2.2	15.2	1.1	-1.1
Income security	86.5	14.6	171.2	12.9	-1.7
Health	23.2	3.9	71.2	5.4	+1.5
Social Security and Medicare	150.6	25.5	373.5	28.2	+2.7
Veterans benefits and services	21.2	3.6	31.3	2.4	-1.2
Education, training, employment, and social services	31.8	5.4	41.5	3.1	-2.3
Commerce and housing credit	9.4	1.6	75.6	5.7	+4.1
Transportation	21.3	3.6	31.5	2.4	- 1.2
Natural resources and environment	13.9	2.4	18.7	1.4	-1.0
Energy	10.2	1.7	1.8	0	-1.7
Community and regional development	11.3	1.9	7.4	0	-1.9
Agriculture	8.8	1.5	14.9	1.1	- .4
Net interest	52.5	8.9	195.0	14.7	+5.8
General science, space, and technology	5.8	1.0	15.9	1.2	+ .2
General government	4.4	.8	11.4	0	- .8
Administration of justice	4.6	.8	12.2	0	- .8
Undistributed offsetting revenues	-19.9	-3.4	-39.4	-3.0	- .4
Total outlays	590.9	100%	1322.6	100%	—
Deficit*	-73.8	14.3*	-268.7	25.5	+11.2

*As a percentage of receipts.

SOURCES: U.S. Department of Commerce, Bureau of the Census. 1992. *Statistical Abstract of the United States.* Washington, D.C.: Government Printing Office; U.S. Department of Treasury. September 1991. *Monthly Treasury Statement.* Washington, D.C.: Government Printing Office.

Medicare ($374 billion), defense spending ($273 billion), and interest on the national debt ($195 billion). From 1980 to 1991, interest on the debt also represented the biggest change in budget priorities, increasing as a percentage of the total outlay by almost 6 percent. The largest budget cuts were in defense, programs for income security (e.g., retirement and disability insurance, housing assistance, food and nutrition assistance, unemployment insurance), education, energy, and community development.

These budget comparisons for 1980 and 1991 provide a clear picture of what the conservative policy involves in financial terms. Increases in government revenues came mainly from increased Social Security payments by individuals, while corporations received the principal tax reductions. On the expenditure side, large increases occurred in Social Security and Medicare, items fixed by previous legislation. The government spent far more than it took in during all these years; the size of the deficit increased rapidly whether expressed as an absolute sum ($44 to $269 billion) or as a percentage of each year's revenues (14 to 26 percent). The difference, of course, is paid for by government borrowing, and the cost of each year's deficit spending shows up in the following year's bill for interest on the debt, which rose from roughly 9 to 15 percent of the annual budget over the period. Moreover, because recent borrowing has been at a higher rate, the interest bill in the immediate future will continue to rise as a percentage unless revenues increase dramatically. The projected federal budget deficit for fiscal year 1992 is $400 billion — a massive increase over 1991's $269 billion — accounted for by a combination of recession and the savings and loan bailout. Revenue increases to offset the rising interest bill have to come from one of the top three sources of taxes, but so far they have concentrated on increases in Social Security payments. This is the fiscal dilemma the state faces in the 1990s: Either Americans, perhaps even corporations, will have to pay more taxes, or an increasing percentage (from 4¢ in 1975 to 15¢ in 1991) of every revenue dollar will have to go toward paying back debts.

In 1991 government at all levels employed 19 million people. A useful indicator of state expansion over time is the proportion of the total labor force employed by government. This figure has climbed steadily from 7 percent in 1900 to a high of 19 percent in 1975. That is, in 1975 nearly one employed person in every five worked for government, most of them at the state and local level. Between 1975 and 1991, a time of recession and government fiscal strain, the percentage of the labor force in government (though not the absolute number) dropped slightly.

If we combine this measure of state expansion with long-term changes in the federal budget, two conclusions are suggested. First, in the mid-twentieth century, state expansion proceeded at an unprecedented peacetime rate. Second, the times of greatest expansion coincided with social and economic unrest and its aftermath. Simplifying a bit, these were the times of depression and the equality revolution — periods followed by vast new state responsibilities that are reflected in the costs and personnel of government.

THE POLITICS OF EXPANSION

Yet the state did not grow in isolation. The same years witnessed an enormous expansion and concentration of the economy. The two went together. A pivotal book on the U.S. economy, by Paul Baran and Paul Sweezey (1966), describes it as *monopoly capitalism.* Those who object to the phrase recognize, nevertheless, that if many branches of industry and commerce are not strictly monopolized by single firms, there has been an inexorable trend toward **oligopoly** (domination by a few) during the twentieth century. This trend, like the growth of the state, has been steady but also punctuated by periods of acceleration and slowdown. One period of growing concentration occurred in the 1890s, another in the 1920s, and the most dramatic shift toward oligopoly came in the post–World War II boom, roughly from the mid-1950s to 1970.

Taking a midpoint of these changes, in 1962 there were about 180,000 manufacturing corporations in the United States. Yet, of that very large number, just the 20 biggest firms owned 25 percent of *all* manufacturing assets; the largest 200 corporations (about one-tenth of 1 percent of the total number) owned 56 percent of all the assets. Concentration took place over a short period of time. In 1925 the 100 largest manufacturing firms held about one-third of all manufacturing assets. That figure rose to 42 percent in the early 1930s when the trend slowed and was even reversed slightly. In the mid-1950s, it began moving upward again such that by 1970 the 100 top corporations held one-half of all manufacturing assets. The same result is obtained by taking the top 500 firms or, say, all those with assets over $10 million. At present more than 80 percent of all corporate wealth is held by about 1 percent of all firms (including banks and large retail firms along with manufacturers). Concentration has grown markedly during the twentieth century, notably in periods of great prosperity.

The figures on growth and concentration indicate something quite the opposite of conventional wisdom or the promises of free enterprise. Two facts are especially striking. First, periods of prosperity (the 1920s and the postwar boom from the late 1940s to the early 1970s) do not lead to greater competition or to enhanced opportunities for small business but to greater concentration by the corporate giants — their invasion of new realms of the economy once reserved for local and family firms. During cycles of depression (the 1930s and late 1970s to early 1980s), the largest firms have trouble, and competition is restored slightly. Second, expansions of state and economy are not opposed to one another; big government does not hamper business expansion. On the contrary, big business and big government foster one another. This is suggested by their coincident historical timing in which state expansion follows closely on periods of economic concentration.

The principal reason for state expansion has been its widening role in the economy. State intervention in the economy is aimed at promoting the conditions of general prosperity, meaning specifically that conditions of stability and continuing economic growth are met. Doubtless, the state does many things that widen its role in society, such as supporting the arts, leisure, or basic research, but its efforts to regulate the economy account for most of the expansion. In the United States and western European countries, "the economy," of course, means an advanced capitalist economy. As Miliband argues, state intervention in these countries has a very specific purpose:

> State intervention in economic life in fact largely *means* intervention for the purpose of helping capitalist enterprise. In no field has the notion of the "welfare state" had a more precise and apposite meaning than here: there are no more persistent and successful applicants for public assistance than the proud giants of the private enterprise system. (1969, 78)

Yet the task of helping capitalist enterprise or ensuring the conditions of prosperity is much more involved than simply providing tax breaks to big business or lavishing a huge chunk of the federal budget on defense industries. The key to understanding the state's expanding role in the economy is in providing the conditions of stability and continuing growth. As we have seen, these conditions include the provision of mass education and a skilled work force, the physical infrastructure of transportation and energy, a population sufficiently prosperous to provide a home market for consumer demand, a global presence that ensures foreign markets and raw-material supplies, support for scientific and technical innovation, labor peace, and mechanisms for dealing with the unfit or the casualties of economic competition (the disabled, unemployed).

The defining feature of advanced capitalist society is that these responsibilities, once shouldered by individual firms and local philanthropies, are now awesomely beyond the capacity of private initiative. The state is the only body big enough and generally accepted for this role. This proposition, which is plausible at face value, is made even more obvious by the history of those periods of accelerated state expansion.

In recent U.S. history, the great eras of state expansion corresponded to periods of social unrest and economic crisis. The modern state began to take shape in response to financial depressions and the Populist and Progressive movements from 1880 to the turn of the century. The New Deal and equality revolution coincided with the Great Depression of the 1930s and its subsequent post–World War II recovery boom. By this I do not mean that protest and reform are caused by economic crises in any simple way or that they are caused by what the state does to solve economic crises (a somewhat more plausible alternative if singular explanations were sought). Rather, I mean that the occasions for a new role of the state begin with crises and the general problem of how to restore social stability and economic growth in a new set of circumstances.

The historical argument cannot be provided in detail here and indeed has been the subject of many excellent books. Concerning the progressive era, Gabriel Kolko's *The Triumph of Conservatism: A Reinterpretation of American History, 1900–1916* (1963) and James Weinstein's *The Corporate Ideal in the Liberal State: 1900–1918* (1968) show that the foundations of the modern U.S. state were laid in the first two decades of this century with measures to regulate business enterprise. Particularly, the Federal Trade Commission and antitrust legislation were designed to restrain ruinous competition, the near-monopolies of Standard Oil and a few railroads, but not to hamstring big business as a whole. On the contrary, the Progressive Era, including the urban-reform movement aimed at big-city machine politics, fashioned a new form of political capitalism in which the general interests of business prevailed. Indeed, the major business associations were the architects of the new regulations designed to bring stable growth and social peace. The state regulated those interests (through interstate commerce rates and franchises) in an undertaking that also significantly expanded its own size and function.

More important from the standpoint of the sheer volume of state activity, if not for the precedent of intervention, was the response to the depression crisis of the 1930s. The New Deal created an important part of contemporary federal and state bureaucracy through the Social Security Act of 1935. The key feature of this legislation was its recognition that private charity could not handle the massive unemployment crisis of the 1930s (12 to 15 million people out of work, representing well over one-third of the labor force) and that the federal government had to take responsibility for relief and for restoring confidence to business and agriculture. The social insurance programs that constitute the largest category of current federal expenditures (and the second largest category of taxes) began in 1935 and were expanded — for example, with Medicare — in the 1960s. The purpose of the New Deal, however, was less to create a "welfare state" than to restart the economy by providing relief, jobs, and consumer demand along with concessions to industry and labor. Once signs of recovery began to appear in the late 1930s, direct relief for the unemployed was converted to work relief, the wages for work relief were reduced so as not to appear more inviting than private employment where it existed, and relief or welfare was steadily redefined as shameful for all but the disabled and poor mothers — and in practice it was even made shameful for them.

The central question in this discussion has been, What explains the growth of the state? One blunt answer is that the state expands in response to the demands of the economy, including both popular protest and financial crises. This is not the only answer, as we will see below, and the relationship between state and economy is not unidirectional. The state is both a source of private economic growth and a problem-solving response to conflicts generated in that growth. The state and economy thus merge in a way that makes it impossible to speak of one thing causing another. As Kolko concludes in his study,

> The crucial factor in the American experience was the nature of economic power which required political tools to rationalize the economic process, and that resulted in a synthesis of politics and economics. This integration is the dominant fact of American society in the twentieth century, although once political capitalism is created a dissection of causes and effects becomes extraordinarily difficult. (1963, 301)

If that much of the argument has been demonstrated, the next question is, What are the implications of state expansion for social organization?

THE STATE AND SOCIETY

State influence is scarcely confined to the economy as the previous illustrations show. The state invades, or at least guards the frontiers of, social life. The question is less whether the state affects everyone's daily affairs than how and to what extent it transforms social organization. This could be answered in countless ways. I will focus on two representative topics that bring the question close to home: family and school.

Invasion of the Family

Family sociology, once a backwater of the discipline, has lately become a lively and quarrelsome arena precisely because of competing claims about how the state has altered this basic institution. The debate joins feminists, conservatives, and conventional family specialists. Many issues are captured in the compound question: Where is the family headed, toward dissolution or adaptation? Conservatives argue that it is in danger of losing its traditional role because of permissive child rearing, two-career parents, or meddling in its province by social service agencies of the state. The last point is the theme of Lasch's popular book *Haven in a Heartless World: The Family Besieged* (1977), a book which argues that "the family has been slowly coming apart for more than a hundred years" owing largely to a "tradition of sociological study" that reflects and shapes "the shattering impact of policy — the impact of the so-called helping professions — on the family" (p. xiv).

Feminists are split on the significance of recent changes, some of which are attributed to their own movement. The new conservative feminists (as Judith Stacey [1983] refers to the latest position of Betty Friedan and others) also believe that the family is threatened, perhaps by too much equality — that its nurturing and emotional support activities are disappearing as family structure is made a public issue by radical feminists. The latter argue that the family should and probably will survive but in a new form free of patriarchy and based on real equality of the sexes.

Conventional family specialists are often aloof to these quarrels. They claim that the basic nuclear family of husband, wife, and children has succeeded the extended family comprised of many relatives living together. But the nuclear family itself has shown remarkable durability, persisting despite divorce, the generation gap, the sexual revolution, and the feminist movement. The family adapts to a changing environment, to the economy in particular, but does not change in the dramatic ways that conservatives bemoan or feminists desire.

As the debate goes on, family structure is changing. The typical family — composed of a married couple, minor children in the home, and one working spouse — in fact is now a rarity. Only about 17 percent of all U.S. families fit this description. By the early 1980s, for example, 62 percent of all married couples had both husband and wife employed, up from 50 percent in 1970 and 40 percent in 1960. Most of the change is due to wives joining the labor force for the first time (Chapter 5) or returning to work after childbearing rather than any appreciable increase in the number of married couples. Between 1975 and 1979, the number of married couples increased by 2 percent while the number of families with employed wives jumped 14 percent. Among married couples, 95 percent of the husbands work, as do 67 percent of the wives. Birth rates edged up in the early 1980s after a low ebb in the mid-1970s, but, at least for the near future, they will remain low by the standards of the immediate postwar decades. Similarly, marriage rates rose in the 1970s and early 1980s after unusual lows during 1955–1965. Divorce rates have risen steadily throughout this century. The pattern, in sum, indicates that contemporary families are marrying, divorcing, remarrying, and working more so than in the past.

The nuclear family clearly persists, as family sociologists claim, but closer inspection reveals that it persists in a form rather different from what was once typical. The special nature of today's nuclear family stands out when compared with earlier forms. If the two-career family is new in the annals of the nuclear family, the nuclear family itself is a product of recent history. In preindustrial societies, the family embraced a wide circle of relatives and household workers. Indeed, the household was the unit of economic activity, the site of the job, combining productive and reproductive functions. Peter Laslett describes the preindustrial English household, using as a prototype a London bakery where fourteen people (including the baker, his wife, children, servants, journeymen, and apprentices) all lived and worked together:

> The only word used at that time to describe such a group of people was "family." The man at the head of the group, the entrepreneur, the employer, or the manager, was then known as the master or the head of the family. He was father to some of its members and in the place of father to the rest. There was no sharp distinction between his domestic and his economic function. His wife was both his partner and his subordinate, a partner because she ran the family,

> took charge of the food and managed the women servants, a subordinate because she was woman and wife, mother and in place of mother to the rest.
>
> The paid servants of both sexes had their specified and familiar position in the family, as much part of it as the children but not quite in the same position. . . . The division of dwelling place and working place was no recognized feature of the social structure of the towns which our ancestors inhabited. The journey to work, the lonely lodger paying his rent out of factory wages or office salary, are the distinguishing marks of our society, not theirs. (1965, 2, 14)

The English preindustrial family closely resembled patterns found throughout the Western world and especially in the colonial United States. In 1790 the average U.S. household contained 5.8 persons; by 1980 the figure had dropped to 2.8. The historical forces that altered that pattern to the nuclear and two-career family residing in a home distinct from the workplace are manifold and conveniently summarized in the master themes of industrialization and modernization. For present purposes, however, the state played an essential role in the transformation — perhaps a leading role.

The transformation involved a shift from large, multipurpose, interdependent family units to small, single-purpose, self-supporting, private ones. In effect, this meant that certain (but by no means all) family responsibilities were lost (or located elsewhere) as the new family responded to incentives to streamline itself. The state facilitated this transformation by creating institutional alternatives to family functions: public schools for the education of children, social insurance for support in old age or unemployment, relief and welfare for needy women and children, asylums for the ill, social workers and probation officers for the errant, public training and employment for the corrigible and able-bodied men.

The state had no conspiratorial aim in fostering policies that had the effect of transforming the more communal, extended family to the nuclear family — no conscious purpose to reduce and privatize the family as an end in itself. Rather, the modern state intended to reform social life in a manner consistent with economic growth and prosperity. The growing demands of the economy — such as mass education and a consumer society — as well as the perils of economic competition — such as recession and unemployment — overwhelmed the capacities of the traditional family. Specialized public institutions were best suited for these new tasks. Conversely, the nuclear family was a more efficient unit for producing and socializing children to assume their place in the social order.

As industry was rationalized for the production of mass-consumption goods, the "privatized" household became the logical market for a standard set of necessary possessions — a set that was duplicated, expanded, and frequently replaced in each home. That fact is commonplace to any observer of family consumption patterns. What may be less obvious is that the state pioneered this materialistic frontier with policies that reshaped the family. The best example in contemporary U.S. history is housing. Through the Federal Housing Authority (FHA), GI loans, and their successors, the state subsidized with mortgage loans the massive suburbanization that made the single-family home a statistical and normative standard (see Chapter 3). In a typical year, the federal government spends on public housing a small fraction of the billions it allows in tax deductions for mortgage interest payments. The nuclear family has become a normative criterion underpinning social policy. As Eli Zaretsky concludes his analysis of this change, "'the family,' in the conventional sense of a private, self-supporting nuclear unit, was to a large extent created, or at least reconstituted, by the modern state" (1982, 192).

Controversy over the nature and plight of the black family provides ironic support for this interpretation. In the fall of 1965, the federal government released a soon-to-become-famous study written by Daniel Patrick Moynihan, entitled *The Negro Family: The Case for National Action* (1965). The Moynihan Report urges, on one hand, true equality for blacks, and reasons, on the other, that the roots of inequality lay in the black family. Beginning with the evils of slavery, the report argues, the black family has been abused in ways that produced a record and a culture of family failure. The record is found in high rates of divorce, desertion, illegitimacy, female-headed households, and welfare dependency. The culture of a mother-centered, or matrifocal, family purportedly causes high rates of delinquency, school dropout, drug use, unemployment, and so forth.

The record cited in the report is accurate and continues in the 1990s. The analysis, however, was sharply disputed, particularly the interpretation that a family culture somehow produced all that pathology. William Ryan, in his memorable phrase, called it "blaming the victim," and critical sociologists showed that there was really no direct connection between having been raised in a matrifocal family and individual delinquency or the rest. Delinquents (etc.), including white youths, come from all sorts of families but mainly from the poor and unemployed. Moreover, many poor families that have not experienced slavery (white and Hispanic) also show the statistical face of a matrifocal family. That is, the matrifocal family appears a pattern and plight of lower classes rather than particular status groups. Many believed that the analysis was wrong.

The lower-class black family has a distinctive profile that is more likely a result of poverty and inequality, which take the form of high male unemployment, low-level occupations, and welfare codes that disqualify families with an able-bodied male in the household. In short, because of discrimination, black families do not enjoy equal access to consumer society or the protections and subsidies of the state. Important from an interpretive standpoint, middle-class black families that have gained such access conform to the nuclear family pattern.

Once again, that is not news. What is revealing, however, is the fact that blacks have developed their own community institutions and culture for coping defensively with a hostile state and environment, *including the extended family.* We have learned, from 1939 onward with the publication of E. Franklin Frazier's *The Negro Family in the United States,* that black families include centrally grandparents, uncles and (especially) aunts, cousins, and friends that have "fictive kin" status — all knitted together in mutual support. With respect to pathology, blacks have a lower suicide rate than do whites because, if Durkheim's theory is right, they have a higher degree of group cohesion. As Frank Riessman says in his criticism of the Moynihan Report, "The Negro has responded to his oppressive conditions by many powerful coping endeavors. He has developed many ways of fighting the system, protecting himself, providing self-help and even joy. One of the most significant forms of his adaptation has been the extended, female-based family" (1966, 475). The lower-class black family is the exception proving the rule of state influence on the nuclear family pattern.

Forming and Reforming the Schools

The social organization of schooling parallels in many ways the economy and family. Beyond intervention, since the late nineteenth century, the state has taken major responsibility for the once private and family chores of education. Beyond responsibility, beginning with the Progressive Era at the turn of the century, the state has reorganized the nature of schooling in ways consistent with its general program of social reform. Raymond Callahan (1962) shows that the key to school reform was in the pressure put on administrators to reorganize educational institutions in the likeness of business corporations — to make them run like factories and produce graduates fitted for their place in corporate and state bureaucracies.

Although school administrators had little sense of how factories operated, they obligingly adopted the familiar organization that seemed to suit the dictates of scientific management: compulsory schooling, regular long hours, required attendance, penalties for absenteeism and tardiness, standardized work (curriculum), time-budgeted instruction, performance ratings (grades), skills analysis (testing), an academic and vocational division of labor (tracking), earned promotion, a chain of command from classroom teacher to principal, dismissal for noncompliance with authority, and the broad ideological defense of these methods founded on the dogma of "achievement" — that is, the certainty that educational performance determined occupational success and future earnings.

Higher education has not escaped these reforms. The state has influenced the character of education in colleges and universities beginning with the creation of the land-grant college system (in 1862), the GI Bill that sent a new kind of student to college in the veterans of World War II, and the vital research and program funds that universities continue to receive from government. Among the consequences of this support, colleges and universities were encouraged, first, to maintain a program of research that served state-suggested public needs. Because the state reflects political pressures, this often meant the needs of private enterprise such as the agribusiness agenda for land-grant colleges and their research documented in Jim Hightower's book *Hard Times, Hard Tomatoes* (1972). Second, it meant that a curriculum was designed to serve the occupational demands of the corporate and state economy — business and public administration, social work, engineering, education itself. The liberal arts took a backseat to the marketable areas of specialization. Today something called general education (humanities, arts) fights a rearguard action against such curiosities as a newly discovered computer science for a place in the curriculum.

Around 1970 a new plateau was reached in higher education as slightly over 50 percent of all high school graduates went on to college. Although an active military draft at the time encouraged the high figure and the percentage slipped back a few points in subsequent years, the coming-of-age of the postwar baby boom marked a permanent change to mass higher education. For twenty years, from the mid-1950s to mid-1970s, community colleges and private and state colleges and universities expanded (and overexpanded) to meet the new demand. With the laudable achievement of mass higher education, however, a crisis set in. The students were unhappy, even rebellious.

Although much of the rebellion was aimed at the Vietnam War and only happened to be staged on the campuses, educators correctly read a lesser theme of educational criticism in the student movement — including university war-related research. Tellingly, students described the university as an "education factory" and (former University of California President) Clark Kerr's approving phrase "multiversity" became an insult in dissident mouths. But the students were not the only ones unhappy. Some educators saw the mass invasion of the university as a threat to its standards, perhaps to its status, and often as the very cause of the unrest (embattled administrators tended to take all this personally and failed to see what a threat Vietnam was to participants in the student movement).

As is usually the case in higher education, the crisis called for a study. From 1967 to 1973, the Carnegie Commission on Higher Education labored over a multivolume answer to university ills. It made many recommendations, but their central thrust was to limit the numbers coming into the universities and to encourage a new division of labor in higher education by segregating ordinary training for the masses from research and advanced instruction for the privileged. In the commission's words,

> "Elite" institutions of all types — colleges and universities — should be protected and encouraged as a source of scholarship and leadership training at the highest levels. They should not be homogenized in the name of egalitarianism. Such institutions, whether public or private, should be given special support for instruction and research, and for the ablest graduate students; they should be protected by policies on differentiation of functions. (*Priorities for Action: Carnegie Commission on Higher Education, Final Report* 1973, 30)

Samuel Bowles and Herbert Gintis (1976), after reviewing the commission's conclusion, ask "Whatever became of equality of educational opportunity?" Obviously, it did not fit with the new demands that the state and economy were placing on the elite universities — demands for new research, marketable technology, and advanced scientific training — least of all when the universities that should be doing this work were in the throes of political dissent. Better to shunt the average student (whom the educational planners may have pictured, mistakenly, as more prone to political action) to the community college or the state college system. Indeed, major state systems began implementing this recommendation. Students judged less promising were diverted to the community colleges or, as some called them, "the high schools with ashtrays." Whatever else it meant, in this case the less promising clearly included the less affluent and minorities, creating a new problem for universities trying to meet their affirmative action goals. The justification for all this, Bowles and Gintis observe, was "starkly reminiscent of the arguments made for high-school stratification [i.e., vocational schools] during the progressive era" (1976, 209).

Reforming the schools seems an unquenchable urge. No sooner had higher education gotten used to the fallout of the Carnegie bomb than a rash of warnings appeared about the deterioration of primary and secondary schools in the early 1980s. Best known, perhaps for its alarming title, is *A Nation at Risk: The Imperative for Educational Reform* prepared by a National Commission on Excellence in Education (three others produced by various task forces at about the same time were *Action for Excellence, Making the Grade,* and *Educating Americans for the 21st Century*). In each case, the condition of public education was described in ominous tones: "the educational foundations of our society are being eroded by a rising tide of mediocrity"; "a real emergency is upon us"; a "disaster." The language suggests that something catastrophic had occurred, a setback on the order of a Russian Sputnik, which prompted similar excitement in 1957.

The reports vary on precisely what had happened. Descriptions of the emergency ranged from falling test scores to apathetic or undisciplined students. The crux of worry, however, jumps out of *A Nation at Risk* in the observation that education must stress excellence if we hope to "keep and improve the slim competitive edge we still retain in *world markets*" (emphasis added). In short, the looming disaster is economic, and the sobering evidence of a narrowing competitive edge is not a Russian satellite but a galaxy of Japanese consumer goods. The connection is made clear in a set of comparisons showing how Japanese children got smarter: They attend school 220 days a year against our 180, they spend three times as many hours studying science in high school, and 95 percent of their high school students graduate versus 73 percent of ours. Accordingly, the recommendations that follow focus on a more demanding regimen of school routines and, particularly, a larger dose of scientific, technical, and computer training.

The critics, mostly public officials, business executives, school administrators, and professors, see another difficulty at the root of the disaster and any practical solution — namely, incompetent teachers. Allegedly, school teachers are recruited from the lower levels of college graduates and lack preparation for math and science teaching. A solution to that problem is envisioned in a master teacher program that rewards good teaching with pay differentials, although there is no clue to how teaching would be evaluated. Teachers become convenient scapegoats. No attention is given to class size, declining salaries, growing workloads, or what actually goes on in the classroom. The responsibilities of parents, school officials, and taxpayers play no part in the problem or its solution. Indeed, there is a striking similarity between the problem conceived materialistically as a disappearing competitive edge in world markets and the solution as monetary incentives for more spirited teaching performance.

In the end, it is the reports themselves that perform badly. Between the hyperbole and scapegoating, the basic analytic questions are not posed. Should a competitive edge in world markets be the aim of education? Are the Japanese doing well because of their schools (or because they make cars that don't fall apart)? If the United States has lost its competitive edge in world markets, how is that the fault of students and poorly paid teachers, rather than the businesspeople who were in charge when it happened? If the schools are doing poorly, why not pay teachers better, give them paid leaves to learn new subjects, reduce class sizes and workloads, rather than giving incentive pay, which would introduce cut-throat competition into the schools? In other words, why blame the victims?

Obviously, the answer to these questions is that the alternatives are costly. In the absence of any probing analysis or unflinching commitment of thought and money to education, the reports provide crass business sense. The recommended dominance of technical subjects in a new curriculum is a modern re-creation of the parody Charles Dickens gave us with Mr. Gradgrind's school of "the facts" in *Hard Times.* At bottom, it is not only an exercise in blaming the schools for circumstances far beyond their control, but it is also one more illustration of how schooling is construed as a servant of the state and the economy.

For good or ill, the state's role in shaping the social organization of the family, school, and other institutional foci of our lives is decisive, if sometimes indirect. Social organization is changing in ways that affect and sometimes make uncomfortable individual lives — families become more isolated and less convivial, schools more bureaucratically regimented in competitive pursuit of material ends. Through these intermediaries, the state weighs on the individual. The root influence is a complex union of state and economy.

These developments summarize in a focused context what grand theories describe as the parallel evolution of bureaucracy and capitalism. This material portrays a pattern of mutual causation in which the economy generates crises due to its own competitive drive and the state responds to those crises with more adaptive policies that influence social organization (e.g., families and schools). But the state also takes the lead by refashioning the institutional rules governing the economy, social life, and the state itself. The direction of these changes is not determined by the requirements of any one institution — not by economic conditions alone, for example. Rather, change is a result of interacting causes: institutional demands and social group conflicts. These principles are demonstrated in our final theme, the direction of change.

THE CRISIS OF THE STATE

Sociological theories have talked about the great transformation from community to society beginning with Durkheim and Tonnies, continuing through the once-fashionable notion of mass society, and resting today with the theory of the state. In that enterprise, the theories have clearly portrayed the historical experience, and a few have explained pieces of it. Marx was certainly right and ahead of his time when he analyzed the trend toward a concentration of capital. Weber added a key element left undeveloped by Marx when he argued that the state expanded "in a complete parallel to the development of capitalist enterprise through gradual expropriation of independent producers. In the end the modern state controls the total means of political organization" (1946, 82).

In one essential respect, however, these grand theories mislead by failing to dwell on the conflicts that attend the process of centralization — conflicts that are important not simply because they may slow the pace of the process but because they define the nature of its outcome. Concretely, this means that the family has not been eliminated by the state (or by the helping professions as Lasch [1977] maintains) or the schools made willing lackeys of the corporation (as Bowles and Gintis [1976] sometimes imply). Instead, families, schools, and many other institutional centers of human endeavor have felt the conflicts and resisted, struggled, acquiesced without enthusiasm, lost grudgingly, and generally responded in ways that also determine the emergent nature of social organization. All these intermediate powers are altered but not eliminated, and their alteration involves a struggle.

This generality takes substance in social organizational forms such as the lower-class and black matrifocal family that illustrates how family and community brace themselves for self-defense and even flourish in a hostile setting. Riessman observes the following:

> In response to the deficiencies of the system the Negro has developed his own informal system and traditions in order to cope and survive. Storefront churches, the extended family, the use of the street as a playground, the block party, the mutual help of siblings, the informal know-how and self-help of the neighborhood, the use of peer learning, hip language, the rent strike and other forms of direct social action are just a few illustrations. (1966, 477–78)

White families cope in other ways such as cooperative schools and day care for two-career couples, new kinds of paid work within the home (the informal or underground economy), house husbands, and experiments in communal living. The two-career family is still a family. It has not been eliminated and may even be facilitated by the much criticized helping professions in matters of child care and extended-day schools, making it durable but different from the traditional nuclear family. Schools resist and adapt with new programs (many teachers have taken their own scarce time to learn the mysteries of computers), alternative schools, and perseverance in teaching old skills like writing, all despite corporate pressures. When the yoke of the state gets heavy, taxpayers revolt. Students revolt, demanding something different from what government and schools have offered, and, in the cases of Vietnam and the "multiversity," they sometimes succeed. None of these responses can reverse the concentration of the state and economy, but together they mediate the impact of those forces on daily life.

Fiscal Crisis

The final irony in this long story of state power is that just as we have come to understand how the state has penetrated our institutions and communities, the state itself has fallen into a crisis of financing and of legitimacy. Its material capacity for intervention is drying up as fast as its conferred right.

The fiscal crisis of the state is elegantly outlined in James O'Connor's (1973) book of the same title. Put simply, the expanded operation of the state, much of it in support of private enterprise, becomes too costly, particularly when profits from government-assisted activities are privately appropriated. That is, the growing costs of big government — new agencies to protect the environment, renew the cities, or finance private health care — are socialized in the sense of being created at taxpayer expense. Meanwhile, the private beneficiaries of these services profit, such as the doctors and hospitals whose social insurance–supported costs have skyrocketed. But there are limits: a political limit set by rebellious taxpayers and a structural limit posed by recession, such as the one that began in 1974 and precipitated the fiscal crisis.

From the late 1970s until the present, the state has found itself faced with the need to cut back on services and personnel, beginning with services to the poor (Social Security payments, school lunches, eligibility under Medicare) and extending to programs for the middle class (education). Much more than a temporary strain, the state began reversing the historical drift of expansion, began to shed some of its responsibilities by reprivatizing services.

Although privatization may please certain business interests, the familiar problem was that the fiscal reorganization of the state has followed lines of class and status privilege. Leafing back to Table 6.1, the effect of the fiscal crisis is evident in federal budget changes from 1980 to 1991. In just eleven years the costs of government more than doubled. The growing tax burden was shifted from corporations to individuals and therefore middle-income individuals who pay a disproportionate share. Expenditures on corporate defense contracting increased, along with interest on the national debt, because the fiscal crisis was dealt with partly by new deficit spending. The other alternative, cutting state services to poor and middle-income groups, also was pursued with great force. Spending on income security (retirement, disability, unemployment insurance, etc.), education, training, employment, and social services were cut most, followed by lesser but cumulatively substantial reductions in programs for the environment, energy, transportation, community and regional development, veterans' benefits, housing, and health (including drug-rehabilitation programs).

A fair summary of the change is that lower-income people paid a greater share of state expenditures — which went disproportionately to the wealthy — and bore the brunt of reduced services. In part, this change reflects the contrasting values of the Carter and Reagan administrations. But the fiscal crisis is a deeper structural problem. It began five years before the Reagan administration (although several years elapsed before it was reflected in the budget), and similar social consequences appear at the state and local levels. Despite partisan charges, particular administrations do not control these economic cycles so much as respond to them with a limited range of stylized policies, which are more or less humane but seldom fair.

Changes in state financing reflect the circumstance of crisis rather than determine its direction. The more important qualitative shift of the 1980s is summarized in the awkward phrase *reprivatization.* That is, a reversal of expanding state responsibility is taking place through actions that return certain functions to the private economy on a profit-making basis. Medical care has been provided mainly through private insurance plans, but recent efforts are aimed at reducing the coverage of Medicare and hospital support in preference to profitable operations. Hospital ownership is rapidly turning into a concentrated industry controlled by three of four major health care–provider corporations. In 1985 there were two dozen major prisons owned and operated by private companies that contract with governments to house their criminals, and the number was expected to double in a year or two. These arrangements may indeed prove more efficient. The issue, however, is whether the individual taxpayer should have to support them and what the equitable responsibility of the state is to its citizens.

Legitimacy Crisis

This raises the other side of the crisis of the state, its declining legitimacy in the eyes of the citizenry. Opinion polls over two decades show an eroding trust in government and in public officials — a trend including but going far beyond Watergate and the Iranian arms–Contra funding exposés. A fiscal crisis is not produced by bloodless structural forces but by social unrest as evidenced in taxpayer revolts and ebbing consumer confidence. In growing unison, conservatives question the propriety of the welfare state, and liberals doubt its equity. Nobody really knows where it is going.

Social scientists do know some important things. The state and economy, with all their effects on social organization, must inevitably restructure themselves in the 1990s. Some of the implications of this fact are discernible.

In the not-distant past, agriculture employed nearly half the population of the developed societies. Today the United States can feed itself and a good many more through the efforts of 3 percent of the population. More important, where industry once grew in a process that steadily incorporated a larger labor force, it now advances with labor-saving technology. The new computer technology requires less labor for its production. The central importance of these facts is that previously the distributional mechanisms of the state and economy were connected directly to the job. That is, the mechanisms that raised money for social programs such as Social Security and unemployment insurance and distributed those benefits according to eligibility criteria were based on the condition of long-term, steady, well-paid jobs throughout the economy.

As industry and profits expanded and concentrated, an improved standard of living *did* trickle down to the general population in the developed countries. State guarantees of social welfare were linked directly to employment or indirectly to the programs maintained by working people and by employers on a per-capita worker basis. Now, however, with a sharp decline in the number of jobs that are needed in the productive segments of the economy, the linchpin of the distributional mechanism falls out. In its absence, the only way to provide for the well-being of increasing numbers of people is through direct transfer payments, or what has been called welfare — a system that was never designed for large numbers and, indeed, was to be supported by a fully employed work force.

The final and noteworthy conclusion is that the welfare state, as it has been conceived, will not work in our technological future. The methods that solved social crises in the recent past are now impractical. This assessment implies several things about the immediate future: (1) State fiscal problems will continue beyond what some saw as only a temporary recession — government faces a long-term revenue squeeze; (2) this, in turn, means a continuing reduction in the services and programs the state has supported — reductions that will continue to spread from the working classes and cost the middle classes in fields such as education and medical care; (3) together, the foregoing imply pitched political struggles over what the state will support; (4) somewhere down the line, a new mechanism for supporting the state will be developed in the political struggle, and, although we cannot envision what it will be, it is safe to say that its terms will command the energies of policymakers in the coming decades.

In the short run, this means that we can anticipate growing inequalities that affect middle-income groups as well as the poor and minorities. The system will not reorganize itself simply through a demonstration of its contradictions. The meaning and occasion of structural crisis are found in the social crisis. The last decade of the twentieth century is going to be a period marked by the same kind of social unrest and wholesale state reorganization as attended the Great Depression and New Deal crisis. When a solution is found, it will involve the state as a more intimate part of social organization.

SELECTED BIBLIOGRAPHY

BARAN, PAUL A., and PAUL M. SWEEZEY. 1966. *Monopoly Capital: An Essay on the American Economic and Social Order.* New York: Monthly Review Press.

BOWLES, SAMUEL, and HERBERT GINTIS. 1976. *Schooling in Capitalist America: Educational Reform and Contradictions of Economic Life.* New York: Basic Books.

BROOM, LEONARD, and PHILLIP SELZNICK. 1968. *Sociology,* 4th ed. New York: Harper & Row.

CALLAHAN, RAYMOND E. 1962. *Education and the Cult of Efficiency: A Study of the Social Forces That Have Shaped the Administration of the Public Schools.* Chicago: University of Chicago Press.

FRAZIER, E. FRANKLIN. 1939. *The Negro Family in the United States.* Chicago: University of Chicago Press.

KOLKO, GABRIEL. 1963. *The Triumph of Conservatism: A Reinterpretation of American History, 1900–1916.* New York: Free Press.

LASCH, CHRISTOPHER. 1977. *Haven in a Heartless World: The Family Besieged.* New York: Basic Books.

LASLETT, PETER. 1965. *The World We Have Lost.* London: Methuen.

MILIBAND, RALPH. 1969. *The State in Capitalist Society.* New York: Basic Books.

O'CONNOR, JAMES. 1973. *The Fiscal Crisis of the State.* New York: St. Martin's Press.

PIVEN, FRANCES FOX, and RICHARD A. CLOWARD. 1971. *Regulating the Poor: The Functions of Public Welfare.* New York: Vintage Books.

Priorities for Action: Carnegie Commission on Higher Education, Final Report. 1973. New York: McGraw-Hill.

RIESSMAN, FRANK. 1966. "In Defense of the Negro Family." Reprint. Pp. 474–78 in *The Moynihan Report and the Politics of Controversy,* edited by Lee Rainwater and William L. Yancey. Cambridge, MA: MIT Press, 1967.

STACEY, JUDITH. 1983. "The New Conservative Feminism." *Feminist Studies* 9:559–83.

WEBER, MAX. 1946. "Class, Status, Party." P. 82 in *From Max Weber: Essays in Sociology,* edited by Hans Gerth and C. Wright Mills. New York: Oxford University Press.

WEINSTEIN, JAMES. 1968. *The Corporate Ideal in the Liberal State, 1900–1918.* Boston: Beacon Press.

WOLFE, ALAN. 1974. "New Directions in the Marxist Theory of Politics." *Politics and Society* 4:131–59.

ZARETSKY, ELI. 1982. "The Place of the Family in the Origins of the Welfare State." Pp. 188–224 in *Rethinking the Family: Some Feminist Questions,* edited by Barrie Thorne and Marlyn Yalom. London: Longman.

7

The World System

Low wages overseas lured away U.S. steel, textiles, electronics, and other industries in the 1970s, creating a "rust bowl" in America that is still plagued with unemployment and uncertainty. (© Bob Adelman/Magnum Photos 1983)

FIVE CONNECTED LIVES

The International Telephone and Telegraph Company (ITT) is a transnational corporation, that distinctively twentieth-century organization that moves money, goods, and information across national boundaries with the impunity of a shark in territorial waters. At its peak in the 1970s, ITT was an empire unto itself, an ensemble of more than 300 companies employing 400,000 people and operating in seventy countries. Yet, unlike IBM and some transnational giants identified with a principal product, ITT is elusive, appearing at some times as a hungry conglomerate and at other times as a disjointed set of firms held together only in a computer's memory. ITT is massive yet formless. It typifies a new way of organizing the world economic system on an axis that cross-cuts national boundaries, political communities, and corporate identities.

Although ITT has suffered from its overextension in the 1970s, in 1985 it was still the twenty-first largest U.S. industrial company with annual sales of $14 billion and 252,000 employees: smaller than Exxon (1), IBM (6), and U.S. Steel (15), but larger than Dow Chemical (25), Westinghouse (30), and Coca-Cola (46). On a world scale of "economic entities" constructed in the 1970s, ITT was the forty-second largest economic entity with annual earnings that exceeded the gross national product of countries like Thailand, New Zealand, Israel, Chile, or Ireland. By any measure, it is still imposing.

ITT was not always such a formidable player in international business. The company was founded by Sosthenes Behn, a resident of the Virgin Islands when the United States bought them from Denmark in 1917. Behn was a sugar broker who acquired a small Puerto Rican telephone company as a bad debt. Quick to see its potential, he expanded the operation by gaining a national telephone and telegraph franchise in Cuba and, soon after, in Spain. Through expansion and acquisition of other companies during the 1930s, Behn came to own major telecommunications systems in other Latin American countries, England, France, Sweden, Belgium, Switzerland, and, in collaboration with the Nazi government, in Germany. The company's shadowy reputation began with the discovery by U.S. intelligence agencies that even after Pearl Harbor, ITT stayed in close touch with Germany through its subsidiaries in Spain and Argentina and actually manufactured communications equipment for the German army. In fairness, ITT made money by helping both sides during World War II — enjoying all the while the protections of a U.S. company and corporate citizen.

Harold Geneen

Despite Behn's intrepid diplomatic style and freebooting business methods, in the 1950s ITT was still principally a telecommunications company ranking just fifty-second among U.S. corporations. From 1959 to 1977, however, the company made radical changes under the leadership of Harold Sidney Geneen, rising to ninth place among U.S. corporations at the height of its expansion and becoming a special kind of transnational. Anthony Sampson describes a meeting with Geneen and the vision behind *The Sovereign State of ITT*:

> He had admired the British Empire and was sorry it had been given up so hastily. Why didn't the British government support the White Rhodesians? . . . He went on to talk about America's difficulties with the rest of the world — how her oil supplies were in danger and how eventually she might need to move into the Arab countries to protect them. As he warmed to his tirade, his whole frame came to life; he began gesturing, pointing, and laughing, his fingers darting around, touching his nose, his ear, his chin, as if weaving some private spell; his greeny-brown eyes twinkled, and he grinned and laughed like a gargoyle. He seemed no longer a dark-suited, owlish accountant, but more like an imp or a genie. I noticed that a clutch of vice presidents were standing around listening, watching him carefully: they laughed when he laughed and nodded when he nodded.
>
> Businessmen, he explained, are the only people who know how to create jobs and make work for people; he was responsible for 400,000 employees, all over the world, and it was his duty to lobby governments on their behalf as effectively as he could. What do governments know about providing jobs? Why does the American government waste time with antitrust questions when it should be supporting the big corporations that are battling with the Japanese and contributing to the balance of payments? . . . I had begun for a moment to get the feel of being inside this amazing corporation, to glimpse it through the eyes of the master and followers. From their camp they looked out on a world benighted with prejudice

> and unreason; where governments were merely obstructing the long march of production and profit; where nations were like backward native tribes, to be placated, converted, and overcome.
>
> Yet on the other side . . . I had the powerful impression that this company was accountable to no nation, anywhere, and held together and inspired by one man, against whom no one cared to argue. A man, moreover, who in spite of his famous accounting skills and discipline yet had the unmistakable style of a buccaneer. ([1973] 1980, 16–18)

The transnational empire that Geneen created retained at its core the telecommunications of a score of European and Latin American countries. But Geneen's strategy for acquiring subsidiaries was designed to simultaneously strengthen ITT's holdings in the United States and to interweave those with a variety of complementary global businesses. Ideally, ITT should be self-sufficient in natural resources, finance, and marketing and invulnerable to the uncertainties of national politics. The takeovers came in rapid succession including many household words in U.S. business: Avis Rent-a-Car, Sheraton Hotels, Continental Baking (makers of Wonder Bread and Hostess Twinkies), Apcoa (parking lots), Howard Sams (owners of the Bobbs-Merrill Publishing Company), Levitt (developers who built Levittown, the New Jersey suburban community that Herbert Gans studied), The Hartford Insurance Group, Aetna Finance, Canteen Corporation of America (a major operator of vending machines), Scott's Lawn Products, Pennsylvania Glass Sand (a manufacturer of ceramics and silicon products), Rayonier (a firm producing rayon, wood cellulose, and paper from its own million-acre timber properties in the United States and Canada), and several hundred more firms devoted to everything from dog food to secretarial schools.

Size, diversity, and geographical spread are vital but alone not what make the transnational firm distinctive as an actor in the global political economy. Rather, those features allow firms like ITT to operate relatively independently of national regulation and to outcompete purely domestic firms. The global conglomerate can buy and sell from itself, or simply pretend to do so, thereby affecting the import–export balances of independent countries through computerized intercompany transfers. The same mechanism allows transnationals to conceal the true value of financial transactions (and taxes owed in one country or another) through advantageous misstatements of the value of goods moving from one country-firm to another (transfer pricing). Moreover, a transnational like ITT can supply its Levitt housing project in Paris with lumber from its Rayonier timber lands in Canada. It can (and does) "leverage" its sales of Rayonier paper pulp to Argentina by indicating that the amount of Argentine grain purchased by Continental Baking depends on reciprocal arrangements. Naturally, wherever ITT executives travel in the empire, they stay at Sheraton hotels, rent Avis cars, and communicate over their own telephone lines.

> On the last Monday of every month, a Boeing 707 takes off from New York to Brussels, with sixty ITT executives aboard, including Geneen or one of his deputies, with a special office rigged up for him to work in. They stay in Brussels for four days, enveloped in their own company capsule, spending most of their time in one of the marathon ITT meetings.
>
> A meeting is a weird spectacle, with more than a hint of Dr. Strangelove. One hundred and twenty people are assembled in the big fourth-floor room, equipped with air-conditioning, soft lighting, and discreet microphones. The curtains are drawn against the daylight, and a big screen displays table after table of statistics. Most of the room is taken up with a horseshoe table, covered in green baize, with blue swivel armchairs and a name in front of each chair, with a bottle of mineral water and a volume of statistics. In the chairs sit the top men of ITT from all over Europe, like diplomats at a conference: in the middle are the senior vice presidents. Among them, swiveling and rocking to and fro in his armchair, surveying faces and gazing at the statistics, is an owlish figure behind a label saying Harold S. Geneen.
>
> The tension of the meetings is not simply the tension of its members as they face the cross-questioning. It is also the inevitable tension of a group of directors from four thousand miles away trying to run an empire. . . . It is in Brussels that the observer can best watch the habits, coloring and nesting habits of the curious new species, multinational man. . . . Defying geography is an essential condition, and an executive may even find himself ordered from Europe to New York just for lunch. . . . In this nomadic existence, telephones become an obsession, not only because ITT makes them but because they abolish distance and provide a reas-

> suring link with home base. The more uprooted the way of life, the more dependent the multinational managers become on their company, which forms the carapace within which they travel. . . . It is a self-contained world. . . . Inside these giant organisms the differences of nationality seem often less important than differences of company. (Sampson [1973] 1980, 95–99)

Oscar the Scavenger

Socially and geographically, it is a very long way from the horseshoe table in Brussels to the garbage dump in Cali, Colombia. Yet Harold Geneen and a Colombian scavenger named Oscar work two sides of the same transnational street. Oscar is in the recycling business. He works the municipal dump in Cali along with 400 others who search incoming loads of trash for rags, bones, glass jars and bottles, sundry reusable household items or industrial scraps, and waste paper. The paper is sold to a buyer at the site who, in turn, sells it in bulk for recycling to local industries, including two large firms that do an international business in cartons, packaging, paper products, and publishing.

> Oscar is 29 years old and came to Cali 5 years ago with his mother, sister, and nephews. Having worked in a variety of jobs in the city, in early 1975 he decided to try his luck as a garbage picker on the dump. . . . Oscar struck up a relationship with Orlando, a fairly enterprising dealer who did not buy at the dump but had a waste paper business very close to it. Orlando gave Oscar the money necessary to buy waste paper on the dump each day, and in addition fixed a price at which he himself would buy the material from Oscar. In many ways Oscar was little more than an employee of Orlando, being paid on a piece work basis but having some room to maneuver. However, after five months Orlando ran into problems owing to a temporary downturn in demand for waste paper in the city. (Birkbeck 1978, 1180–81)

Later, the waste-paper market revived, and Oscar managed to become a buyer himself, although his economic fate was uncertain when Chris Birkbeck wrote this account. Oscar and Orlando are in one sense small-business owners, buying and selling recyclable material at a small profit. Because the only market for their main product is the local paper industry, in another sense they are disguised wageworkers, but workers who lack any employee benefits or protections against risks such as unemployment and who are thrown on their own resources when the demand for paper falls. The flexibility to use or dispense with these informal workers is a great convenience to local firms, insulating them from the costs of maintaining a regular work force in slack times and thereby producing a cost saving that enables local firms to remain competitive with the transnationals.

In Cali, Colombia, local industries that purchase recyclable paper are even more directly connected to transnational corporations including ITT. The ITT subsidiary Rayonier sells pulp paper, and, depending upon its price, demand for waste paper in Cali will fluctuate. Moreover, one of Cali's most prominent firms, Carvajal, besides manufacturing paper products, also publishes children's books for companies based in the United States. Hallmark and Price/Stern/Sloan Publishers both commission Carvajal to print their pop-up book versions of *Snow White, The Night Before Christmas,* and other classics (capitalizing on the lower costs of publishing these intricately designed works abroad). The Bobbs-Merrill subsidiary of ITT is also a major producer of children's books for the U.S. market. Presumably, the success of Bobbs-Merrill in domestic competition with firms like Hallmark would adversely affect the publishing business in Cali and, accordingly, the demand for recyclable paper from the dump. Such are the intricacies of the global economy.

Wiza Leitao

Following the web of international business to another example, Brazil is an industrial power in the region and major coffee exporter, competing with Colombia in the world commodity market. In recent years, however, the Brazilian economy has gone from a self-proclaimed miracle to the world's most aggravated case of debt crisis. Brazil presently owes $110 billion to international banks. Interest on the debt alone, if it could be paid, would cost Brazil one-half the annual value of all its exports or 40 percent of its gross national product — sources of income that are vitally needed in programs for the nation's poor majority.

In part, Brazil's debt crisis resulted from costly imports including petroleum, basic parts for automobile assembly (by Volkswagen, Ford, General Motors, Fiat, and Chrysler), high technology, and luxuries. Key exports from Brazilian industry such as steel and shoes did not balance this trade deficit. As the crisis developed, government spending increased in an effort to stimulate the economy in the long run, although its short-run effect only aggravated matters by reducing state revenues that might be applied to the debt. International lenders, including many U.S. banks, imagined Brazil a good risk and were happy to fill the gap, until the possibility began to loom that the country would default on its loans. At that point, the U.S. and Brazilian governments, on the advice of the International Monetary Fund (IMF), began to consider the gloomy and unpromising solutions.

In broad outline, the recommended austerity program contained two provisions. While the debt crisis was eased with a renegotiated payment schedule, the Brazilian government would have to reduce substantially its domestic spending, and the trade balance would have to be shifted. The latter condition, of course, meant that Brazil would have to sell more than it bought from countries such as the United States. This involves many social repercussions on both sides of the equator. To the south, "the poor are still suffering, but new tax rises, designed to reduce Brazil's budget deficit, have hit middle-income taxpayers hardest. Cutbacks in cost-of-living increases have squeezed white-collar salaries."

> Wiza Leitao, a 38-year-old schoolteacher, is typical of white-collar workers here. A year ago Mrs. Leitao was struggling but was able to continue her middle-class ways. In recent months she has had to give up some of these trappings. First went language classes, swimming lessons and therapy sessions for her children. She had to sell her car to pay the rent. More recently, she pawned her jewelry and silverware and moved into a less-expensive apartment. Her weekly menu has far less variety than it used to. And she has taken out a loan. . . . [Ms. Leitao sees her plight in political terms]: "To change the economic situation in our country, you have to change the president, change the ministries and change the system. They've

made a disaster out of the economy." (*Wall Street Journal*, June 22, 1984)

Stanley Barr

But, for the moment, the bankers are breathing a bit easier as austerity measures have taken hold and Brazilian exports have closed the trade deficit. The crisis has been forestalled, however, at the expense of the real income of the Brazilian lower and middle classes and to the detriment of certain U.S. manufacturers.

> Welpro Inc. of Seabrook, N.H., a 50-year-old maker of women's dress shoes, will cease operating July 1, Stanley Barr, president and son of the company's founder, told the 250 employees at noon today.
>
> Mr. Barr said the decision was made in response to Wednesday's decision by the United States International Trade Commission not to give tariff protection to American shoe manufacturers. He said he and his brother were going out of business "before we run out of cash" to meet the payroll.
>
> [Another shoe manufacturer] complained that while imports were flooding the United States, American producers were rebuffed abroad by embargoes and duties foreign countries placed on American-made footwear. Brazil, the biggest supplier of the American market, will not allow American shoes to be imported, while Taiwan and South Korea, the second- and third-largest suppliers to the United States, levy duties of more than 50 percent on American-made shoes. (*New York Times*, June 8, 1984)

Steelworkers in Youngstown, Ohio, echo the same complaint, pointing to rising steel imports from Japan and Brazil as a major cause of their town's demise (see Chapter 3). Meanwhile, transnational companies like ITT contribute to the problem from the standpoint of domestic manufacturers and unions by going abroad with new production facilities in order to take advantage of lower labor costs.

Chen Choi Ling

Coming full circle, actions taken to protect U.S. jobs against undermining by cheap imports have feedback effects on people and jobs in the Third World. Restrictions on imported textiles, for example, would bar apparel assembled by stages in two countries. Millions of sweaters sold in the United States are affected because they are first knitted into "panels" in China and later finished in Hong Kong.

> Chen Choi Ling runs a knitting machine. She works six days a week, earns pretty good money and hopes to visit Hong Kong one day. Miss Chen never heard of protectionism or import regulations, and she does not see how she could figure in any trade dispute. . . . She lives with her parents and two sisters on a small farm, and earns enough — about $60 a month — to support them all.
>
> But Miss Chen and 500 other women at the Shenzhen [China] knitting factory here just across the border with Hong Kong stand to lose their jobs when the United States imposes new textile import restrictions. (*New York Times,* September 3, 1984)

Harold Geneen, Oscar the scavenger, Wiza Leitao, Stanley Barr, and Chen Choi Ling begin to personify the interrelated constituents of the modern world system. The ambitious concept "world system" correctly assumes that the global economy is increasingly interdependent — a system in which the political and economic interests of nations, citizens, and transnational corporations intertwine like snakes in a sack.

THE MODERN WORLD SYSTEM

A system means an ensemble of parts whose action can be explained only in their relation to other parts and a whole governed by specifiable interrelationships. The physical environment, for example, is an ecological system. The modern world system suggests and the illustrations demonstrate that the advanced and developing nations are interdependent, that old-fashioned rivalries between relatively independent nations or empires are now replaced by cross-national class conflicts (international bankers vs. schoolteachers in Brazil and steelworkers in Ohio) or contests between transnational corporations and national states. The modern world system demonstrates, moreover, that there can no longer be a national, or otherwise parochial, sociology. Classes, status groups, the state, all the fundamental social categories take on a distinctive aspect by virtue of their inclusion in a changing world system. Modern sociology is necessarily international sociology.

That point has been made most forcefully in Immanuel Wallerstein's work, *The Modern World System: Capitalist Agriculture and the Origins of the European World Economy in the Sixteenth Century* (1974). Wallerstein's fine volume, the first in a series of four, offers a model that is inclusive and intuitively appealing:

> A world system is a social system, one that has boundaries, structures, member groups, rules of legitimation, and coherence. Its life is made up of the conflicting forces which hold it together by tension, and tear it apart as each group seeks eternally to remold it to its advantage. . . . It is the peculiarity of the modern world-system that a world-economy has survived for 500 years and yet has not come to be transformed into a world-empire — a peculiarity that is the secret of its strength. . . . Capitalism as an economic mode is based on the fact that the economic factors operate within an arena larger than that which any political entity can totally control. This gives capitalists a freedom to maneuver that is structurally based. It has made possible the constant economic expansion of the world-system, albeit a very skewed distribution of its rewards. (1974, Vol. 1, 347, 348)

The world system is composed of a core of advanced and dominant states, a periphery of dependent and mainly exploited states, and a semiperiphery of buffer states that deflect the political pressures that would otherwise be directed at the core. Since the rise of the modern world system in the sixteenth century, the nations occupying these positions have shifted, but the system has persisted. Today's core includes the United States, western Europe, Japan, and other advanced countries. The periphery, of course, is composed of the poor nations of Africa, Asia, and Latin America. Semiperipheral or buffer states include industrialized countries in the developing world such as Brazil and Israel, as well as poor countries within the developed regions such as Ireland and Spain. Although these states are closely interconnected as an economic system, they are not politically integrated; no central authority governs their competition and conflict. Far from a defect, however, the absence of political integration is a necessary condition of its economic expansion.

Wallerstein's work has done much to restore the classical tradition and Karl Marx's global perspective to modern sociology. Wallerstein adopts Marx's position that "the capitalist era dates from the sixteenth century" ([1867] 1977, 876). The development of capitalism depended on many vital conditions such as the creation of wage labor, but, in its fully developed, industrial form, it was fundamentally an international system. Marx described how handicraft production in India was destroyed by British colonial policy to create a market for English textile factories. In the new system, India was assigned the role of producing cotton "just as Australia, for example, was converted into a colony for growing wool."

> A new and international division of labour springs up, one suited to the requirements of the main industrial countries, and it converts one part of the globe into a chiefly agricultural field of production for supplying the other part, which remains a pre-eminently industrial field. (Marx [1867] 1977, 579–580)

In the early years of North American sociology, with neglect of the classical tradition (see Chapter 1), the study of development disappeared. The result, as Anthony Giddens notes, "essentially severed social theory from the concerns which originally inspired all of the most prominent social thinkers — the nature of the transformation which destroyed 'traditional' society and created a new 'modern' order." After World War II, and with the emergence of new nations in Asia and Africa, sociologists returned to the study of development. At first, however, the developing nations were cast in the interpretive terms of "modernization theory" — itself largely a product of social evolutionary theories.

Modernization was ahistorical, taking for granted the idea of a nation itself, rather than understanding it as a historically created form of society — in the manner, for example, that the new nations were created out of former European colonies. Modernization theory held that nations develop according to a single model defined by such (Western) experiences as capital formation and investment, innovation, entrepreneurship, and, eventually, industrialization. In the metaphors of the time, nations takeoff at a point when those conditions are secured and then move along a Western developmental continuum — the stages of which are established by the history of the now-industrialized countries. New nations can imitate that success if only they could get themselves together: Abandon their traditional ways, acquire a dose of the achievement motivation that supposedly did so much for U.S. growth, educate themselves, have fewer children, and take on the mental outlook of modern people.

Most of these modernization formulas were nonsense. If traditionalism sometimes deterred economic development, in many cases, such as family-based saving and enterprise, it promoted it. Moreover, the theory never explained how once-prosperous societies became *underdeveloped* and how others were prevented from developing by the imperial designs of the advanced societies. In that sense, the social science of the more developed countries itself tended to be imperialistic, blaming the poor nations for their own plight. In the sharpest statement of this general idea, André Gunder Frank claimed that

> in contrast to the development of the world metropolis which is no one's satellite, the development of the national and other subordinate metropolises is limited by their satellite status. . . . The regions which are the most underdeveloped and feudal-seeming today are the ones which had the closest ties to the metropolis in the past. (1969, 9, 13)

Commercialism

Historically, the world system took shape as the dominant European states acquired the transportation technology and the political machinery necessary to sustain colonialism. With navigation skills taken over from the Portuguese, Spain became the first core power of the world system in the early 1500s. Spain colonized South America and portions of North America, the Caribbean, and the Philippines, all in the interests of its empire. Spanish conquistadors and friars took their Christian civilization around the world less for the salvation of heathens than for the commercial enhancement of the Crown. The greatest prize of the early expeditions was the gold and silver bullion taken from mines in Mexico, Peru, and Colombia or from Indian shrines. In Latin America and the Caribbean where precious metals were not found, the Spaniards took up agricultural production of sugar, indigo, cacao, beef, and sundry products, which were in demand in Europe. Typically, production depended on enslaving Indian labor, and the native populations died off at alarming rates from overwork or European diseases. The slave trade conducted by various European countries was designed to provide or replenish the labor supply in the colonies. In other places, such as the Philippines, Spanish colonialism was devoted chiefly to the transshipment of goods (spices and Chinese porcelain) from the East via the *Manila Galleon* which docked in Acapulco.

In this initial stage, little thought was given to any authentic or self-sufficient development of the colonies themselves. Their function was to supply the empire with what it could not get as cheaply elsewhere, rather than to produce what would meet local needs. Cities were located in places of strategic control over the extractive economy and were discouraged from trading with one another in preference to funnelling their surplus to the center. In these ways, the colonies were misdeveloped from the start.

Ironically, Spain did not develop either, in the sense of profiting from the empire or reinvesting in the economy at home. As a result, the Spanish Empire began to collapse in competition with the Dutch and English and with the eventual independence movement in Latin America. Although commercial power waned and Spain later lost its place in the core, the nature of the world system remained the same under the trade policies of England, Holland, and other European states. Just as in Spanish America, the colonies of these new core powers (India, Egypt, Macao, Curaçao, etc.) did not develop in any autonomous fashion.

Imperialism

The first real transformation of the world system came 300 years after its creation with the development of industrial production in the nineteenth century. Led by England, industrializing nations such as France, Germany, and the United States began looking to actual and potential colonies for raw materials and markets. What was new and critical at this stage was that the core powers needed a dependable supply of certain raw materials (cotton, minerals, timber, etc.) and foreign markets for *industrialized* goods. The colonies had to be developed now as something more than passive suppliers or commercial enclaves. They had to supply raw materials needed by industry in the core and to do so in ways that generated a sufficiently strong market in the periphery for the purchase of manufactured (consumer and, especially, capital) goods. Historian E. A. Brett, discussing British–East Africa trade, put the problem clearly:

> At the colonial end this system required that each territory find one or more primary products in demand on world markets. . . . At the British end it required export-oriented industry capable of taking advantage of the opportunities created by new colonial markets. . . . The problem was one of creating demand by expanding colonial primary production. (1973, 73–74)

The industrializing core powers acquired a new and more aggressive interest in both the size of their colonial possessions and in the development of each in the very restricted sense of a lively market for the manufactured exports of the core. This meant, first, a new rush to acquire colonies: the "scramble for Africa" among England, France, Germany, Belgium, Holland, Spain, and Portugal, as well as the U.S. military campaign to replace Spain in the Philippines and Caribbean. Second, it meant a more active involvement of core states in the domestic economies of the colonies — a colonial service to oversee local political affairs and production policies, and managed investments such as in railroads that would consume core capital goods and market local products more efficiently. Above all, it meant a growing and critical *inter*dependence, core rivalries with a potential for global conflict (as World War I is sometimes interpreted), and a more tightly integrated world system. All these features defined a second stage, after the long era of trade and commerce, in the development of the world system — a stage based on industry and empire.

Multinationalism

Since the mid-twentieth century, the world system has been undergoing another major transformation. Outwardly, the change is represented by the end of formal colonialism and the emergence of the new nations in Africa and Asia, the world dominance of the United States from 1945 to 1975, and the appearance of the transnational corporation. Beneath these appearances, however, is the more fundamental fact of a *global reorganization of production* — the shift of economic power away from nations and national capital and toward vertically integrated transnational firms whose operations and choices about production and marketing are based on a global financial strategy. It is the world of international companies such as ITT, but also a world in which U.S. firms export their production facilities to manufacture goods abroad in cheap labor markets for reimportation and sale in U.S. markets. Capital is internationally mobile, as is the labor that migrates from the Third World to the corporate farms and factories of the advanced countries. This is the era of multinationalism.

The new stage is no more favorable to the underdeveloped nations of the periphery than the last two. On the contrary, to the extent that transnational firms operate in a large number of countries, they may be less obliging in particular ones. Where England, for example, once had a heavy stake in Kenya's domestic prosperity as a supplier and market, the transnational firm manufacturing color television sets in Taiwan for export to the United States and Europe may find that labor is now cheaper in Malaysia or Ireland is offering a tax haven from which shipping costs are lower. Taiwan, which faithfully provided a stable political environment and enthusiastic workers, is dispensable when those exist elsewhere at cost savings. Taiwan supplies nothing unique, and its internal market is inconsequential from the transnational standpoint.

With the exception of politically important semiperipheral states and the petroleum exporters, the underdeveloped countries face enormous competition and protected markets in the advanced states when they endeavor to develop their own export industries. Typically, industrialization in the underdeveloped countries is limited, dependent on costly imported components (machinery or parts), and competitive neither in home markets nor as exports. Peripheral states are usually stuck with just their primary products for export (food or minerals). Faced with no other alternative, the poor countries try to produce as much of their one or two export crops as possible, meaning that domestic food production drops and starvation increases. Moreover, even the demand for these primary products tends to deteriorate as the cost of manufactured goods increases. The problem, called deteriorating terms of trade, is summarized, using hypothetical figures, with the situation in which a tractor that could be acquired in 1960 for 50 bags of coffee now requires 1000.

The world cannot be neatly divided into geographical zones like core, semiperiphery, and periphery with distinctive functions assigned to member countries. Rather, those categories highlight processes that govern the relations among states. In the concrete world system, there are countries such as China that deliberately keep their distance from international capitalism. Their socialist economies based on state ownership of production and central planning, nevertheless, make cooperating with the world capitalist system convenient on many occasions — for example, when China wants to acquire Western technology. Many states practice nonalignment and proclaim economic nationalism, but their leverage varies from the independent stance of oil exporters to the diplomatic persuasion of important yet dependent hemispheric powers such as India or Mexico. And some pure client states — from the former Soviet satellites of Eastern Europe to Taiwan and South Korea — depend on major powers such as Russia or the United States for their political stability and economic survival.

In summary, since its beginning in the sixteenth century, the world system has evolved from trade and commercialism based on state and merchant capital, to imperialism resting on national industrial capital, and, presently, to multinationalism centered on international production and finance capital. The world system of the late twentieth century is characterized by (1) concentration of capital in the giant transnational corporations, the majority of which are U.S.-based, although others are increasingly European, Japanese, and merged hybrids; (2) growing monopolies and government supervision of trade; (3) a more active role of the state in international investment; and (4) the steady development of an internationally determined class structure.

INTERDEPENDENCE OF TODAY'S WORLD SYSTEM: THE U.S. CASE

When the modern world system enters into social and economic analyses, it is usually for the purpose of explaining why the Third World remains poor and exploited at the hands of imperialism and dependency. This line of argument is useful and has explained a good deal that modernization left obscure. Yet, oddly, the same worldly analyses are seldom applied to the advanced nations. In the United States, for example, recessionary cycles, inflation, and unemployment are more often understood as the growing pains of a postindustrial society than as the fruits of a world system changing at its core.

If the essential premise of a world-system analysis is that development is a global process and if this requires, in part, a closer examination of how the practices of the advanced countries have contributed to the underdevelopment of the periphery, it also requires that equal attention be given the structural transformations of the advanced societies that stem from their participation in an interdependent world economy.

Because participation in the world system can mean a great many things, let us focus the question and introduce some evidence about the magnitude of foreign investment, using the United States as the leading example among the advanced countries. Table 7.1 shows the growth in U.S. foreign investment since 1950 by world regions. The most impressive feature of these data is the enormous increase shown across the top line. The value of U.S. corporate investment more than doubles with each decade and by 1990 had increased thirty-six times since the end of World War II. Second, and contrary to some popular notions, the great bulk of this investment is in the other advanced nations (especially Canada and western Europe) rather than the less developed regions of Africa, Asia, and Latin America. In fact, the proportion of total direct investment in the "developing" (poor) nations has been declining from about one-half of the total in 1950 to one-third in 1990. Third, the *rate* of increase has slowed (from approximately 200 percent in the 1950s to 100 percent in the 1980s).

A full appreciation of the changing character of the modern world system requires that we look behind these figures. Specif-

TABLE 7.1 U.S. Direct Investment Position Abroad at Year-End (in Millions of Dollars): 1950–1990

Country	*1950*	*1960*	*1970*	*1980*	*1990*
All countries	11,788	32,788	78,179	219,737	431,321
Developed countries	5,697	19,328	53,145	157,084	312,186
Canada	3,579	11,196	22,790	44,640	68,431
Europe	1,733	6,681	24,516	95,686	204,204
Japan	19	254	1,483	6,274	20,994
Australia, New Zealand, and South Africa	366	1,195	4,356	10,484	18,557
Developing countries	5,735	12,032	21,448	52,684	105,721
Latin America and other Western Hemisphere countries	4,576	9,271	14,760	38,275	72,467
Africa (excluding South Africa)	147	639	2,614	3,703	3,780
Middle East	692	1,139	1,617	2,281	4,755
Asia and Pacific (excluding Japan, Australia, and New Zealand)	320	983	2,457	8,397	24,719
International and unallocated	356	1,418	3,586	9,969	13,414

SOURCE: U.S. Department of Commerce. August 1962, August 1977, August 1991. *Survey of Current Business.* Washington, D.C.: Government Printing Office.

ically, three major interpretive points help explain what is going on here.

1. The particular U.S. firms that are overwhelmingly responsible for increases in foreign investment are the corporate giants rather than any cross section of domestic business. Fewer than one-tenth of 1 percent of all U.S. corporations receive one-half of all business profits earned abroad. The same tiny fraction (1/1000) claims 90 percent of all foreign tax credits provided by the U.S. government. When these facts are combined with the trend toward a growing concentration of corporate ownership within the United States (see Chapter 6), where

most of the world's largest firms are based, the obvious result is a growing international concentration of capital.

2. In the 1950–1990 period covered by Table 7.1, the composition of U.S. foreign investment has changed. In the earlier years, the investment was diversified in the sense of a balance among portfolio investments (stocks, shares in foreign enterprises, public bonds, etc.) and the direct ownership or control of firms. By the mid-1970s, nearly 80 percent of all investment was direct, that is, made up of companies that operated in other countries but were totally owned or effectively controlled by U.S. parent firms. The significance of this is that U.S. investment has expanded rapidly by buying out firms that were operating successfully under national ownership in their own countries. This means that U.S. investment abroad has not gone into new or complementary ventures requiring large capital or technological expertise. On the contrary, it has invaded branches of national economies that domestic capital has already successfully developed — facts that contradict the conventional justifications of U.S. foreign investment, for example, that it transfers technology or complements local efforts.
3. The data in Table 7.1 underestimate both the amount and the impact of foreign investment in the developing countries. On one hand, for tax purposes U.S. firms understate the value of their holdings and conceal them through intracorporate transfers. On the other, U.S. companies operating abroad often raise much of their capital from local sources, meaning that the amount of money that the firm itself has invested is far less than the value of the enterprise controlled. From the standpoint of Third World (public or private) corporations and states, the U.S.-based transnationals present an imposing presence. As we have seen, an ITT or General Motors may be richer and more powerful politically than the host nation, circumstances that severely compromise national sovereignty and the ability of developing countries to regulate the transnationals in their midsts. With their ability to shift production and capital from one country to another

or seek protection from U.S. laws as a last resort, international corporations elude host-country authority over their operations.

From the standpoint of the U.S. economy, international business for many years proved highly advantageous. There is no doubt that much domestic prosperity, though unequally shared by the corporate rich and the middle and working classes, derived from business abroad. For many years, automobiles, steel, a host of consumer goods, and, preeminently, agricultural commodities produced in the United States or in U.S. subsidiaries abroad provided through foreign sales a great deal of employment and income at home. Harry Magdoff (1966) argues that upward of 20 percent of corporate income derives from foreign sales and the figure rises to 50 percent with the addition of military purchases by the U.S. and foreign governments.

The reported earnings on foreign investment by U.S. firms has climbed from about 10 percent of the total after-tax profits of all domestic nonfinancial corporations in 1950 to 20 percent, a figure that has remained steady from the mid-1960s until the present. Reported figures, however, again understate true earnings abroad because they include only income that is sent home (or repatriated), not profits that are reinvested abroad, moved to other-country subsidiaries, or concealed. How much national income really originates abroad is therefore guesswork, but it is probably 25 percent of what U.S. citizens and corporations earn from all sources.

Pursuing the matter of interdependence, the other side of involvement in the world system is investment in the United States coming from foreign countries. Although the general public may lack the analytical understanding of how our world is changing, the world system is nevertheless a matter of daily experience. When the U.S. dollar is strong, for example, prices are cheaper if we travel abroad. But U.S. exports, including the leading category of agricultural products, are more expensive in world markets, thus reducing demand, profits, and jobs. A weak dollar makes exports more competitive abroad, but it raises the price of gasoline, coffee, and other imports. Until the mid-1980s, the United States had a very strong dollar and high interest rates, which attracted a great deal of foreign investment to the country. Table 7.2 shows the trend in this cross penetration of international capital over twenty-eight years.

Cross penetration of international capital has increased rapidly in recent years. Comparing Tables 7.1 and 7.2, in 1970 foreign investment in the United States ($13 billion) was just 17 percent of U.S. investment abroad ($78 billion). By 1990 they were roughly equal (at just over $400 billion). Moreover, as U.S. investment abroad slowed in the 1980s, foreign capital literally rushed into this country, increasing over threefold in just six years. The trend reflects foreign acquisition of U.S. land, businesses, and securities (including treasury bills that finance the deficit) and a growing number of joint ventures by the leading transnational firms based in the United States, Japan, and Europe. Toyota-General Motors and Fiat-American Motors are manufacturing automobiles for the North American market, while U.S. and Japanese high-technology firms are cooperating in some areas as they compete fiercely in others.

Behind the facades of Main Street America, foreign investors have taken over a variety of domestic firms. The Korvettes retail chain was purchased by a French group of investors, while Canadians were trying to buy Woolworth's. U.S. banks have been a favorite acquisition for European, Japanese, and Middle

TABLE 7.2 Foreign Direct Investment Position in the United States at Year-End (in Millions of Dollars): 1962–1990

Country	*1962*	*1970*	*1973*	*1980*	*1990*
All countries	7,612	13,269	20,556	65,483	403,735
Canada	2,064	3,117	4,203	9,810	27,733
Europe	5,247	9,554	13,937	43,467	256,496
United Kingdom	2,474	4,127	5,403	11,342	108,055
Europe (excluding the United Kingdom)	2,773	5,427	8,535	32,126	148,441
Japan	112	229	152	4,219	83,498
Other	189	369	2,264	7,987	36,008

SOURCE: U.S. Department of Commerce. February 1973, October 1977, August 1981, August 1991. *Survey of Current Business.* Washington, D.C.: Government Printing Office.

Eastern capital needing a base of operations. Office buildings and urban real estate are common speculative ventures for foreign investors, but their entry into farming and agricultural property has caused the greatest concern:

> Such household staples as Alka-Seltzer, Baskin-Robbins ice cream and Libby's canned food are produced by foreign-owned companies in the United States. In recent years foreign conglomerates have bought such American institutions as Saks Fifth Avenue, A & P grocery stores and Howard Johnson's restaurants and hotels.
>
> Farmers have complained that well-heeled foreigners have bid up the price of agricultural land to the point where it becomes too expensive for Americans.
>
> And even as foreign-held deposits in U.S. banks have grown, foreigners have also turned to buying the banks themselves. State [of California] Banking Department officials estimate that when Midland Bank Ltd. of London completes its takeover of Crocker National Corp., parent of Crocker National Bank, almost one-quarter of California banking assets will be under foreign ownership. (*Sacramento Bee*, December 7, 1981)

CONSEQUENCES

The United States

International capital has been lucrative for all classes within the United States until recently. Since the 1970s, however, a steady reversal of the advantages that accrue to the advanced countries from the global economy has set in. The new winners in this game are not the developing countries but the transnationals themselves, or what some observers call an international capitalist class. Barry Bluestone and Bennett Harrison describe the change as *The Deindustrialization of America* (1982) — the export of U.S. industrial production to offshore plants or runaway shops located in enclave economies of the semiperiphery or in cheap labor markets of the Third World.

The observant U.S. consumer knows that there are no radios and few if any television sets "made in America" anymore — although large numbers are sold there. Similarly, color television sets, shoes, shirts, jeans, sweaters, toys, glassware, cassettes, calculators, all manner of electronic products, and increasing numbers of automobiles, steel, and countless other items sold domestically are manufactured abroad by U.S. transnational corporations. These firms go abroad with production plants, in the first instance, to reduce their wage bills from, say, the \$25–40 per day plus employee benefits paid in the United States to the \$3–5 daily wage without benefits paid to foreign workers. Second, host countries are happy to attract these sources of employment and typically offer tax breaks to the transnational firms. Common offshore production sites are in Latin America (Mexico, Puerto Rico, Jamaica), the European semiperiphery (Spain, Ireland, Scotland, Portugal), and, especially, Asia (Hong Kong, Taiwan, India, Singapore, Malaysia, South Korea, Philippines, Thailand). As if these benefits were not enough, the U.S. government waives most of the duties on the products reimported for sale and even provides nominal insurance against loss of property, for U.S. businesses operating abroad.

As the list of reimported products suggests, it is not just any industry or subdivision of a larger industrial firm that will profit from moving abroad. First, the more easily and cheaply the product can be transported, the more likely that its production will be exported (clothing, footwear, electronic components). Bulky products, like the Ford Motor Company's Escort "world car" made in Spain, may be manufactured abroad when the intended market includes a variety of equidistant, third-country sites. Generally, industries producing for the mass-consumption market typify the runaways. Second, it is the actual production activities (fabrication, assembly, testing) that are more apt to be exported, rather than management, clerical and financial activities, or research and development.

The significance of these two points for the domestic labor force is that most of the jobs being exported are ones that traditionally define the U.S. working class — the factory jobs of blue-collar, ethnic, and minority workers. Over the last two decades, reports of a growing number of plant closings, runaway shops, and abandoned towns have been chronicled in the news media. The devastating effects of multinational production on U.S. domestic economy are not confined to outmoded industry or even to the rust bowl.

Sioux City Still Suffers After Its Top Employer Moves Business Abroad

Ever since the Zenith Radio Corp. pulled out of here, the local economic picture has been dark. The unemployment rate has soared to the highest level in Iowa. The city's biggest auto dealer says repossessions are up sharply. Home building has come to a halt. "I've shut down and laid everybody off," says Jon Winkel, a local builder. "Everybody I know is in the same situation."

Zenith was the biggest employer in this manufacturing and meatpacking town of 80,000 population. When business was going strong, nearly 2,000 workers made parts for stereo and television sets in a huge plant on the edge of town. But Zenith closed the plant a year ago. It moved its business to Mexico and Taiwan to compete better with Japanese and other imports to the U.S., and the 1,400 remaining workers were out on the street.

"I think it could have been avoided," complains Frances Sorensen, who worked at Zenith for 22 years. "I have no ill feeling toward Zenith. It's the federal government's fault. They aren't putting high enough duties on imports."

Zenith and Sioux City aren't unique in having problems with import competition. In industries as diverse as steel, shoes, textiles and electronics, the same competition is taking place, and communities across the U.S. are losing jobs. Television makers alone say import competition has claimed 60,000 jobs in more than a dozen cities over the past few years. During the past 18 months, Zenith has cut back operations in Glenview and Paris, Ill., Springfield, Mo., and Watsontown, Pa., in addition to closing the Sioux City Plant.

Besides affecting the laid-off workers, the closing dealt a blow to small-business men who supplied the plant with parts. "It really hurt me," says Harry Sachau, whose storefront company in South Sioux, Neb., makes antennas. "Zenith was two-thirds of my sales," he says. "I ran about $250,000 of business with them a year." When the plant closed, he adds, "I went from 32 employees to eight. . . ." Some prominent people have been caught in the rubble. Three of the biggest home builders in town have gone out of business. Longstanding real-estate firms have been financially shaken, too. (*Wall Street Journal*, April 5, 1979)

Warner Communications, Inc., is a transnational firm that grew from the Warner Brothers movie company to a conglomerate with publishing, home video, and a computer branch, which included Atari — the once profitable creator of Pac-Man.

Although Warner eventually sold Atari, in early 1983 it was attempting to revive the ailing firm by moving abroad.

> Atari to Idle 1,700 at California Site, Move Jobs to Asia
>
> Atari, Inc. said it is permanently eliminating 1,700 of its 7,000 jobs in Santa Clara, Calif., so it can move manufacturing plants to Hong Kong and Taiwan, where it already has facilities.
>
> The employees are nonunion workers involved in making home computers and video games. Production of essentially all of Atari's home-computer products and home video games will be moved.
>
> "This will enable us to keep prices at attractive levels." . . . While declining to compare specifically costs in the Far East with those in Santa Clara, the spokesman said costs in Asia will be "substantially less. Wages are an important factor in the equation, and so are taxes, real estate and other costs." (*Wall Street Journal,* February 23, 1983)

The stunned reaction to Atari's announcement arose from the belief that computers and California's Silicon Valley together symbolized the wave of future employment, the place where outmoded blue collar workers would be absorbed.

> Politically, Atari symbolized the notion that high-technology employment could replace jobs lost in U.S. basic industries — that gleaming factories in the Silicon Valley of California could absorb workers from the devastated industrial "rust bowl." . . . Firing a quarter of its U.S. workforce and shifting production overseas seemed an ironic blow to the notion that high-tech jobs might be waiting for laid-off auto and steel workers. (*Sacramento Bee,* March 7, 1983)

The trend continues in the 1990s. Invented in the United States, the typewriter is no longer manufactured domestically. Smith Corona, the last U.S. typewriter maker announced in July 1992 that it would move its Cortland, New York, factory which employed 800 workers, to Mexico. "Smith Corona has already set up factories in Malaysia and Singapore, so that its employment in Cortland has slumped from a high of 4,200 in the late 1970s to only 1,245 now" (*New York Times,* July 22, 1992). Smith Corona was reluctant to move the last of its manufacturing capability offshore and blamed the federal government for failing to protect U.S. firms against unfair trade practices. It accused Japan of "dumping" or selling typewriters in the United States below the cost of production, in order to ruin domestic firms and gain control of the market.

As these illustrations show, there are at least two sides to the problem of exported jobs. Organized labor argues that profit-hungry transnationals have unpatriotically abandoned U.S. workers for Asian sweatshops, although they still want privileged access to the home market. The federal government, it is charged, shows its probusiness colors by enabling the runaway shop with special provisions in the tariff laws. The result is growing unemployment and degradation of the labor force, while transnational profits continue to mount.

International firms, on the other hand, argue that global competition forces these moves; without access to cheap labor "offshore," Japanese imports into the United States and other third-country sites would undersell the U.S. transnationals, damaging the entire domestic economy in the long run. Moreover, they allege, business abroad actually creates new jobs at home in the areas of management and sales that counterbalance losses in production. In many senses, both sides are right because they speak from different positions in what is ultimately a contradictory global political economy, contradictory in that the aims of one group can be attained only at the expense of the other.

This is not to say that every claim of the protagonists is correct. On several key issues, drawing on available research, it is possible to evaluate competing claims. A central question is the matter of job loss. Does exported production produce a *net* loss of employment in the United States or do new headquarter activities of the transnationals generate an equal or greater number of compensatory jobs?

The initial studies of this question using data from the 1960s were crude in methods and preconceived about conclusions. One conducted under AFL-CIO auspices found that about 500,000 jobs were lost to the domestic economy from 1966 to 1969, based on what *would have been* had U.S. firms produced at home what they reimported from their offshore plants. Conversely, a group from the Harvard Business School polled firms about their operations at home and abroad, reporting that

600,000 new domestic jobs were created in connection with improved export business. These studies talked past one another by posing hypothetical conditions that ignored the question of net gains or losses. The U.S. Senate Finance Committee ordered a more rigorous analysis by Robert Frank and Richard Freeman who concluded that "the net impact of foreign investment by U.S. multinationals is a substantial domestic employment demand reduction" (1978, 156). Whatever compensating trends exist (and they are real), the net loss has accelerated in the late 1970s and 1980s due to the global recession. Bluestone and Harrison (1982) estimate the net loss at 2.5 million jobs per year.

Given the controversy between researchers in sympathy with business or labor over whether there was really any net job loss, it is significant that no one argues the substantial *displacement* effect on the labor force due to exported production. That is, everyone agrees that the jobs that are lost come from the categories of the skilled blue-collar trades and semiskilled factory work. Exported jobs tend to be in the more labor-intensive, traditional manufacturing fields that formerly absorbed the labor of less affluent workers and ethnic minorities. Although some new clerical and managerial positions replace these losses, it is doubtful on an individual basis that displaced workers can be retrained to qualify for those jobs involving completely different skills. Recall the unemployed Youngstown steelworker who lamented being unable to "play Atari" (Chapter 3). To the extent that the displaced blue-collar industrial workers are reabsorbed at all, it appears to be in low-paying and nonunion jobs in the service economy — in fast-food chains, as clerks in liquor stores, and the like.

This massive shift in the work force implies that the share of national income going to labor declines as the share going to capital increases; and a study for the U.S. Senate by Peggy Musgrave (1975) confirms that reasoning. This shift toward greater inequality is a fairly straightforward result of there being fewer well-paid, unionized industrial workers and thus a downward movement in average wages, yet no long-term slackening of corporate profits — particularly for transnational firms.

To summarize these specific developments in broader sociological terms, the internationalization of capital has transformed the class structure of the United States and other advanced countries. The evidence shows how corporate power and concentration are strengthened while organized labor is penalized by the overseas flight of production. On one hand, the domestic labor force is professionalized with the creation of a few new positions in the clerical, sales, management, and engineering fields. On the other hand, it is degraded with a great deal of job loss and displacement to low-paying service-sector positions. The overall result is a growing polarization of the class structure in advanced societies.

The Third World

It is impossible in a short space to deal with all of the ways in which Third World development is affected by incorporation into the world system. Based on material discussed in previous chapters, however, it is possible to take one approach to the problem that most effectively reveals a general pattern. I argue that Third World urbanization provides a wide window showing connections between developed and underdeveloped countries, as well as a case study of how the expanding world system can retard the progress of poor nations.

Recall that the world city (Chapter 3) includes as a hinterland and economic extension parts of the developing world in Africa, Asia, and Latin America. This was true of previous city forms, but today's links are more direct and essential. If the pessimistic Youngstown steelworker actually looked for a job with the Atari Company, he would find it had moved. As previously noted, the company shifted its production to Taiwan and Hong Kong at the expense of 1700 jobs in Santa Clara, California — which retains, nevertheless, the corporate headquarters. This contemporary example continues a long record of foreign participation in the economies and urban development of Third World cities.

We have seen that urbanization in the developing areas, particularly Latin America, has increased in the last sixty years at a rate exceeding that of the more developed regions. A larger proportion of the world's total urban population now lives in the less developed countries. The great world cities of New York, Moscow, London, and Tokyo are all smaller than Shanghai, China; Calcutta, India; São Paulo, Brazil; and Seoul, South Korea. Among those likely contenders, the world's largest urban area soon may be Mexico City.

Yet the appearance of cities in the Third World is unprepossessing at best. At worst, it is one of seldom-relieved poverty, makeshift shacks, open sewers, bleeding sores, and idle humanity. If we look beneath that appearance, a more livable, if still poverty-stricken, picture emerges. In broad terms, however, cities of the Third World have not developed along a trajectory established in Europe and North America. In a sociological sense, there is no universal rural–urban continuum or generally applicable Western model of urbanization. Specifically, Third World cities have not passed from a commercial to industrial

TABLE 7.3 Labor-Force Distribution in Selected Countries

		Primary: Agriculture, Fishing, Forestry (%)	*Secondary: Industry, Construction, Mining (%)*	*Tertiary: Services, Commerce, Administration, Transport (%)*
United States	1860	59	20	21
	1900	38	30	32
	1950	12	33	55
	1980	3	30	67
Argentina	1925	32	20	48
	1960	22	21	57
Brazil	1925	68	12	20
	1960	52	13	35
Chile	1925	37	21	42
	1960	25	17	58
Colombia	1925	65	17	18
	1960	49	15	36
Mexico	1925	70	11	19
	1960	53	17	30
Peru	1925	61	18	21
	1960	54	15	31
Venezuela	1925	63	10	27
	1960	32	12	56

SOURCES: U.S. Department of Commerce, Bureau of the Census. 1982. *Statistical Abstract of the United States, 1980.* Washington, D.C.: Government Printing Office. Fernando H. Cardoso and Jose Luis Reyna. 1968. "Industrialization, Occupational Structure and Social Stratification in Latin America." In *Constructive Change in Latin America.* edited by Cole Blasier. Pittsburgh: University of Pittsburgh Press.

stage and beyond. Their development has followed a very different path — one that needs to be described and explained.

Table 7.3 contrasts the labor-force distribution in the United States and major Latin American countries in various years that reflect early and, by national standards, mature industrialization. This comparison shows dramatically a pattern of difference between the advanced industrial countries and the developing world in general. The pattern is this: In the United States, the industrial era beginning around 1860 leads to a steady decline in the (primary) agricultural labor force with a parallel rise in the (secondary) industrial and the (tertiary) commercial–services sector. In the midperiod of industrial maturation (say, around 1900), employment across the three sectors is relatively even, although by 1950 the industrial labor force stops growing and services grow alone.

In the developing countries of Latin America, which raised their industrial production significantly from 1925 to 1960, however, industrial employment scarcely grows at all. Reductions in agricultural employment owing to mechanization and commercialization are taken up almost exclusively by rapid expansion of the service sector (and by unemployment). Put simply, the developing countries do not experience a three-stage process of modernization, from agrarian to industrial to services, in employment terms. Instead, their labor forces shift from mainly agricultural to mainly commercial and service-based. Combining this pattern with simultaneous rapid urbanization, the result is that Third World cities are dominated by a bloated or overcrowded service sector and its twin, high rates of underemployment.

Table 7.3 portrays what observers see in the streets of Third World cities — a great many seemingly idle people, a profusion of petty traders (lottery-ticket sellers, shoeshine boys, fruit sellers, vendors of clothing and toys), people with only their bodies to sell, and others up to no good. The question then is, Where do these urban masses come from, or, more precisely, why did the industrial sector fail to grow as it did in the more advanced countries?

The answer is the subject of many volumes devoted to development economics and the sociology of underdevelopment. It is summarized briefly in the notion of Third World dependency in the world system. The basic fact about the Third World is that it was incorporated into the world economy by imperialism. The Spaniards in Latin America; the French, Belgians, Germans, Portuguese, and English in Africa; the English and French in Asia; the North Americans in the Philippines, Caribbean, and Central America — all went to those societies in search of plunder, raw materials, captive markets, and client states. They did not go to develop local resources or industry for the purposes of internal development. On the contrary, their intervention typically destroyed local artisan industry where it existed and substituted, through coercive colonial policies, local consumption of European and North American manufactured goods in exchange for exported primary products from the mines and fields. Third World countries developed as appendages or enclaves of the core nations of the world system, and in that sense they were typically misdeveloped from the standpoint of a healthy or balanced local economy.

The pattern of dependent urbanization continued after the formal end of colonialism. Industries were created in the developing countries, but often under the ownership or control of foreign interests seeking now to dominate world markets without the costly political apparatus of colonialism. The methods by which this ambition was pursued varied greatly around the world. Some developing countries succeeded in avoiding its worst effects and even in gaining limited autonomy. The dominant pattern, however, was market control by the advanced nations through exportation of their own industrial goods or through offshore plants that manufactured products for local consumption in the Third World.

Foreign-controlled industry did three things to misdevelop Third World economies: (1) It used advanced industrial production methods, capital-intensive or inappropriate technologies that employed very few people; (2) it concentrated on high-profit and luxury markets, failing to industrialize mass-consumption goods; and (3) it competed with potential national industrialists, discouraging local investment while taking its own profit out of the countries. A national industrial class did emerge in many countries, often in joint ventures with foreign firms. But that class was generally subservient to the wishes of foreign capital, forced to adopt the same, capital-intensive methods. This prevented the developing countries from generating industrial employment as a source of national development, which is reflected in Table 7.3.

Foreign capital also went heavily into rural areas where subsistence farming by peasant families was converted to commercial agriculture for the urban and export market. Typically, this meant that small parcels of land were centralized for large-scale operations, the peasants were bought off or kicked off the land, production was mechanized, and, generally, the rural areas no longer provided a livelihood for poor people. Indeed, their massive migration during the mid-twentieth century explains the rapid urbanization of the Third World.

The so-called overurbanization of Third World cities now makes sense. It involved, on one hand, too many people forced off the land by commercially mechanized agriculture and flooding into the urban economy and, on the other hand, too many people for the dependent city to productively employ.

The physical results of this process are the extensive slums and squatter settlements observed from Lagos to Lima. Unlike U.S. cities, where poverty and unemployment are concentrated in the decaying inner cities, urban areas in the Third World display vast "peripheral" belts of unserviced, self-constructed housing. Worldwide patterns vary, but, in the more urbanized Third World such as Latin America, central cities have been maintained as the locus of government, business, and elite residence. With overcrowding, suburbs have also grown for the rich and poor. In major metropolitan areas such as Mexico City, however, the downtown core has been renovated and poor neighborhoods bulldozed to make room for high-rise apartments and offices. Central plazas built in the 1530s are still administratively and commercially central places in most Latin American cities, despite the growth of suburbs and even shopping centers. The city has not been abandoned, and the old charm endures alongside the plaintive poverty.

When we trace back the characteristics of urban life in Third World cities to the economic and political root, their causes are found in a distinctive pattern of development within an interdependent world system — specifically in the historical development of core and periphery countries. The urban poor live the way they do because rural society provides few opportunities for economic survival and the cities present few industrial or solid working-class jobs. Instead, the cities offer many small niches for enterprising souls in the overcrowded economy of commerce and services.

In summary, the decisive forces shaping social classes in the United States and in the Third World stem from the world system. In developed countries, their present effects include the elimination of industrial jobs and polarization of classes. In the Third World, cities grow to historically unprecedented sizes while their dependent economies fail to employ, feed, or house their masses. The global economy is becoming less a series of national economies than a joint venture dominated by the richest nations and the largest transnational corporations, themselves increasingly independent of nations.

THE STATE AND THE WORLD SYSTEM

Yet, the modern world system is based on contradiction between national and transnational interests. If offshore production benefits the U.S. consumer, foreign worker, and transnational firm, it penalizes the domestic manufacturer, U.S. worker, and local community. If U.S. and Japanese firms cooperate to produce for home markets with domestic labor, then French and German manufacturers suffer. At bottom, there is no roundly beneficial solution to these contradictions, short, perhaps, of a global political authority that might regulate conflicts in the best possible compromise. But it is precisely with this dilemma that modern states struggle — protecting home markets (which Japan does extensively) or enforcing international agreements that benefit the advanced countries (as the United States does through its influence over international agencies).

The most far-reaching current example of state efforts to develop a strategy for competition in the world system is the regionally based alliance for a North American Free Trade Association (NAFTA). The agreement was negotiated in response to the establishment of a European Trading Community in 1992 and proposes to create a single, duty-free market composed of 370 million consumers in the United States, Canada, and Mexico. The respective governments hope to benefit from one another's markets and open investment, limiting at the same time Japanese and European Community penetration of the regional bloc. What used to be done by states is now being done by cross-national alliances.

Although the free-trade agreement promises benefits to each of its members, there are potential costs for certain groups and interests within each country. Under the free-trade agreement, goods would pass between countries without tariffs (after a fifteen-year period of phased reduction), much as they do between U.S. states. Similarly, firms and jobs could move freely to more advantageous sites, and some employees of these firms could also move without the usual immigration procedures. The perceived advantages for the United States and Canada include unlimited access to Mexico's consumer market, banking and insurance sectors, and cheap labor pool. The United States expects a major boost in exports with its two largest trading partners. Mexico, in turn, expects that more U.S. industries like Smith Corona will move south, creating jobs; the U.S. and Canadian markets for agricultural produce will expand; and Mexicans will enjoy freer movement across the border.

All of this sounds persuasive, and there is no doubt that some interests will benefit. Nevertheless, there are certain zero-sum conflicts. If Mexican agricultural exports to the United States increase, it is either because Americans suddenly start eating more or because Mexican goods undersell and so displace U.S. farm products. When U.S. firms go to Mexico to produce, say, typewriters for the international market, U.S. jobs are lost; when U.S. firms go to Mexico to produce, say, construction materials, Mexican firms and jobs are lost (in part because the more efficient or "capital-intensive" U.S. firm , even though it employs Mexican workers, needs fewer of them). Free trade is likely to magnify the problems that already exist with job flight from the United States, the *net loss* of traditional blue-collar jobs, and the reduction of U.S. wages, benefits, and safety standards as workers are forced into competition with their Mexican counterparts. Defenders of the agreement say that that will not happen because U.S. workers are more productive and thus preferred by many firms that will stay at home. The argument is not true, however. Given comparable technology, workers in the Third World rapidly attain levels of productivity comparable to the same sectors of the industrialized world.

Critics of the free-trade agreement worry that it will drive down standards of environmental protection in a manner parallel to labor standards. If it is difficult to stop pollution and enforce environmental protection laws within the United States, it would be impossible to do so in Mexico where the country is desperate for jobs and law enforcement is corrupted by commonplace bribery. Even a casual examination of growing pollution throughout Mexico (in unlivable Mexico City, of course, but also in other cities, on the beaches, in the rivers and lakes) makes the idea of environmental protection there implausible, despite the assumptions of the agreement. The danger, accordingly, is that U.S. and Canadian firms burdened by the costs of environmental protection will either go to Mexico to pollute the planet or win concessions on the measures they are required to take at home.

The winners in North American free trade, of course, are the multinational firms capable of operating across borders and mounting investments that drive out competing national capital. Banks and insurance companies in the United States hope to gain control of those sectors that were formerly protected from foreign investment in Mexico. Although wealth is very unevenly distributed in Mexico, it is a wealthy country in many ways. The consumer market of 80 million people, already dominated by U.S. products, is expanding. There is probably a higher percentage of American cars sold in Mexico than in the United States. But the new opportunities that attract investors are in commerce and services: banking, insurance, advertising, management and computer consulting, investment services, and so on. And a few large Mexican firms will also succeed in the multinational competition, sometimes in joint ventures with U.S. companies.

> The 660 workers of the automated Corning Inc. factory [in Martinsburg, West Virginia] can expect greater job security than they have now, and new jobs may be added. Their employer, like scores of other big American companies that have become more international in the last decade, has formed an alliance with an overseas competitor — in this case Vitro S.A., the giant glass company in Mexico. Corning is expected to gain substantially from rising sales as Mexico cuts its tariffs. But, about 2,000 workers in small handmade glass factories 160 miles to the west, on the other side of the Appalachians [in Ellenboro, West Virginia] face possible financial ruin. (*New York Times*, July 21, 1992)

The small factories in Ellenboro and Doddridge County — the seventh poorest in West Virginia, with an unemployment rate of 11.5 percent — would lose its market in high-quality flower vases and drinking glasses to the Mexican giant.

The North American Free Trade Agreement is a major intercountry initiative to realign the modern world system and the position of the United States, Canada, and Mexico as a regional bloc with respect to other regions, especially Europe — that is unifying and expanding into the former Soviet satellite countries. The world system of the twenty-first century is taking a new shape and a structure of privilege determined in important part by state management of the change.

Political Management of Conflict

Until recently, the U.S. Congress regularly approved an annual foreign-aid bill after much conservative handwringing and liberal self-congratulation. Conventional wisdom supposes that foreign aid is a gift from the taxpayers to the less fortunate people of the world. The first part is right — but the gift, or about 90 percent of it as congressional studies show, is aid for the purchase of U.S. products by foreign countries. In the language of foreign affairs, it is "tied aid" — funds (including loans) that recipient countries are required to spend on such developmental needs as can be met with U.S. goods and services. Foreign aid is provided to developing countries to build dams with U.S. construction equipment and engineering firms, to start factories with U.S. machinery, and to feed the hungry with U.S. grain. Some of these projects may have constructive effects. Many do not since they preempt local production or make the countries dependent on imports. The point here is that the projects also provide a captive market and indirect subsidy for U.S. business.

A number of public programs complement the subsidies based on foreign aid. Under provisions of the Tariff Schedules of the United States, commodities manufactured in offshore plants and reimported for sale in the country have all tariffs waived, and duties are paid only on the "value added" in foreign assembly. The transnationals are awarded a decided edge over domestic or foreign firms that produce for the U.S. market.

Moreover, the federal government insures U.S. firms operating abroad against loss through unprovoked nationalization by foreign governments. The federal government supports a Border Industrialization Program with Mexico that encourages offshore production conditions contiguous to national territory. Free Trade Zones are supported in friendly countries and within U.S. borders. In short, public programs for assisting international capital are abundant.

A program of Worker Adjustment Assistance extends the benefits of government protection to people who lose their jobs as a direct result of foreign competition. Workers who qualify receive extended unemployment benefits and opportunities for job retraining. This assistance, however, is ineffective, short-term, difficult to qualify for, and funded at a small fraction of the costs of subsidies to capital. In a related vein, organized labor has pressed for more protection of jobs, but favorable action in selected areas seems to come only when the forces of business and political circumstance combine with labor. For example, in mid-1984 the shoe and steel industries both applied to the government's International Trade Commission (ITC) for import curbs on their foreign competitors. The shoemakers' request was rejected, while the jointly submitted petition of the United Steelworkers of America and Bethlehem Steel Corporation resulted in a favorable recommendation to the president and a strongly worded suggestion that foreign steel producers voluntarily reduce imports into the United States.

State influence spreads far beyond domestic issues. A number of nominally international agencies set policies, which serve the special interests of the advanced countries, for the global economy. One of the most important of these is the International Monetary Fund (IMF), created in 1944 to foster economic exchange and stability among the North Atlantic member nations. The organization, which began as a source of technical advice and "drawing rights" to meet balance-of-payment deficits, was soon converted into an arbitrator of domestic policies in the Third World. The transformation was accomplished by the IMF assuming a greater role in evaluating the credit worthiness of countries applying for loans to its sister institution, the World Bank. During its first thirty years, the IMF fostered policies for developing nations that stressed economic (and political) stability and the consumption of exports from the advanced countries. IMF ideology suggested that free trade was the answer to underdevelopment, even when that trade was coerced by discretionary World Bank loans and even when it stifled autonomous development.

Beginning in the early 1970s, the IMF was forced to alter these policies in the face of a mounting Third World debt crisis. Worldwide recession and substantial petroleum price increases forced the developing countries to borrow heavily from the World Bank and from international private banks. As the debt and interest rates rose, it became clear that the banks had a crisis on their hands. The developing countries could not borrow fast enough to pay just the debt service on past loans. The danger in this situation was that defaults or long-delayed interest payments on the loans threatened the stability of the world financial system and the prosperity of many of the largest U.S. banks.

The IMF swung into action with a series of renegotiated debt-payment plans and new short-term loans. More important, it instituted austerity programs for the debtor countries, intended as long-term solutions. Governments were advised to reduce expensive imports and domestic spending and to apply their cash savings to debt payments. Hardships were imposed on the poor and middle income as Third World governments were obliged to cut food and petroleum subsidies to consumers, reduce wages, and eliminate public projects that provided employment. But the amount that could be generated from those savings alone would not significantly reduce debts running to $50 (Argentina) or $120 (Brazil) billion. The IMF answer was that Third World countries had to expand their exports.

As sound economic advice, the IMF prescription had only one flaw. More than forty developing countries were on the IMF's critical list. While Third World exports were being urged as a practical solution, the advanced countries were suffering recessions and protecting home markets from cheap imports or trying to expand their own exports. The contradictions loom once more. The IMF and U.S. banks want Third World countries to export so they can pay their debts, but that means countries such as the United States have to import goods that compete with domestic manufacturers and jobs. In effect, the IMF recommends that everyone sell more to everyone else.

Obviously, it is not sound economic advice that is being offered but a political response to the contradiction that the IMF has not been able to conceal — that prosperity for some in the capitalist world economy comes at the expense of others. The structure of the world system is revealed in such economic policies and state actions.

A choice example illustrates the political complications that surround seemingly efficient solutions. Brazil is the world's leading debtor country and a producer of high-grade steel but no oil. The United States has been importing some Brazilian steel to the disadvantage of its domestic industry. In the summer of 1984, the Reagan administration was looking for needed electoral support from labor. A solution of sorts appeared in the Middle East:

> The United States government guaranteed $425 million of loans to help Iraq build an oil pipeline through Jordan only after Iraq revised the proposal to include large quantities of American steel pipe. . . .
>
> The Export-Import bank (which provides direct credits and credit guarantees for overseas purchasers of American products) was initially opposed to giving financial guarantees for the pipeline because it believed that the Iran-Iraq war posed excessive financial risks. . . . Those reservations were overcome when Iraq agreed to order about $100 million worth of American steel pipe. . . .
>
> The episode shows the role that steel politics is playing this election year. "You can bet this order is going to be played hard up and down the steel country. . . . That's a very substantial piece of business." It represents about 10 percent of last year's total steel exports.
>
> Iraq's willingness to buy American steel rather than cheaper steel from West Germany, Japan, or third world exporters was taken as an indication of its desire for a stronger American participation in the project.
>
> Iraq had approached the London subsidiary of Bechtel Group, Inc. last spring. The San Francisco-based engineering and consulting concern, which has strong links with the Reagan administration, is the prime contractor in the $1 billion pipeline project.
>
> Bechtel had been headed by Secretary of State George P. Schultz before he came to Washington. Defense Secretary Caspar W. Weinberger was vice president and general counsel for Bechtel before coming to the Pentagon. . . . The Iraqis wanted a substantial American participation in the pipeline principally as an "insurance

> policy" against any future action by Israel against the pipeline. In 1981, Israel bombed an Iraqi nuclear reactor. (*New York Times*, July 16, 1984)

In short, a combination of domestic political considerations (needed labor support and corporate influence) dovetailed with foreign policy considerations (Middle East tensions) to override the possibility of debt relief from Third World steel exports. Ironically, this case also shows that the United States was happy to do business with Iraq's Saddam Hussein during his war with Iran. Within a few years, the United States was at war with Iraq, which had built up its military strength using U.S. agricultural credits for weapons, and ready to explore better relations with Iran.

Several conclusions about the politics of multinationalism and the changing state may be drawn from these examples. The domestic policy of states is increasingly affected by the global economy. The state in developing countries is forced to compromise the interests of its own working classes and national manufacturers in policies that benefit transnational firms and international agencies and banks. In concrete instances, the state in advanced countries must take sides over the conflicting interests of transnational capital and domestic capital and labor. In the Iraq deal where these seem to coincide, U.S. banks with outstanding loans to Brazil would suffer. A parallel contradiction prevails internationally where advanced countries want to revitalize debt-ridden economies and to expand their own markets at the same time. An arrangement in which Brazil swapped steel pipe for Iraqi oil would have helped the debt crisis but not the U.S. administration's political problems. There are no ready solutions to such dilemmas short of an effective world political authority. Yet the privileged nations resist that alternative with grudging support for the United Nations and self-interested insistence on the IMF's short-term crisis management. Meanwhile, the world system drifts toward deeper crises.

THE GLOBAL FUTURE

The epic change in the world system witnessed in the 1980s and 1990s has been the collapse of Soviet socialism and a realignment of Central and Eastern Europe with the West. We will discuss the causes of this revolution in Chapter 9, but some of the obvious consequences deserve attention here.

As the former socialist states are transformed to electoral democracies, state-managed economies are converting to free-market principles. Major enterprises once owned by governments are being privatized in sales of firms to national and Western investors — and sometimes to their own employees. Western firms have acquired some successful enterprises, such as the manufacturers of light bulbs and a recording company in Hungary, and consumer goods are beginning to replace rationing under controlled prices. At the same time, however, the introduction of capitalist methods is generating new inequalities, which may or may not be addressed with economic revitalization. Unemployment is a new and extensive problem. Wages are too low for most people to participate in the new consumer economy. And the expected rush of foreign investment has not arrived, partly because of a recession in the West and partly because the market simply is not very attractive. The prosperity that many East Europeans expected to appear suddenly with the removal of communism is nowhere in sight. Indeed, the more likely prospect is that new class divisions will grow within the formerly socialist countries and that the countries themselves will either fragment (as in the former states of Yugoslavia, Czechoslovakia, USSR) or realign along a new North–South axis of rich and poor countries that supersedes the East–West axis of the Cold War period.

To grapple with these questions, the conceptual framework based on a world system has several distinct advantages over conventional theories of development and modernization. Two such advantages have already been elaborated. First, the world system provides a rich and holistic scheme for describing the modern world, a framework that is capable of incorporating and connecting the economic circumstances of people, firms, and countries around the world. Second, it explains the origins of unequally distributed prosperity and poverty worldwide and explains those through the workings of global capitalism as a historically evolved system. A third advantage is that the framework provides concrete solutions for an equitable form of international development. The gap between rich and poor nations widens with time as an economic and political consequence of the present structure of the world system. That trend, however, is not inevitable. Practical proposals exist that would change matters.

In 1980 a prestigious Independent Commission on International Development headed by former West German Chancellor Willy Brandt issued a report entitled *North-South: A Program for Survival.* The report's key premise was that world peace and survival in the face of ecological and nuclear threats depended on a new pattern of international cooperation. The report addresses the "North-South divide as broadly synonymous with 'rich' and 'poor,' 'developed' and 'developing' " (p.31) — after some arbitrary rearranging that puts China with the South and Australia with the North. The North–South language is intentionally designed to supersede presumably outmoded notions of East–West conflict and stress that the socialist countries must be a part of any new world order:

> The crisis through which international relations and the world economy are now passing presents great dangers, and they appear to be growing more serious. We believe that the gap which separates rich and poor countries — a gap so wide that at the extremes people seem to live in different worlds — has not been sufficiently recognized as a major factor in this crisis. It is a great contradiction of our age that these disparities exist — and are in some respects widening — just when human society is beginning to have a clearer perception of how it is interrelated and of how North and South depend on each other in a single world economy. Yet all the efforts of international organizations and meetings of the major powers have not been able to give hope to developing countries of escaping from poverty, or to reshape and revive the international economy to make it more responsive to the needs of both developing and industrialized countries.
>
> The nations of the South see themselves as sharing a common predicament. Their solidarity in global negotiations stems from the awareness of being dependent on the North, and unequal with it; and a great many of them are bound together by their colonial experience. The North including Eastern Europe has a quarter of the world's population and four-fifths of its income; the South including China has three billion people — three-quarters of the world's population but living on one-fifth of the world's income. In the North, the average person can expect to live for more than seventy years; he or she will rarely be hungry, and will be educated at least up to secondary level. In countries of the South the great

> majority of people have a life expectancy of closer to fifty years; in the poorest countries one out of every four children dies before the age of five; one-fifth or more of all the people in the South suffer from hunger and malnutrition; fifty percent have no chance to become literate.
>
> Behind these differences lies the fundamental inequality of economic strength. It is not just that the North is so much richer than the South. Over 90 percent of the world's manufacturing industry is in the North. Most patents and new technology are the property of multinational corporations of the North, which conduct a large share of world investment and world trade in raw materials and manufacturers. Because of this economic power northern countries dominate the international economic system — its rules and regulations, and its institutions of trade, money and finance.
>
> The South has called for a new regime to protect the commodities which they export against price-falls and fluctuations. The North has only slowly moved towards this. . . . Exports of manufacturers are important for developing countries' industrialization, but the North is raising obstacles against these too. . . . Current trends point to a somber future for the world economy and international relations . . . poverty and hunger . . . mounting debts and deficits . . . major tensions between countries competing for energy, food and raw materials. . . . And overshadowing everything the menacing arms race. (Independent Commission on International Development 1980, 30–47)

The balance of the 300-page report is devoted to solutions. Although the distinguished authors are far from naive, many of their recommendations contrast sharply with current practice. They suggest, for example, a large-scale transfer of resources to developing countries, elimination of tied aid, IMF reforms that respect the domestic social and political objectives of developing countries, greater access to markets in the industrialized countries for exports from the South, priority for the needs of the poorest countries, and better prices for raw materials — in short, a New International Economic Order.

In many senses, a new order is emerging, although so far it does not resemble the hopes of the Brandt Commission. The unrivaled power of the United States in world affairs from 1945 to 1975 is certainly diminishing in the face of industrial competition from Japan and western Europe. Producer associations such as the Organization of Petroleum Exporting Countries (OPEC) gained influence in world affairs, although it has not been used in the interest of other developing nations. The high-technology future is an arena of intense competition as France, England, and Germany attempt to make inroads on Japan and the United States.

The old mechanisms of domination are faltering. Third World debt has become so enormous, and implicated so many financial institutions in the advanced countries, that the threat of default forces compromise. Brazil, for example, has refused to accept IMF terms that threaten its own fledgling democracy. A new sense of autonomy and needed reform exists in the developing countries and parts of Europe. While the United States waged war with a social revolution in El Salvador and postrevolutionary mixed economy in Nicaragua, socialist governments have come to power in five countries of southern Europe: France, Spain, Portugal, Italy, and Greece. The world system is undergoing fundamental realignment. If a New International Economic Order does not promise rapid equalization of world prosperity, it suggests an unprecedented challenge.

ANOTHER LIFE

Meanwhile, the core responds in its own ways. Increasingly, the placeless transnational corporation is the seat of global economic power rather than the nation state. Transnational corporations like ITT and Atari take full advantage of the protections offered by their nominal citizenship but show little allegiance when it comes to moving their capital or production around the world.

The class structure that emerges with multinationalism includes menial workers in the Third World enclave economies, displaced workers in the core, a new ethnic immigration to cheap-labor jobs in world cities of the core (see Chapter 3), and a new class of international capitalists. Yet as the fortunes of the transnationals flourish, the relative prosperity of the advanced nations declines. New regional pacts such as the North American Free Trade Agreement are likely to unify interests that operate across countries while they widen inequalities within. The danger is that a lowest common denominator of wage and environmental standards will become irresistible to profit-seeking firms. The late twentieth century presents a global society with a global class structure and globally patterned inequalities of wealth and poverty. Alone that might suggest equalization and cause for some optimism. But the new inequalities are also more intense, dividing the developing countries into a smaller privileged elite and a larger impoverished citizenry, just as they invade the developed nations.

The commendable Brandt Commission report has yet to redirect international development policy. Indeed, the drift is in another direction. Contemporary policies on international development, encouraged by powerful agencies such as the IMF, stress government frugality with little mention of its human consequences. Meanwhile people persevere living in the conditions that result:

> The government economic policies that have drawn such praise from Mexico's international creditors were not on Jesus Sanchez's mind today as he sifted through the municipal garbage dump here, scavenging scrap to sell to feed his family.
>
> Mr. Sanchez says that on an average day he can gather enough old bottles, cardboard and newspapers in his scuffed canvas sack to earn about 400 pesos, the equivalent of $2.16. It is enough to feed his family of eight with tortillas and chili, he said, but not much more. Last year there was milk nearly every week for the children; this year, with the price nearly two-thirds higher, it is a luxury.
>
> Reduced government spending on items such as salaries, subsidies and public works translate into immediate hardship here.
>
> Mr. Sanchez, who is 60 years old, said he once worked in construction and other odd jobs, and his deeply lined face has a leathery look. With construction in the city at a virtual standstill and other jobs scarce, the dump here is now his only source of income. (*New York Times*, July 7, 1984)

The world system is a grand abstraction, yet one that vividly connects Mr. Sanchez, Mexico, and the poor countries with international banks and corporations. When these connections are laid bare, solutions for world development readily appear to participants in the North–South struggle. And with those solutions, the bedrock obstacles also appear in an international politics of inequality. This, at least, focuses the conflict and the agenda for a New International Economic Order, although it cannot forecast the outcome of the struggle.

SELECTED BIBLIOGRAPHY

BIRKBECK, CHRIS. 1978. "Self-Employed Proletarians in an Informal Factory: The Case of Cali's Garbage Dump." *World Development* 6(9–10):1173–85.

BLUESTONE, BARRY and BENNETT HARRISON. 1982. *The Deindustrialization of America.* New York: Basic Books.

BRETT, E. A. 1973. *Colonialism and Underdevelopment in East Africa: The Politics of Economic Change, 1919–1939.* London: Heinemann.

FRANK, ANDRÉ GUNDER. 1969. "The Development of Underdevelopment." Pp. 3–17 in *Latin America: Underdevelopment or Revolution.* New York: Monthly Review.

FRANK, ROBERT H., and RICHARD T. FREEMAN. 1978. "The Distributional Consequences of Direct Foreign Investment." In *The Impact of International Trade and Investment on Employment,* edited by William G. Dewald. Washington, D.C.: Government Printing Office.

GIDDENS, ANTHONY. 1973. *The Class Structures of the Advanced Societies.* New York: Barnes and Noble.

INDEPENDENT COMMISSION ON INTERNATIONAL DEVELOPMENT. 1980. *North-South: A Program for Survival.* Cambridge, MA: MIT Press.

MAGDOFF, HARRY. 1966. *The Age of Imperialism.* New York: Monthly Review.

MARX, KARL. 1867. *Capital,* Vol. 1. Reprint, translated by Ben Fowkes. New York: Vintage Books, 1977.

MUSGRAVE, PEGGY B. 1975. *Direct Investment Abroad and the Multinational: Effects on the United States Economy.* U.S. Senate Committee on Foreign Relations, Subcommittee on Multinationals. Washington, D.C.: Government Printing Office.

PORTES, ALEJANDRO, and JOHN WALTON. 1981. *Labor, Class, and the International System.* New York: Academic Press.

SAMPSON, ANTHONY. 1973. *The Sovereign State of ITT.* Reprint. New York: Stein and Day, 1980.

WALLERSTEIN, IMMANUEL. 1974. *The Modern World System: Capitalist Agriculture and the Origins of the European World-Economy in the Sixteenth Century,* Vol. 1. New York: Academic Press.

8

Social Control, Deviance, and Rebellion

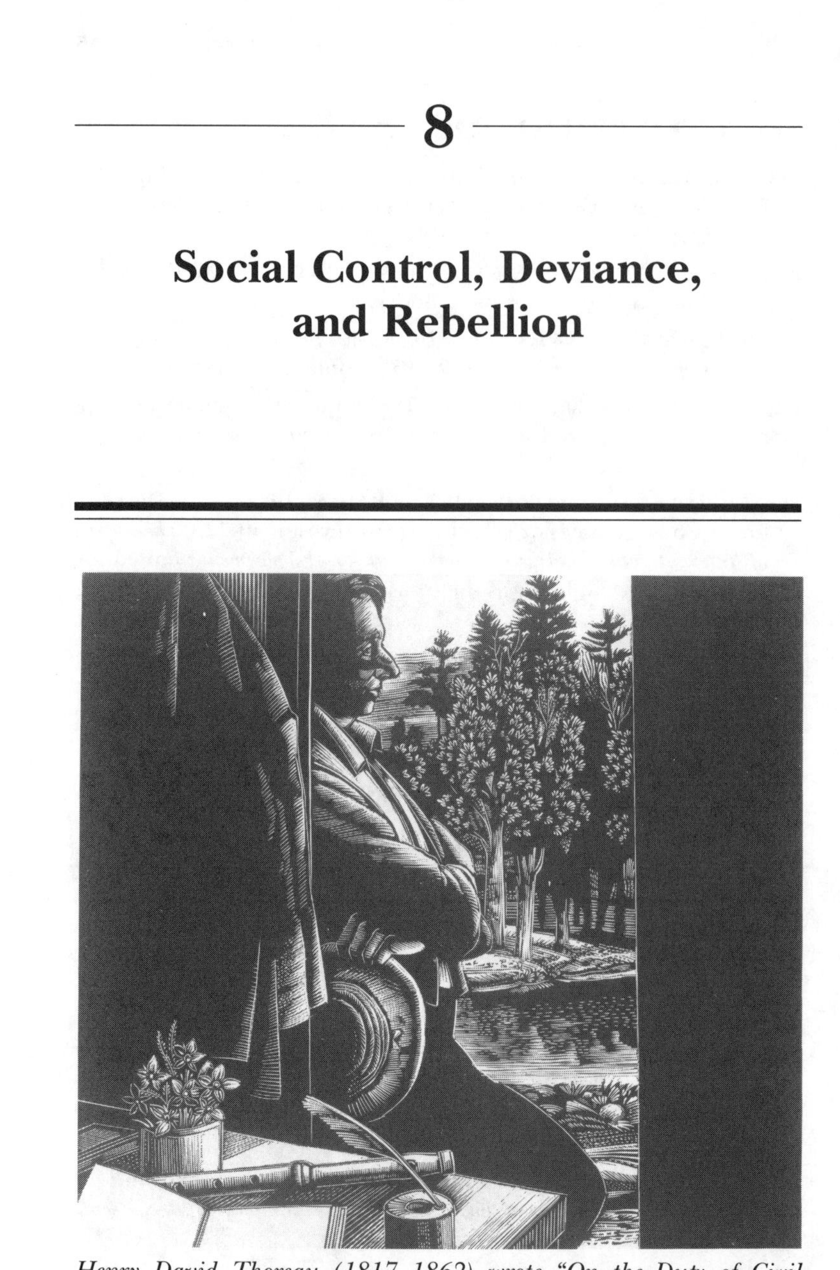

Henry David Thoreau (1817–1862) wrote "On the Duty of Civil Disobedience" in 1849. The essay, which called for free exercise of "moral sense" in public affairs, was read around the world, and its principles inspired the passive resistance movements of Gandhi and Martin Luther King, Jr. (Wood engraving by Michael McCurdy)

SOCIAL CONTROL

Today, abortion is a major issue, a conflict over social control in the most intimate way and a battleground on which political fortunes are being decided. Twenty or thirty years ago, abortion, although practiced with some regularity, was morally taboo — a deviant act. With the development of the women's movement and pressures for population control, however, values began to change in some quarters. A new women's right to control over their own bodies was asserted by some and disputed by others. Doubtless, those favoring abortion were affected by changes in the economy, increased levels of female employment, the trend toward smaller families, and liberalizing sexual mores. Those opposing abortion were equally affected by status threats to the traditional family and homemaker role and by a resurgence of religious commitment that followed on rapid social change.

In 1973 the U.S. Supreme Court ruled that states could legalize abortions on demand. Pro-life and pro-choice forces entered into a conflict that involved churches, feminists, men, state agencies, medical and insurance businesses, and, not least, rival political factions. Twenty years later, the conflict has intensified. Massive pro and con political rallies mark the anniversary of the court's decision. Candidates for the highest offices make direct appeals to abortion opponents or women's political groups. Antiabortion vigilantes have bombed and burned many clinics in defiance of the Supreme Court and its secular law. Women and men anguish over what to do about an unwanted pregnancy, finding little comfort in the cross fire of irreconcilable values and moralistic injunctions. How can we make sense of examples like abortion? Why, despite claims about common values or the authority of bodies like the Supreme Court, is society ridden with conflict?

Social control is a useful concept for answering those questions and a good summary of many processes discussed previously. In a general sense, social control refers to the means by which groups influence or direct individual members. In light of earlier chapters therefore, social control deals with the means by which class privilege, status group honor, and the authority of the state, family, church, or association are maintained and sometimes altered. Social control focuses on the circumstances under which people conform to group expectations, the bases of that consensus, and how it is enforced, resulting in conduct judged as deviant in the individual and rebellious in the group. This takes in much territory.

Concretely, every group, association, and institution in society exercises social control or tries to influence its members to some degree: Families teach and sanction moral standards, schools do the same while encouraging achievement, churches preach, jobs prescribe and reward particular types of behavior, professional groups define their own ethics and expel rule breakers, political groups rally their followers around causes, and the law and its enforcing agencies define the ultimate boundaries of acceptable conduct. Individuals, in turn, gear their action to all sorts of influences that groups control: rewards, values, customs, esteem, punishment, obligations, recognition, and fear. Although each of these roles of the social-control relationship seems so diverse as to defy any organized analysis, this chapter proposes to illuminate the nature of social control by linking it to the structures of class, status, and the state.

A fruitful way to begin is to consider how we think about social control or, more precisely, what people believe society has a right to demand of individuals. Most people agree that a certain amount of social control is essential for the efficient and harmonious operation of social life. We need laws that protect us from scoundrels, agreed-on rules for operating an automobile or a government, and shared goals that govern business or education. We agree, in short, to a social contract that guarantees individuals certain satisfactions provided they play by the rules.

Yet, at the same time, we are ambivalent about the appropriate subjects and limits of social control. There ought to be more law and order, we say, but less restraint on individual liberty; more vigilance of the public interest, but less bureaucracy; civilization without regimentation. Among social critics worried about the proper balance, modern society is alternately portrayed as verging on the barbarous struggle of children that William Golding described in *The Lord of the Flies* or George Orwell's oppressive *1984*.

Ironically, these opposing views err in the same fashion by posing the problem as one of more or less social control, with dangers at each hypothetical extreme. That is, they mistakenly assume that society possesses a coherent set of master values or compatible interests that are implemented to a greater (regimented) or lesser (anarchic) extent. This misconception is promoted by certain social theories that overstress the potent effects of an integrated value system or a unified power elite to the neglect of conflict and competition. Although folklore, fiction, and theory all illuminate different surfaces of social control, none highlights the whole crystal. Social control involves more than a unified set of expectations dictated from the top down. It also includes resistance and conflict.

The abortion controversy is not unique in demonstrating this point. Major social issues such as civil rights, war, and social welfare exhibit the same profile. All sorts of groups in society attempt to influence or control people according to group interests. Social control comes from groups at the top, bottom, and around all sides of society. Coming from powerful groups, it has a better chance of success, but that does not avoid stalemates, as in the case of abortion, or fragile agreements. The value of this conception of social control is that it avoids deterministic interpretations of social life, appreciates the diversity of social motivation, and reckons with the struggles people endure as they labor to define and act on moral grounds. Social control is no more or less than the public expression of right and wrong — the politics of morality. Values and power affect social control but in ways that are worked out in varied particular situations (rape, gross fetal defects, or pregnancy-related threats to the mother's life, e.g., are contingencies that change the meaning of abortion).

The argument that follows specifies the outlines of social-control situations (the constraints of class, status, and the state), as well as the main kinds of variation that govern those situations. *Social control* is defined here as the negotiated and reciprocal obligations among participants in collective actions that derive from group support and influence conduct broadly understood as moral. It results in a social contract that, depending on shifting circumstances, may vary from quite oppressive conditions to liberal compromises and even stalemate. The kind of theory that applies here is one that accepts the role of values and power but emphasizes, with the sociologist Herbert Blumer, that collective action arises from the "fitting together of individual lines of action" (1962, 184). The appropriate imagery is neither Golding's war of all against all nor Orwell's big brother state but Leo Tolstoy's vision in *War and Peace* that history unfolds "in accord with the will of the hundreds of thousands of individuals who [take] part in the common action."

This chapter develops the argument and its implications through four straightforward propositions: (1) Social groups attempt to set standards for acceptable and deviant conduct, although these efforts vary widely in their success; (2) the state is the fundamental but not exclusive agent of social control; (3) social control exercised mainly by the state follows lines of class and status group privilege; and (4) attempts at control commonly meet with noncompliance, resistance, or outright rebellion.

DEVIANCE AND THE POLITICS OF SOCIAL CONTROL

Common sense would tell us that deviant behavior involves violations of the demands of social control. Deviants include criminals who break the law, homosexuals or eccentrics who flout convention, the mentally ill evidently incapable of understanding or acting rationally, people who abuse drugs or children, and even ideologues who preach Nazism or communism. In each instance, there are **social-control agencies**, from police to the mental health center, specially commissioned to identify, admonish, treat, or punish deviants.

The question of what constitutes deviant behavior, however, is quickly complicated by the unsystematic application of social control. Eccentrics and, increasingly, homosexuals enjoy some tolerance, particularly if they occupy positions of social honor — celebrities or artists, for example. Vigilantes in the abortion clinic or New York subway are not always considered criminals by sympathizers and courts. Drunkenness is acceptable in an entertainer whose act features it as humor, but not in the family doctor or accountant. Oddballs with money or with caretakers are just odd; unprotected or on the streets, they are mentally ill persons requiring confinement when that is available. Moreover, some people who act deviantly escape notice and are not treated as deviants, whereas others who do not engage in those acts but are suspected of them are treated as deviants.

Two critical implications follow from these observations. First, social control is a public response to behavior or anticipated behavior that varies widely in its severity or the sanctions it carries. Second, deviance is a socially conferred label applied only to some of those who engage in deviant acts and to others who do not engage in those acts. Deviance is defined by the social reaction to conduct rather than conduct as such. The implications require elaboration.

Social control can be passive in the manner of settled standards that prescribe appropriate and deviant conduct or active in the sense of consciously applied rules. The first form belongs to culture as represented in social groups. The second lies in the domain of social-control agencies such as police, courts, legislatures, and a variety of nongovernmental bodies — churches, reform groups, guardians of public morals, advocates of legalizing this or banning that.

Social control, conceived in this way, is characteristically an interactive and contingent process. Certain social groups and institutions, for many reasons, endeavor to set standards for the rest of society — to play the role of "moral entrepreneurs." To the extent that they are legitimate rule makers (e.g., legislatures) or rule enforcers (courts), or to the extent that they enjoy status and power (ministers, psychiatrists, bosses), they are more likely to successfully define and sanction deviant conduct. In this process, groups articulate norms or group-defined expectations of individual behavior. The limits of acceptable conduct are suggested, thereby sketching the boundaries of the moral order. But the actual effect of prescribed rules, the real moral universe, depends on how seriously the rules are taken by people, what power backs them, what other rights compromise them, and how systematically they are enforced. **Deviance**, according to Howard Becker, is defined in the following terms:

> Social groups create deviance by making rules whose infraction constitutes deviance, and by applying those rules to particular people and labeling them as outsiders. . . . Deviance is not a quality of the act the person commits, but rather a consequence of the application by others of rules and sanctions to an "offender." The deviant is one to whom that label has successfully been applied; deviant behavior is behavior that people so label. (1975, 9)

This fertile statement gives a new meaning to concrete events that would otherwise appear as morally inconsistent. Homicide, for example, is often considered a moral absolute — universally recognized as the ultimate crime and severely sanctioned in all cases. Yet mercy killings are sometimes considered justified. And the rules change. Recently, several battered wives have been exonerated in the killings of their husbands. Tax evasion, drunk driving, and dope peddling are formally serious crimes. Yet how serious they are in practice, as measured by the actual punishment they draw, depends on whether the tax evader is a big corporation or a religious fanatic such as the jailed Reverend Moon; whether the alleged drunk driver is a laborer, the county sheriff, or a famous lawyer such as F. Lee Bailey; whether the dope peddler is a black slum youth, a white suburban Dr. Feelgood, or a colorful industrialist such as John DeLorean — who was indicted for cocaine trafficking in a well-publicized FBI campaign against celebrity dopers and found innocent by a jury persuaded that FBI entrapment was the more grievous deviant act. These examples only begin to suggest the kinds of contingencies that determine whether the deviant label is successfully applied.

Now, if the study of individual deviance concludes that society's reaction determines what is deviant behavior, then the important questions are: How does certain conduct come to be regarded as deviant? When do courts and control agencies punish? Why don't those things happen on other occasions that involve similar individual acts? If deviance is no longer a quality of conduct as such but a social reaction to certain conduct under certain circumstances, then the social process of rule making and rule enforcement becomes key. That is a big step and a far-reaching implication. John Kitsuse provides the most cogent summary of how sociologists understand deviance:

> In modern society it is difficult or impossible to derive theoretically a set of specific behavioral presuppositions which will in fact be normatively supported, universally practiced, and socially enforced by more than a segment of the total population. Under such conditions it is not the fact that individuals engage in behaviors which diverge from some theoretically posited "institutional expectations," or even that such behaviors are defined as deviant by the conventional and conforming members of society. A sociological theory of deviance must focus specifically on the interactions which not only define the behavior as deviant but also organize and activate the applications of sanctions by individuals, groups and agencies. For in modern society the socially significant differentiation of deviants from non-deviants is increasingly contingent upon circumstances of situation, place, social and personal biography and bureaucratically organized agencies of social control. (1962, 256)

This insight is enormously helpful in sorting out inconsistencies in who is and who is not considered a deviant. Its implications, however, are much broader. The idea can be extended beyond the usual confines of social-control agencies and individual deviance to the general relations of state and society by recasting the relations between social-control agencies and individuals in terms of the state and such major social groupings as classes and status groups. The state is the chief arbiter of social control, maintaining social order through manifold agencies, both public and private (see Chapter 6). Yet the state does not monopolize social control or successfully contain dissent and rebellion. It is, however, in the interplay of state power and dissent that the most general features of social control are found. Some examples of the parallels between individual and political deviance will clarify this avenue of generalization.

Deviance Generalized to Politics

The societal reaction view of social control suggests that political deviance, like its "pathological" cousins (crime, mental illness, etc.) arises in a process that defines certain behaviors as a threat to established order. Reaction to the threat takes different forms such as identifying and prosecuting political deviants (traitors, bribe takers, pacifists), suppressing dissent, or reorganizing state power depending, once again, on situational conditions. Illustratively, high officials involved in the Watergate scandal, for a number of reasons including their ambitious rivals and a growing distrust of President Nixon, came to be defined as criminals rather than overzealous campaign aides. Watergate justice never held the president publicly accountable, and a minor figure, the intrepidly defiant ex-spy Gordon Liddy, drew the longest prison sentence. But, once the president had felt obliged to say "I am not a crook," it was clear that the social reaction had turned against the highest officials of the state and that many would pay with a deviant label. Conditions such as status and remorse affected the link between proven guilt and sanctioned deviance. The discretion that enters into judging wrongdoing, scornfully called "politics," was the same as that accorded so-called behavioral disorders. This suggests, again, that deviance and social control pervade all of social life, rather than operating only at the bizarre fringes.

In an earlier epoch, partly through the energies of Congressman Nixon, screen writers, professors, and State Department officials were blacklisted for alleged (and unproven) loyalties to the Communist party. In both of these cases, what explained the official proceedings was less the gravity of political corruption or of dissent (as when real Communists were politically active in the 1930s) than national crises that demanded the reform of state power. In the Watergate case, for example, the crises included an unwinnable war in Vietnam, landslide Democratic party defeats, and the first of the 1970s recessions.

Yet states are not always so successful in prosecuting a new class of deviants and thereby reaffirming organizational principles. Under other circumstances, alleged culprits successfully redefine charges of wrongdoing. They revolt, and the state teeters on the edge of suppressing rebellion, negotiating peace, or suffering revolutionary transformation. The American revolutionaries of 1776 said they were not treasonous subjects of George III or delinquent taxpayers but free people who would have effective representation or self-government — and, through skill and good fortune, they made it stick. Civil rights activists in the 1960s said they were not Communists (as repeatedly alleged) or mean agitators but also free people who would have freedom now — and slowly they got some. Agrarian populists, the International Workers of the World (IWW), the American Indian Movement (AIM), and others, for a variety of reasons, all suffered suppression and a deviant label that stuck for actions no more intrinsically radical than those of George Washington and Martin Luther King, Jr.

Tolstoy mused about the "movement of nations" and events unexplained by the will of rulers or the actions of individuals. Asking "what does this signify?" he concluded "that the force which decides the fate of peoples lies not in conquerors, nor even in armies and battles, but in something else . . . the activity of all the people who participate in the event." So, in the case of attempts at social control that may ensure order or tumble into rebellion, the "something else" that causes the difference is the social condition that makes individual and institutional lines of action dovetail with state power or veer off in another direction. Our task therefore is to examine the conditions of state and society that explain social control.

THE TECHNOLOGY OF STATE POWER

The notion of "political technology" is more than a metaphor. The state fashions techniques for maintaining public order that are similar to the methods industry uses to produce commodities. In each case, the technology includes ideas that define the task (the principles of mass production or of equal responsibility under the law) and physical implements for doing the job (machines or prison cells). The technology of state power appears in many related forms: in laws, courts, prisons, parole, welfare agencies, schools, congressional committees, licensing boards, riot commissions, and hospitals that anguish over the fate of a Baby Jane Doe. Understood in this way, social-control agencies are linked to the broader social institutions from which they receive the authority to impose sanctions. The diverse actions of social-control agencies begin to make coherent sense when they are seen as originating from the common root of state efforts to maintain a particular kind of social order.

Michel Foucault, the late French philosopher and historian, elegantly developed the idea in his book *Discipline and Punish: The Birth of the Prison* (1979). Foucault shows how, over a short time beginning in the late eighteenth century, the gruesome spectacle of public torture vanished as a technique of punishing crime for the benefit of onlookers and was replaced by the rehabilitative methods of the prison and asylum. The dramatic shift, in Europe and North America, coincided with the demise of absolute monarchies, the rise of modern democratic states, and coincident legal reform movements. In Foucault's terms, new states adopted a more subtle, perhaps insidious, technology of power, one that avoided public violence — which had turned dangerously against the kings themselves — and attempted instead to restore deviants to the moral order through institutional discipline. This technology attacked the soul rather than the body.

The idea of political technology alerts us to how social-control methods change over time in response to economic and political developments. The role of punishment in society, as Foucault notes, goes far beyond an effort to reduce crime or deviance and includes "a whole series of positive and useful effects." One of these is an economic effect, as shown in historical studies, that

> relate the different systems of punishment with the systems of production within which they operate: thus, in a slave economy, punitive mechanisms serve to provide an additional labour force — and to constitute a body of "civil" slaves in addition to those provided by war or trading; with feudalism, at a time when money and production were still at an early stage of development, we find a sudden increase in corporal punishments — the body being in most cases the only property accessible; the penitentiary . . . forced labour and the prison factory appear with the development of the mercantile economy. But the industrial system requires a free market in labour and, in the nineteenth century, the role of forced labour in the mechanisms of punishment diminishes accordingly and "corrective" detention takes its place. (Foucault 1979, 24–25)

A neat example of this point is provided by English vagrancy laws. Vagrancy as a crime punishable by forced labor first appeared in England in 1349 when the plague, European wars, the crusades, and town migration all combined to deplete the supply of rural labor. Vigorous enforcement of the vagrancy statutes helped keep scarce laborers on the job. In time, these problems were overcome, and vagrancy disappeared as a common crime. Miraculously, however, it reappears as a capital crime in the sixteenth century at the same time expanding commerce is endangered by bandits and highwaymen. Later, when the poor house came into use as a method for encouraging industrial labor, vagrancy was once more demoted to a public nuisance, and the laws were used to clear the streets of misdemeanor bums. The example should not suggest any simple link between deviance and the economy, but it does show how the state intervenes in the relationship.

Crime and the State

English historian E. P. Thompson (1975) has written a remarkable book about the Black Act of 1723 that made poaching a crime punishable by death. The question Thompson asks himself is, Why in the world would the British government resort to a drastic law that proposed hanging country folk engaged in deer hunting (with faces blackened for stealth) in the forests and game preserves set aside for the aristocracy? Why, moreover, was this offense of the Blacks soon joined by several hundred more capital crimes of a similar nature, ranging from breaking fish ponds to cutting sod for use in fireplaces? What purpose was the Black Act designed to serve?

The conflict had a clear history. Civil wars and the commonwealth government that temporarily replaced the British monarchy in the previous century had greatly reduced royal amenities, including the hunt and exclusive use of the deer parks. Correspondingly, common rights to these preserves for agricultural purposes had grown. Fish ponds that flooded areas of sod cutting and the deer that ate crops were nuisances to the agrarian economy. The Blacks were no mere poachers. They hunted deer, cut trees and sod, broke dams, and committed other "offenses" in their efforts to farm the forest lands, an activity that they had come to consider as their right.

Yet these facts alone hardly explain the massive official reaction in the new law. Why did the state choose to make a dramatic example of the seemingly petty crimes of the foresters? On one hand, a new royal house (the Hanovers), who realistically feared popular opposition such as the Blacks expressed by their defiant use of the forests, had come to power in 1714. On the other hand, in the interests of consolidating its power, the government had begun to award new privileges of forest use to its own aristocratic supporters. Licenses for hunting game and patronage positions such as gamekeeper were distributed in a blatant bid for support. The government used "the deer as a screen behind which to advance their own interests."

When the Blacks resisted these politically motivated incursions by poaching, now intended and understood as an act of political defiance upheld by common rights, a true emergency was created because the Blacks won community support. "It was the displacement of authority, and not the ancient offense of deer-stealing, which constituted, in the eyes of the government, an emergency" (Thompson 1975, 191). The interests of agrarian society met head on an effort to reorganize the state along modern lines that gave more power to a new royal bureaucracy at the expense of the king. The Black Act therefore was just one expression of a realignment of state power. Enforcement of the act, moreover, was "seized upon by Walpole [the new prime minister] as a heaven-sent opportunity to consolidate his power" (p. 201). The prime minister and his supporters were creating a new technology of power, a state that would be independent of the king and based on new bureaucratic forms and a law that put aristocratic privilege ahead of the common people's interests. In the process, the reorganized state created a "forest bureaucracy" to share in the privileges of social control and a new brand of criminal.

The previous points take advantage of circumstances in which a strong state was in the process of formation with the consolidated assistance of powerful social classes. In that sense, they may give the impression that states unfailingly develop potent mechanisms of social control — an observation that is subject to wide variation. Prior to the American Revolution, the British colonies in North America were governed by fragile local authority, menaced by sectionalism and religious controversy. As England struggled to establish a new state and parliamentary form, the Massachusetts Bay Colony in North America faced a more elusive threat to public order — witches.

Kai Erikson's study reveals that the Massachusetts Bay Colony underwent three different outbreaks of religious heresy, crime waves that saw public prosecutions rise significantly. Each outbreak was connected with political uncertainty about the colony's mandate and future. The most serious episode occurred around 1692 when political dissension, land disputes, and commercial competition began to plague the colony: "It was while the people of the colony were preoccupied with these matters that the witches decided to strike" (1966, 141).

The famous witch trials of Salem Village began as an effort to stamp out a small cult of teenage girls gathered around a household slave from Barbados steeped in the practice of voodoo. The hysteria spread until hundreds of citizens were accused and held for trial as witches. The freewheeling accusations touched even the president of Harvard College and the legendary pilgrim Captain John Alden. Erikson explains these events as a reflection of deeper social troubles, of unspoken threats to social order that call forth a renewed sense of solidarity and a reaffirmation of society's boundaries:

> Whenever a community is confronted by a significant relocation of boundaries, a shift in its territorial position, it is likely to experience a change in the kinds of behavior handled by its agencies of social control. The occasion which triggers this boundary crisis may take several forms — a realignment of power within the group, for example, or the appearance of new adversaries outside it — but in any case the crisis itself will be reflected in altered patterns of deviation and perceived by the people of the group as something akin to what we now call a crime wave. These waves dramatize the issues at stake when a given boundary becomes blurred in the drift of passing events, and the encounters which follow between the new deviants and the older agents of control provide a forum, as it were, in which the issues can be articulated more clearly, a stage on which it can be portrayed in sharper relief. (1966, 68–69)

A social order menaced by an invasion of witches may seem bizarre today. Yet modern societies continue to turn up diabolical threats to their own values and stability just when established boundaries are shifting. The Salem Witch Trials have been instructively compared with the fanatical anticommunism crusade that swept the United States in the late 1940s and early 1950s — and has erupted periodically before and since. The comparison goes deeper than the mean and spooky tactics of witch-hunting. Erikson's observation about witchcraft in the seventeenth century — that "no other form of crime in history has been a better index to social disruption and change" (1966,

153) — applies as well to the periodic "red" scares of the twentieth century. The red menace began cropping up in the United States even before the first Russian Revolution in 1905. Dissenters once labeled revolutionaries or anarchists now became associated with communism, even though the issues involved in their dissent were not regularly joined by the American Communist party — itself not formed until 1919.

For nearly three decades, U.S. communism drew only occasional condemnation and seldom posed a real subversive threat to national security. Indeed, in 1936 Earl Browder ran for president on the Communist party ticket, receiving nearly 75,000 votes. From its high point, when the Communist party had perhaps 200,000 sympathizers, it declined steadily, particularly after Stalin's cruelty in the Soviet Union during the 1930s. The Committee on Un-American Activities of the House of Representatives was formed in 1938 but achieved celebrity only a decade later. Ironically, the great anti-Communist crusade and McCarthyism in the early 1950s came long after the movement had lost a significant following. What then explains the American Inquisition, the period of U.S. history that David Caute (1978) calls the "great fear" when thousands of citizens from the schools to Hollywood were labeled agents of the Communist conspiracy?

Consistent with Erikson's interpretation, postwar U.S. society plunged into a political crisis compounded of domestic strife and international challenge. For sixteen years of depression and war, the country had been ruled by the Roosevelt administration. With Republicans and conservative Democrats on the sidelines, fundamental social reforms had been instituted from Social Security to unemployment insurance and labor legislation (see Chapters 5 and 6). Under emergency conditions, the rudiments of a welfare state had been built that irked the political opposition all the more because they correctly saw themselves as unable to regain the White House during the war and immediate aftermath. Growing domestic irritation was compounded by the postwar international scene. European states were in ruin, and a newly empowered United States looked across the wreckage at the Russian bear.

By 1948 when Harry Truman ran for reelection the political crisis was ripe for a Cold War against the Soviet Union and any of its alleged domestic sympathizers: "bleeding hearts" like Eleanor Roosevelt who had defended the civil rights of Communists before a congressional committee; appeasers like the late President Roosevelt, who had been duped by Stalin, "communistically inclined" (in Joseph McCarthy's phrase); Democrats who had supported labor, internationalism, and related left-wing causes — in short, a large segment of the Democratic party. Red-baiting came into fashion in the 1946 congressional elections and paid off handsomely with major Republican party gains. Truman knew he was in trouble, although he had two years before the next presidential election in which to mount his own campaign against the red menace. And so he did with a flourish. To steal Republican thunder, Truman began a purge of purported Communists in the federal civil service. An attorney general's list was issued naming fancied subversive organizations. Deportations of suspect immigrants took place. Communist party stalwarts were indicted for alleged plots to overthrow the government. And the Truman Doctrine abroad pledged U.S. support for countries (no matter how repressive) engaged in a struggle with communism.

Judged by the result of Truman's reelection, the campaign worked. But it also let loose years of mean-spirited intimidation and a political technology of official suspicion (committees, subversives lists, kangaroo hearings, investigations, loyalty oaths) that ruined thousands of lives, uncovering in the process no red plot to overthrow the government. As Caute (1978) observes, Truman's use of the Cold War was crucial in sowing the seeds of McCarthyism. By 1952, with the election of General Dwight D. Eisenhower to the presidency, the political crisis faded. The witch-hunting Senator Joseph McCarthy was disgraced, and the Communist conspiracy suddenly looked less menacing, although the climate of fear retreated only slowly.

Three interpretive points summarize the historical evidence on the state and social control. First, there is a literal technology of state power embodied in ideas about and mechanisms for implementing social control. Second, technology changes with economic and political development, most notably during times of social disruption or crisis. Finally, the state is a critical source of social control but not the only one. State and society — government and church, for example — influence one another. At bottom, the process of defining and sanctioning deviance is a political process — whether in specific cases the deviance concerns political activity, crime, insanity, or some other malady. Social control is political because it involves exercising power for the social order.

CLASS, STATUS, AND SOCIAL CONTROL

So far we have dealt with the mechanisms of social control, where they come from and when they are brought to bear. The next, more practical matter is how those mechanisms are actually employed and how their application is *socially mediated* by the influence of class and status. The previous discussion demonstrates how the zealousness or perceived demand for social control varies with periodic crises and shifting social boundaries. But social control is routinely affected by the power of social classes and status groups. Analyzed along these dimensions, social control tends to follow lines of privilege rather than any objective needs of social order.

Class

The connections of class and crime appear in different forms from one historical period or society to another. In the modern United States, the issue has been analyzed most completely in connection with what Edwin Sutherland (1940) called "white collar crime." The phrase emphasizes that legally defined crime is hardly confined to a shadowy underworld or desperate working class.

A long list of "upperworld" crimes compiled by Albert Morris in 1935 included stock fraud and embezzlement, bank manipulations of investment trusts and securities, manufacturers' price-fixing, graft in business and politics, employers' exploitation of unorganized labor, and many more. Today, the list could be vastly expanded to include, for example, industrial dumping of toxic waste, violation of product safety laws from children's toys to nuclear reactors, computerized investment swindles, corporate manipulation of stock values and sales, fraud in government health programs and defense contracting, and multinational bribes to foreign government officials for the purchase of U.S. exports.

Sutherland's (1940) original research examined decisions by courts and commissions against the seventy largest U.S. industrial and commercial firms, with respect to four types of laws: antitrust, false advertising, labor relations, and patent infringement. He found a total of 547 violations; each corporation had at least 1, and the average was 7.8 per firm. Revealingly, however, only 9 percent of the violations were treated as criminal matters, the rest being dismissed as misdemeanors, or minor infractions. Noting this tendency to coddle corporations, Sutherland went on to examine the 547 incidents to determine whether, by the standards of social injury and legal provision of a penalty, these were true criminal acts. In virtually all cases, the two standards were met, and the first — social injury — applied in all 547. That is, in 100 percent of the cases, corporations were guilty of having committed socially injurious crimes, although those crimes were punished as such in just 9 percent of the cases.

What were the corporate criminals up to that Sutherland considered socially injurious? In the false-advertising category, he discovered the now-familiar practice of gross misrepresentation: cigarettes that claimed to cure throat irritation, facial creams billed as "skin food" that removed wrinkles, and tea supposedly made from tender leaves for one firm but actually bought in bulk from the same source used by competitors. In the field of labor relations, corporations tried to prevent union organizing by employing private goon squads to break up, with clubs and gas guns, peaceful assemblies of workers — over a four-year period, one steel corporation purchased 143 such guns while the entire Chicago police department found need of only 13. Corporate officials embezzled from their own stockholders in the form of paying themselves excessive and fraudulently calculated salaries and bonuses.

Finally, the large corporations were undermining the free-enterprise system and attempting to replace it with their own "private collectivism" through the restraint of trade. In the major trade associations, manufacturers conspired to limit production and fix prices. Companies protected themselves by applying for patents on "every possible modification of procedure, bringing [patent infringement] suits on every possible pretext. . . . The Good Humor Corporation engaged in patent litigation for more than a decade with the Popsicle Company and other manufacturers of ice-cream bars to determine which firm had invented this contribution to science and the arts" (Sutherland [1956] 1968, 69). Today, roughly forty years since this incident, the same practices are routine in the business world, down to a protracted court battle over who "invented" and holds patent rights to the frozen pizza.

In a strongly worded conclusion, Sutherland claims "that the ideal businessman and the large corporation *are* very much like the professional thief" ([1956] 1968, 69). Like common criminals, 97 percent of corporations are repeat offenders. Illegal behavior is much more common than prosecutions suggest because few corporations are punished for these industry wide practices. Violators do not lose status in the business community for their deeds, and business holds contempt for legislators, courts, and government snoopers. "Businessmen, being like professional thieves in these four respects, are participants in organized crime." Although Sutherland's original work was done over fifty years ago, his results, if not his vigorous analogies, are replicated in recent research, by Marshall Clinard and Peter Yeager (1978), for example.

Class bias reflected in the lax prosecution of white-collar crime is further demonstrated by the enthusiasm with which social-control agencies go after violations by labor unions. Sutherland notes that "of the [Department of Justice] actions against business firms and associations of business firms, 27 percent were criminal prosecutions; while of the actions against trade unions, 71 percent were criminal prosecutions" (1949, 137). Official scorn is heaped on "welfare chiselers" but spares firms that defraud the government by billing for unnecessary or unperformed medical services under Medicare or selling defective and overpriced merchandise to the Department of Defense. In recent cases, faulty computer chips were knowingly sold to the government and installed in an undetermined number of missiles. A ten-cup coffee maker for the air force C-5 transport plane came at a cost of $7622 to the taxpayer.

Selectivity in prosecution is compounded in the rare instances when corporate crime is actually proven in courts or agencies, yet receives light to suspended sentences. A Canadian study, by Laureen Snider (1982), of ordinary and corporate thieves shows that common criminals are much more likely to go to jail and to stay there for long terms.

The relationship between class and crime is more involved than the old exposés of white-collar skulduggery would suggest. To say it is more complicated, however, does not mean that today's corporate outlaws appear in a more favorable light. Rather, some things have changed in the public and official attitude toward these crimes. New laws aimed at upperworld crime have been fashioned and applied with perhaps a broader deterrent effect than the ordinary criminal statutes. Moreover, the national political mood has shifted back-and-forth in support for prosecution of those who violate the public trust.

An insider's account of the U.S. Department of Justice by Jack Katz (1979) shows how a "social movement against white-collar crime" arose in the 1970s within the federal government. The timing was crucial because the Watergate scandal had stripped away the protective cover of normal corruption. Young prosecutors promoting their own careers exploited the general mood of indignation. Katz argues that the Justice Department is highly selective about which corporate crimes it will prosecute because they are typically complicated and "tough to make" (prove) — not least because of the elaborate legal defense that the firms can afford. But when the government does decide to go after a particular offender, to put in all that work with smaller staffs and budgets, it pursues the case more vigorously. White-collar crime is not casually or systematically ignored, although it is selectively prosecuted with the result that it is less risky an undertaking.

Among other saviors, corporations can usually count on changes in political administrations and the cooling fires of reformism. New administrations suspend the overzealousness of their predecessors. Katz, for example, predicted: "As a practical matter, the White House [after Watergate] temporarily lost its considerable ability to protect lesser political centers from moral attack. Accordingly, the movement against white-collar crime should recede as the Presidency is restored to the insulation of presumptive legitimacy" (1979, 178).

Lately a shift back to leniency has confirmed these predictions. In the early 1980s, investigations and prosecutions of industrial polluters dropped sharply. At the same time, the

Justice Department's active cases under the Foreign Corrupt Practices Act of 1978 dropped from 400 actions employing fifteen full-time lawyers to 14 cases that now occupied just three prosecutors. The widespread crime of bribing foreign officials to purchase U.S. corporate products (such as airplanes) and services (oil drilling) was effectively shelved as the new administration turned instead to "fraud involving military supplies, food stamps and drug money." The "multinational fraud branch" of the Justice Department was quietly eliminated.

Corporate crime is treated gingerly for a variety of reasons. Legislators and agency officials fear the political power of big business. Cases involving sophisticated financial dealings and resourceful concealment are difficult to make. Government agencies are understaffed for the demanding purpose. Highly paid corporate lawyers can defeat, stall, and trivialize even well-prepared cases. Violators can play on political sympathies, claiming that honest business is persecuted while muggers and rapists go free — despite the fact that these crimes are prosecuted and punished more vigorously. Typically, these methods succeed in limiting the number of white-collar crimes that are brought before the law and minimizing the sanctions imposed on those where guilt is fixed.

Social-control agencies in general, and those responsible for the definition and enforcement of crime in particular, reflect the broader social pattern of class interest. It is not the seriousness of crimes, weighed on some scale of social injury — the lives lost due to unsafe automobiles and toxic pollutants compared to the toll from street crime, for example — that determines their official treatment. Rather, it is the consequences of those crimes for a complex and sometimes conflicting set of class interests that mediate definitions of seriousness and prescribed treatments. The state and its social-control agencies are deeply involved in the same structure of class privilege — often defending it and sometimes reforming it.

Status

The state and class interests, although basic, are by no means the only sources of social control. Diverse values carry prescriptions about social control, particularly values that distinguish cultural and status groups from one another. Conduct considered shameful or illegal by middle-aged, white, Yankee Protestant women may be seen as necessary, resourceful, or admirable by Latin immigrant young men trying to survive and better themselves in the urban slum.

Certain forms of crime and deviance are explained by the practices, even the normal ambitions, of status groups. Daniel Bell (1961) once described "crime as an American way of life," suggesting that successive generations of immigrants have provided, in small numbers, recruits for the rackets. The romanticized notion of organized crime is better understood as a status group response to poverty and ambition. The infamous Italian Mafia, for example, was neither distinct nor permanent but an effort by particular members of one status group to achieve wealth and prominence through illicit trades, especially prostitution, bootlegging, and gambling.

Proof for this argument comes from the fact that Mafia business was previously run by the Jews and Irish, until success allowed them to move on to more respectable enterprises. As new immigrants, each of these status groups in succession responded to a situation in which their social mobility was impeded along conventional routes but facilitated in illicit pursuits by urban political machines anxious to attract ethnic bloc support.

> The desires satisfied in extra-legal fashion were more than a hunger for the "forbidden fruits" of conventional morality. They also involved, in the complex and ever shifting structure of group, class, and ethnic stratification, which is the warp and woof of America's "open society," such "normal" goals as independence through a business of one's own, and such "moral" aspirations as the desire for social advancement and social prestige. For crime, in the language of the sociologists, has a "functional" role in the society, and the urban rackets — the illicit activity organized for continuing profit, rather than individual illegal acts — is one of the queer ladders of social mobility in American life. (Bell 1961, 129)

A fair summary is that new forms of organized crime appear as successive waves of ethnic migrants are confronted with the norm of material success, restricted opportunities in legitimate enterprise, and shifting opportunity structures for illicit trades (such as the shift from machine politics and a tacit acceptance of gambling and prostitution for certain Italians in the 1930s to widespread recreational drug use today). The hypothesis suggests, moreover, that today's organized crime should assume a new look as Italians join their Irish and Jewish predecessors in respectability and new immigrant groups, such as Latin Americans, come along. This too seems consistent with the evidence of a South American connection in drug trafficking — a connection between status and crime that whets the appetite of social-control agencies.

The relationship between status groups and social-control agencies is analogous in many ways to the previous revelations about social class. The intricate ways in which status mediates punishment parallels the general argument, just as it taps some engaging research on criminal sentencing. This is another area in which common sense is easily deceived. For example, conventional wisdom might assume that racial inequality in the United States would punish minority groups more severely than the majority for violations of the same law. Critics of racial justice sometimes ask rhetorically, Why is a white youth caught stealing sent home with a warning, while a black is sent to prison? There is some truth in the implied answer, but there is also much error. Race affects social-control practices including criminal punishment, but it does so in special ways that reflect the social meaning of race.

A sociological classic on prison sentencing done in Texas could scarcely be criticized for choosing a legal system that was soft on blacks. Indeed, Texas juries were unstinting in penalties meted out to black and white convicted felons alike. But they made significant distinctions according to the type of crime involved. Contrary to expectations, Henry Bullock's (1961) research showed that "juries tended to give Negro prisoners committed for murder [and rape] shorter sentences than they gave whites who were committed for the same offense. They gave Negroes committed for burglary longer sentences than they gave whites committed for this offense." What might explain this pattern? Why were blacks actually treated more leniently in some instances than were whites?

The answer was that murder and rape were mainly *intra*racial crimes, burglary an *inter*racial crime. Blacks and whites tend to murder and rape within their own racial groups. Such crimes committed by blacks victimize primarily other blacks and therefore were not deemed as serious or deserving of severe punishment as the murder or rape of whites (by other whites). The tendency for juries to be lenient with these black offenders was interpreted as a more fundamental *devaluation* of black people. The relationship was reversed in the case of burglary where typically propertyless blacks steal from whites and draw longer prison sentences than do white burglars. This patent racism, manifested in the devaluation of black victims, has the ironic consequence of leniency for black offenders in some cases.

This provocative finding has impressed the field of race relations as a whole, just as it has been replicated in the study of social control and victimless crimes associated with drugs. Drug-related crimes have the additional advantage of showing how the social definition and perceived threat of drugs has changed over time, and with it the methods of punishment.

For many years, the recreational use of nonprescribed drugs, with the exception of alcohol, was primarily a minority group practice of blacks, Mexican Americans, Native Americans, and occupational groups such as musicians and physicians. Given the status of most of these users, drugs were treated relatively unambiguously as criminal. In the late 1960s, drug use became more common among white and middle-class youth, provoking something of a crisis in the social definition of drug crime. As more white youth flagrantly committed this crime, agents of social control, faced with the unhappy choice of prosecuting their own children or ignoring their laws, conveniently chose a third course. Drug offenses were redefined in the federal Comprehensive Drug Abuse Prevention and Control Act of 1970, notably with a key distinction between the "real" drug-pusher criminals and the innocent drug-user victims. In short, one-time offenders became officially defined as victims because the status and intertwined class backgrounds of the drug-using population changed to coincide more with majority groups in society.

All this created an unusual situation because a fair number of drug users were still minorities and most drug dealers were white. That would suggest a contradiction with earlier patterns of enforcement and punishment that focused on minorities who were the principal users. Nevertheless, an imaginative study by Ruth Peterson and John Hagan (1984) reasons that the new definition, so urgently arrived at, would primarily dictate enforcement patterns (interacting with racial factors in exceptional cases). This is essentially what they discovered in a study of New York prosecutions. Beginning in the early 1970s, sanctions against drug use declined sharply, while drug dealing became a more severely punished crime — meaning, in this case, that the majority of white dealers were punished more harshly than before. Black drug users actually enjoyed dual qualification as victims, of racism (in the liberal climate of the time) and of mercenary dealers, and received the most lenient sentences, followed by white users. The exception to this pattern, however, was the rare black dealer who was punished even more severely than the white dealer on the premise that it was especially contemptible to victimize fellow victims.

The specific point demonstrated in all this work concerns the situationally prescribed social meaning of race, that a "convincing explanation of differential lenience *and* severity in sentencing is to be found in race-related conceptions of offender-victim relationships, conceptions that are specific to the contexts in which they operate" (Peterson and Hagan 1984, 57). Generally, status groups are associated with patterns of the origin and treatment of socially controlled conduct.

Summary

The interplay of status and social control ranges over a much broader field. Social status affects the likelihood, for example, that a person perceived as a troublemaker by his or her associates will be committed to a mental institution and for how long a period. An absorbing subject of social-control research concerns "status offenses," or rule infractions, defined by *who* does what: minors who purchase alcoholic beverages or parolees who carry a firearm, for example. Status offenses often arise from social-reform movements such as the nineteenth-century efforts that sought to protect children from criminal careers and adult punishments, although, by creating the special legal status of juvenile delinquent, they also left children assigned to that status without formal procedural guarantees in law. Suffice it to say that these issues have distinctive origins and effects, much like the parallel developments discussed in connection with the links between race and punishment and the changing definition of property crime.

Social class and status affect social control in blatant, subtle, and intricate ways. Separately and in combination, they help determine what actions are defined as criminal or otherwise deviant. These designations change over time; innocent vagrancy or hunting at one time are later punishable by death. And they change as a result of emergent class and status interests. Beyond their influence on the definition of *what* is deviant, class and status interests play key roles in determining *how* the deviant is treated; whether corporate criminals are tolerated or drug pushers rapidly become the most dangerous of public enemies, as their clients become victims. Implementing all these changes and inconsistencies is the state that at one time fronts for class interests and at another — in an era of political crisis, reform, and official career making — suddenly turns on corrupt practices in politics and business. We are understandably fascinated by the exceptions to common sense, the circumstances in which deer hunting becomes a hanging crime or blacks ironically get lighter sentences. Conversely, we may find tiresome yet another sensationalistic exposé of class privilege in the very machinery designed for social justice. Yet all these tendencies coexist and join in a pattern that changes but is far from incomprehensible at any moment. Social control is control by the powerful; but it is also limited, replete with irony, lenient when its subjects are not valued, forgiving when it needs their loyalty, and always susceptible to challenge. Social control, in short, is politics.

THE LIMITS OF CONTROL: SAYING NO TO AUTHORITY

Social control has decided limits. People do not always follow the dictates of authority, nor can they be coerced without end. On occasion they withdraw, decline cooperation, resist, rebel, or simply move apart from their leaders and social-control agents. Foremost among the conditions of social order and the limits of social control are what people are willing to obey.

Social control is challenged in active rebellion and in passive, nonviolent ways. The latter takes two additional forms that differ in assertiveness: noncompliance and resistance. Let us begin with noncompliance and resistance, the circumstance in which laws or expectations put forth by social-control agencies are ignored and subverted. With that foundation, active rebellion will be considered in the following, final section.

Noncompliance

Instances of noncompliance are commonplace and provide handy ammunition for critics of society's hypocrisy. Standard examples include widespread use of alcohol during prohibition, ever-present gambling and prostitution, recreational drug use, income-tax cheating, abortion when and where it is illegal, evasion of the military draft, and so forth. The accusation of hypocrisy, however, mistakenly assumes that some kind of moral consensus is possible at all and misreads deeper sentiments of value conflict that prompt people not to comply. Noncompliance on a broad scale does not arise from perverseness but from rational, material, and ethical motives in their own right. Two kinds of noncompliance demonstrate the point: one arising in public opinion or societywide disaffection from the norms of particular social-control agents and a second originating within the apparatus of social control through conflict and stalemate in official circles. In concrete cases, these two overlap, but the distinction is useful for identifying origins of noncompliance.

Occasions of societywide disaffection from the state's agenda are rare but decisive. A splendid example is the way broad public support for the Vietnam War reversed itself, becoming widespread opposition within the calendar year of 1968. In January 1968, public opinion polls showed that Vietnam "hawks," favoring a stepped-up war effort, outnumbered "doves," supporting peace negotiations, by two to one (56 percent to 28 percent, with 16 percent having no opinion). Within just two months, however, the percentage of hawks fell by 15 points (to 41 percent), the doves gained about as much (to 42 percent), and that trend continued: By the end of the same year, doves outnumbered hawks by nearly two to one.

Sociologist Howard Schuman (1972) asked, "Why the massive shift? What prompted the public to reverse its stand on such a basic issue in so short a time?" Two facts informed a sensitive interpretation. First, the campus-based antiwar movement had peaked earlier and did not appear to spill over to the general public. Indeed, the public disapproved of the tone and methods of student dissent. Second, the deterioration of support for the war coincided with a major (Tet) offensive of U.S. foes in Vietnam, coming at a time when the administration and military were announcing imminent victory. In short, the public rather abruptly became disenchanted with the war, got sick of hearing about U.S. casualties in an indecisive fight over vague objectives, decided it was a mistake on pragmatic (rather than the students' moral) grounds, and withdrew its support despite the continuing efforts of leaders to rally patriotic zeal.

Vietnam produced a collapse in the moral authority of the state — a temporary and circumscribed collapse to be sure, but one that is not without precedents and antecedents. Not only did the people desert leaders, but social-control agencies also found themselves without customary power to command and enforce purported deviance. Summonses from draft boards went unheeded; prosecutions for draft evasion were contested and discredited. Ultimately, the military draft was reformed, and an amnesty of sorts offered evaders. Massive civil disobedience, when it led to arrests, was treated lightly by the courts. The state had exceeded its limits of control and was forced to backtrack.

In less dramatic ways, state power and the limits of social control are continually renegotiated. One reason for this is simply that the times, circumstances, and attitudes change in ways that sap the authority of earlier definitions of deviance. Specific instances have their own causes. Taboos and social-control practices surrounding homosexuality, for example, fell away rapidly from the 1960s to the 1980s. This, no doubt, had to do with changing sexual mores, an expansionary period for civil rights, and the organized political power of the Gay Liberation Movement. Recently, the AIDS epidemic has revived homophobia (principally against males and with less hostile attention to the intravenous drug users who spread the disease), now with a new public health argument.

A second kind of noncompliance originates in conflicts within diverse interests for and agencies of social control. Systems sometimes break down under the sheer weight of competing demands over the definition and enforcement of nonconforming conduct. A lucid example is Pamela Roby's (1969) analysis of what was intended as a straightforward revision of the New York State Criminal Code on prostitution. Proposed reforms unexpectedly brought forth legions of interest groups at each stage of planning and implementation.

Initially, a state commission proposed reforming old laws in a manner that would reduce the penalties for prostitution and, in the interests of equality, make patronizing a prostitute a minor offense also. At public hearings on the proposed reform that followed, the American Social Health Association objected to reduced penalties in the interests of containing venereal disease, and police hesitated arresting patrons who were usually their only witnesses against the prostitutes. When the new law was implemented, particularly with police sweeps of New York City centers of action, hotels and businesses complained about harassment of their patrons, police and politicians lauded their own clean-up campaign, the Civil Liberties Union protested over innocent women caught in the dragnet only because of their stylish dress, judges and police quarreled over the politically mandated number of arrests that crowded busy courts and jails, the Legal Aid Society objected to racial and sex bias in prostitute arrests and lax enforcement of the law for patrons, and classy prostitutes remained indifferent because they never worked the streets where enforcement was focused. Negotiation that attempted to accommodate all those interests capable of exerting power at some stage in the process led to stalemate. In the end, the reforms were retained but enforcement lapsed.

The lesson to be drawn from New York's engagement with prostitution is that social control (and law) is a highly politicized process. Theoretical limits to the definition and sanction of deviance are drawn by wide-ranging interests, and very practical limits are set by those charged with enforcement in an atmosphere of conflicting pressures. The usual result in such cases of ambiguous moral authority is a deterioration of control. Moreover, what is demonstrated here in a minor instance of vice control applies broadly to other victimless crimes and controversial policies whether they concern the environment or social welfare.

Noncompliance with the designs of social control therefore may originate with the people, as in the Vietnam experience, or with the controllers themselves. These two overlap or converge. Vietnam certainly produced deep splits between Congress and the administration, just as competing social-control agencies in New York responded to different organized segments of the public. The distinction is useful, however, for illuminating the diverse sources and pervasiveness of noncompliance.

Resistance

Resistance is another matter, differing from noncompliance in deliberate efforts to escape or repudiate the clutches of social control. I use the words *escape* or *repudiate* advisedly because resistance, too, comes in many shades of passivity and activity, with the former doubtless more common. During World War II, the French maintained an extensive underground resistance movement to Nazi occupation. More recently, James Scott (1985) has described how Asian peasants practice various forms of "everyday resistance" to landlord exploitation in the form of foot dragging, sabotage, gossip, and ridicule. Resistance as it is used here includes all these forms and is characterized by persistent opposition short of open rebellion.

Passive resistance to social control as a group phenomenon generally takes the form of some alternative society — religious communities such as Quakers, farming cooperatives, hippie communes, Bohemian villages, and secret societies (from Freemasons to Communist party cells). It is no accident that wartime resistance movements, such as the French in World War II, employ and re-create these types of secret organization. The underground during Nazi occupation found moral and organizational support in alternative structures such as the radical political parties, secret societies, the church, and the independent peasantry. The same is true with more contemporary underground networks of U.S. political radicals (Weathermen) and humanitarians assisting Central American refugees (Sanctuary).

Although alternative societies have a long pedigree, in most historical periods they embrace a minute segment of the population. Judged by such numbers, passive resistance to social control would be rare indeed. Yet there is a much larger buffer zone between alternative and straight societies. Large numbers of relatively conventional people engage in denatured versions of radical nonconformity. David Matza (1961) makes this point in a memorable discussion of the "subterranean traditions of youth." Pure forms practiced by extremist minorities include delinquency, radicalism, and bohemianism, but each is paralleled by "conventional versions of these traditions . . . experienced by broad segments of the youthful population" (p. 105).

> Conventional versions are reasonable facsimiles of subterranean traditions in which their most offensive features are stripped away or tempered. . . . this is not by design, but as a result of emergent syntheses of conventional and rebellious sentiments or as a consequence of the fortuitous existence of independent traditions. A conventional version of the delinquency tradition is what has come to be called teen-age culture. . . . A conventional version of the radical tradition may be found in the long-standing American posture of "doing good." (Matza 1961, 105, 116–17)

Matza's point is that youthful resistance to the dominant culture comes in several forms and that together they signify much more disaffection than would be indicated by open rebellion alone. Events of the last twenty years seem to confirm and extend the argument. From the late 1960s onward, a counterculture attracted a broad following to the clash between an efficient yet value-barren technocratic society and its growing youthful opposition. Theodore Roszak observes that "no analysis seems to make sense of the major political upheavals of the [1960s] decade other than that which pits a militant minority of dissenting youth against the sluggish consensus-and-coalition politics of their middle-class elders" (1969, 2). Illuminating as it was, the counterculture beacon began to fade in the 1970s with the end of the Vietnam War (and military draft) and a return to normalcy. The generational revolt that Roszak expected to peak about 1984 looked very different when the fateful date arrived. Yet the "emergent synthesis of conventional and rebellious traditions" (p. 116) that Matza spoke of continues today in a deeply divided world of youth and adulthood.

The very word *youth* has replaced the transitional notion of adolescence and come to signify a distinct subculture with its own values, troubles, and entertainments. The world of rock, heavy metal, punk, and new wave music, although tightly controlled by the major recording companies (as subsidiaries of other major broadcasting and entertainment industries), is nevertheless foreign to most of adult society. Problems of "discipline in the classroom" may stem less from the possibility that today's youth is more perverse than usual than from an increasingly competitive and uncertain occupational future. As the connection between education and available good jobs becomes murky, instruments of social control once relied on by the schools lose their potency. If Émile Durkheim was right, and it seems he was, the increasing rate of teenage suicide points to a heightened sense of normlessness that is troubling to many young people.

Finally, resistance to social control takes active forms that go far beyond subcultural alternatives. Protest movements once focused on the Vietnam War, civil rights, and the environment persist, addressing new objectives such as nuclear disarmament and peace in Central America. Antiabortion forces have grown to a sprawling social movement engaged in actions as diverse as prayer vigils and abortion clinic bombings. Today's technocratic society spawns a new kind of computerized rebellion: break-ins using personal computers to tap the secrets of U.S. defense establishments, research hospitals, and banks. Although innocent in many cases, even careerism in others, some young rebels also recognize the new vulnerability of a society increasingly dependent on information that can be freely appropriated. As a one-time political activist turned software engineer explains,

> Information is becoming our most valuable commodity, and yet electronic information is being housed in a technology that makes free dispersal of it very easy. So while the establishment is pouring millions into perfecting its computer technology, it is undermining its centralizing power. I hope the establishment realizes it is being revolutionized. (*Wall Street Journal,* October 1, 1983)

Just as outlaws once enjoyed public sympathy for robbing the rate-gouging railroads or mortgage-foreclosing banks, certain indisputable crimes against powerful institutions are considered just or clever today. Donald Black draws an even broader lesson from the past: "Much of the conduct classified as crime in modern societies such as the United States is similar to . . . traditional modes of social control and may properly be understood as self-help" (1983, 34). Black has in mind a set of defensive and rebellious acts intended to redress wrongs: robbery as a form of debt collection, assault in self-defense, homicide in response to adultery or seduction into drug addiction, vandalism (spray painting) as protest, looting as retribution for merchant exploitation, and so forth. Whatever the consensus on whether these acts are justified, a good deal of nonpredatory crime is at least intended to right wrongs. It is social control in Black's view, or certainly social control from below. In many instances, it is also active resistance to institutions that monopolize legitimate social control.

In summary, noncompliance and resistance to social control are far more extensive than normally appreciated. They appear in unexpected forms: behind common instances of public opinion about war, in lax law enforcement, youth disaffection, and a variety of justifiable or mitigated crimes. Whatever we may think about these actions, their considerable analytic importance lies in demonstrating real limits on the scope of official social control. Viewed in this light, the degree of legal and moral conformity (or value integration) in society is far less than conventional social theories claim.

Finally, noncompliance and resistance are based on more than public opinion, conflicting social-control agencies, subcultural disaffection, or any other set of restraints on social control. The limits of social control are a matter of more than imperfect coverage or incomplete value integration. There is also a positive right to dissent, a cultural and political basis precisely for questioning the legitimate limits and the moral authority of social control and its official agents. The right to dispute social control is also a value.

Henry David Thoreau wrote,

> I think that we should be men first, and subjects afterward. It is not desirable to cultivate a respect for the law, so much as for the right.

> The only obligation which I have a right to assume, is to do at any time what I think is right. . . . How does it become a man to behave toward this American government to-day? I answer that he cannot without disgrace be associated with it. I cannot for an instant recognize that political organization as *my* government which is the *slave's* government also. ([1849] 1960, 223, 224)

Thoreau's essay later became a basic text for Ghandi's anticolonial struggle in India as well as for the civil rights and antiwar movements in the United States of the 1960s. It inspired people to claim a right that superseded the existing premises of social control. In that sense, of course, Thoreau and other writers succeed by articulating deeper cultural values.

Yet the connection between moral acts of civil disobedience and social movements of resistance is complex. Resistance movements have broader and more pragmatic causes and may even embrace sentiments hostile to the lofty principles of civil disobedience. Thoreau's objections to slavery were not among the social forces leading to a civil war that freed the slaves. Civil disobedience helps spark rebellions and provide them with important symbols. Whether they catch fire and how hot they get depends on the social kindling. That is the final consideration.

RIGHTS AND REBELLION

Rebellions are organized, sustained, and violent struggles by groups that refuse to continue living under their present conditions. The discussion that follows uses rebellion and revolt interchangeably. Organization, persistence, and a clear purpose distinguish rebellions from riots, which are usually short and spontaneous. Unlike protests, which are typically nonviolent and brief, rebellions are sustained and usually result in violent conflict, even when that is not part of the plan — for example, as a result of official retaliation.

Rebellion issues from ordinary forms of dissent rather than from rare and monumental causes or sudden traumatic changes. The causes of rebellion lie in normal group politics. More precisely, they arise in conflicts between injured groups and social-control agencies that become irreconcilable within routine politics. Indeed, rebellions usually begin when dominant groups and their social-control agencies — their guards, police, or armies — take some coercive action against dissident groups in an effort to finally solve a problem or transform a situation troublesome to the dominant groups themselves. Contrary to conventional thinking, rebellions seldom originate with heroic rebels who decide to take on the established order or to make a revolution. Insurgents get such ideas only when concrete exploitative conditions in some sense force them to do battle. Nobody starts a rebellion for the romance of it all, or even to create a just society as opposed to eliminating injustices.

This interpretation of rebellion is drawn from a variety of sociological studies devoted to protest movements, strikes, revolutions, and, generally, to the subject of collective violence. It comes also from attempts to test popular theories about when and why social groups rebel. Some of those are based on common sense. Simplifying somewhat, these theories may be divided into three categories claiming, respectively, that rebellion is caused by (1) the commonsense of **immiseration**, the idea that desperation and rebelliousness grow apace of misery and, particularly, economic hardship; (2) the counterintuitive hypothesis of **relative deprivation**, or rising expectations; and (3) the recent synthesis of **moral and political economy** characterized in the preceding paragraph.

Immiseration, or hardship interpretations, are based on the seemingly obvious proposition that rebelliousness is directly correlated with mounting inequalities. People endure exploitation up to a point at which they can't take it anymore. Rebellion occurs when conditions become so bad that starvation, slavery, a way of life, or life itself is in the balance. The best that can be said about this idea is that it receives little factual support. The worst of times are often the most socially calm.

Second, and conversely, relative deprivation reasons that rebellion occurs when formerly miserable conditions are improving. Alexis de Tocqueville studied life under the old regime overthrown by the French Revolution and concluded that "the French found their position the more intolerable the better it became" (quoted in Davies 1962, 6). The plausible idea here was that as bad conditions improved, people acquired both a sense of what they had suffered and a hope that all suffering might be eliminated. Karl Marx advanced a similar, if grimmer, prediction that the revolution will come with the "relative immiseration" of the working class — not necessarily a growing absolute poverty but a declining share in the total wealth of society, even though that wealth may be increasing. Growing inequality attacks even the higher classes, reduces many to the ranks of workers, and spreads a revolutionary spirit to the majority.

The ideas of de Tocqueville and Marx have been reformulated to fit a variety of modern forms of social protest according to the theory of relative deprivation, or rising expectations. People rebel — according to one exponent of this idea, James Davies — when an intolerable gap appears between what they want and what they get, "when a prolonged period of objective economic and social development is followed by a short period of sharp reversal. People then subjectively fear that ground gained with great effort will be quite lost; their mood becomes revolutionary" (1962, 5). Although, once more, the idea is intuitively appealing, it receives little research support.

The third view is broader and more historical. It rejects the simple idea that revolt and collective violence can be explained solely or even principally by material conditions. Causes must be sought in the moral order that justifies rebellion and in the political order where protesting groups and the state interact. Protest and rebellion require the interaction of at least these two parties, each with its own motives for pursuing the conflict. David Snyder and Charles Tilly (1972), for example, find no historical connection between economic hardship and social protest. Collective violence does correlate, however, with major periods of political activity including occasions of government repression. Economic conditions clearly play a contributing part in rebellion, along with the structures of class and status in which they are embedded. Particularly, they mobilize groups when an alternative to hardship appears. But alone they are insufficient to explain when and how protest makes its appearance. The most promising alternatives lie in the analysis of struggles for power. In other words, rebellion arises in struggles over social control and represents the extreme at which conflicting values defy coexistence or negotiated compromise. At bottom, rebellions are always about claims of injustice, sharply opposing values, and political tactics — about the moral and political economy.

Some examples help clarify these points in two ways. They allow, first, an evaluation of the theories presented. Second, they make the thematic point of this discussion by showing how rebellion occurs, under specifiable conditions, as an extreme recourse for solving persistent conflicts over social control. In that sense, rebellion does not differ from noncompliance and resistance because of the issues that prompt it but in the circumstances that prevent a simpler solution. To provide some scope for this interpretation, the following illustrations are chosen from such diverse settings as the total institution, community, and society.

Rebellion in the Institution

The sociological laboratory is a useful tool for analyzing elementary relationships such as that between rebellion and social control. We may examine, for example, a setting in which social control seems so complete that revolt is impossible. If, despite all odds, rebellion does occur in that place, it probably reveals something quite essential. This is the elegant point of Gresham Sykes's (1965) book that argues that even in the most formidable institution of social control, the maximum security prison, ruling groups do not enjoy total power or defenseless subjects.

Prisons have this in common with any society: They must maintain order and get work done, if only to complete peacefully their daily routines. To the extent that prisoners can withdraw their cooperation by refusing to work or to act peacefully, they have some power because their custodians' success depends on their compliance. This implies that the prison must provide inmates with certain incentives to cooperate, what Sykes calls the institutional reward system: freedom from close supervision, immorality, a blind eye to such practices as the contraband trade in narcotics, gambling, and home brewing. It is mainly by manipulating these small privileges, rather than by force, that the power of custodial rulers is peacefully maintained.

The only hitch in this delicate balance of power is that it relies on corruption from the standpoint of strict prison management — the official rules that are bent precisely to give bargaining power to the rulers and the ruled. The prison runs smoothly only by creating incentives for cooperation based on *negotiated rules*, including clear violations of official rules. Sykes insightfully labels this fact the "defects of total power" (1965, 40ff.). Thus, the New Jersey State Prison functioned, corrupt but peacefully efficient, until reformers came along intent on restoring the system to sound management. The result was a destructive prison riot that had all the characteristics of rebellion — that is, duration, organization, and purpose. The rebellion, following on reforms aimed at eliminating corruption, revealed

> a basic paradox: The system breeds rebellions by attempting to enforce the system's rules. . . . The effort of the custodians to "tighten up" the prison undermines the cohesive forces at work in the inmate population and it is these forces which play a critical part in keeping the society of the prison on an even keel. (Sykes 1965, 124)

To place this interpretation in a broader context, the prisoners did not rebel because they were put away and suffered the pains of imprisonment. Instead, from necessity they developed a social order and a culture for managing their unhappy plight. They did rebel when their means for coping with life were threatened, when the "rights" they had struggled to secure were taken away. They rebelled because the moral economy was upset. Conversely, the "technology of state power"— of which Foucault makes so much in the precise case of the prison — also has its defects, its material and cultural limits beyond which control crumbles into revolt. Social control and rebellion exist in interaction, the potential, nature, and limits of each defining the other.

Community Revolt

What is true of social order and the roots of rebellion in prison applies as well to communities and societies. Another modern sociological classic dealing with this question is Liston Pope's classic study *Millhands and Preachers: A Study of Gastonia* (1942). Here in the North Carolina town of Gastonia, a textile-manufacturing center was developed in the late nineteenth century in an industrious, God-fearing, and notably paternalistic setting. Workers came out of the southern mountains to take places in the factories, mill-sympathizing churches, and company-provided houses, acquiring in the transition an appreciation for their menial jobs and patrons' favors. "Up to 1929 the political and moral rights of the manufacturer to dominate the entire community had hardly been questioned" (p. 214). Yet on April 1, 1929, a bitter, "communist-led strike" erupted that shattered the moral order of the industrial community:

> Gastonia was catapulted from obscurity into headlines around the world. In rapid succession metropolitan newspapers told their readers of a communist uprising in the little textile town, of mass parades and the use of troops, of mob scenes and the destruction of life and property. Outbreaks of violence flared for several weeks, culminating in the death of the chief of police and the wounding of several of his officers in an armed skirmish with communists and strikers. Scores of arrests followed, and two dramatic trials focused the spotlight of publicity on a tense courtroom for additional weeks. At last seven men were convicted of murder and sentenced to prison, only to escape their sentences by flight to Russia. (1942, 3)

What happened? Pope reviews the standard explanations such as low wages and worker grievances, industrial competition, and outside agitators but finds none of those satisfactory. What had been happening in Gastonia over a period of years was a reorganization of the community, amounting to a social and cultural crisis. As prosperity in the textile industry began to decline after World War I, the companies looked for ways to reduce labor costs. Wages were cut. A stretch-out system of mechanization was adopted in the mills, requiring workers to operate several machines at once and deliver more labor for less pay than previously. Lower wages, however, did not incite the strike "nor was the fact of a stretch-out of supreme importance" (1942, 231). Rather, the arbitrary and impersonal methods of management were a basic cause, particularly as cost-cutting methods unwittingly attacked the tissue of community order.

Industry's social-welfare paternalism was changed from a system designed to promote loyalty to one preoccupied with labor discipline. "Established residents of the community and leading church members were dismissed under the stretch-out . . . and their places were taken by cheaper workers from outside" (1942, 232). A community integrated around traditional factory methods and their reflection in a company town was converted to a collection of disconnected individuals to whom the appeals of militant unionism made more sense as a substitute:

> The deeper causes of strife were not simply economic but cultural in a broader sense. Alien methods of industrial relations had been introduced into an industrial situation which had come to depend, for unity and peace, on paternal relations, and to expect emphasis on community welfare rather than on productive efficiency alone. (1942, 232–33)

National Revolts

These insights about rebellion gleaned from total institutions and communities may be generalized to societies in a manner that brings together many of the themes developed in previous chapters.

Revolutionary movements in the Third World arise in conditions of economic underdevelopment and social inequality: commercial agriculture that dispossesses peasants, urban migration to an overcrowded service economy, multinational exploitation of labor, and growing unemployment (see Chapter 7). Yet, again, these conditions provide the backdrop to revolutionary movements that spring more immediately from social dislocations, cultural judgments of injustice, and political openings. The social revolution in El Salvador began in 1980 when the U.S. government and local elites instituted a series of reform measures in the agricultural and export economy. The reforms were designed to avoid "another Nicaragua" in this neighboring country. The effect of the reforms, however, was to impoverish more peasants and antagonize the right-wing coffee planters and industrialists. Moderate politicians were driven out of a coalitional government, by the army and its death squads, and into a revolutionary movement. The rebels never started out to make a revolution but, on the contrary, pursued legitimate methods for achieving equality that aroused official violence.

The successful revolution of 1979 in Nicaragua was the culmination of similar processes begun much earlier. The United States occupied Nicaragua with a military force in 1912 and remained there almost continuously until 1932 when it installed the dictatorship of Anastasio Samoza and trained his national guard. Peasant rebels who had fought the U.S. occupation under the leadership of Augustin Sandino were suppressed by the new regime. Indeed, Samoza repressed all elements of the society with the exception of his private army and close business associates. He turned the country into his own "family state" (G. Black 1981, 34–38). Between Anastasio, Sr., and his son who succeeded him, the Samoza family came to own much of the national economy: fifty-one cattle ranches, forty-six coffee farms, eight sugar plantations, the national railroad and airlines, the principal port facilities, the most lucrative import franchises (e.g., Mercedes Benz), mines and timber industries, banks, radio and television stations, and a host of businesses from insurance companies to houses of prostitution — some of this empire financed with U.S. aid money.

Beginning in the early 1960s, a band of student revolutionaries renewed Sandino's struggle for national liberation but met with little success as a guerilla force in the mountains against Samoza's national guard. By the 1970s, however, and particularly with the Carter administration that withdrew support from the dictator, the regime began to falter. It was not the new Sandinistas alone who made the difference. The middle class and business interests outside Samoza's group had suffered and made common cause with the revolutionary movement. A major earthquake in the capital city of Managua had ruined small businesses, and the dictator refused to rebuild the city, preferring to construct a new one on land he owned and could profit from in speculation. By the summer of 1979 when the revolution succeeded, estimates suggest that perhaps 90 percent of the population was opposed to the regime. The Sandinistas provided much of the moral energy in the movement's goal to regain national sovereignty from a regime maintained by Yankee imperialism. Hard times helped mobilize this broad front of national opposition. But the turning point came with the political opening that was a joint product of the new revolutionary coalition and fading support for Samoza from the United States.

The road to successful revolution is a treacherous one. Political movements that begin with reformist aims, only to meet state repression, do not regularly thrive until a political opening comes along. South Africa's African National Congress (ANC) is a choice example. Beginning in 1912 as a moderate civil rights organization, the ANC grew steadily through the years as the Dutch colonial republics formed a Union of South Africa independent of Great Britain (1931) and the Afrikaans Nationalist Party gained control of the government (1948), which it retains today. From 1948 onward, the present system of apartheid (an Afrikaans word meaning *separatehood*) was implemented with the objective of making South Africa "a white man's land" in which the white 15 percent of the population has complete political domination over the noncitizen majority of black (70 percent), mixed-race "colored," and Asian population. The ANC opposed this deepening racism, particularly its laws that confined nonwhites to segregated Native Lands and required that special passes be carried by those who were working or traveling outside their restricted areas. In the 1950s, growing ANC militance demanded an end to segregation with demonstrations, strikes, boycotts, and legal challenges. The first great violent conflict of the modern era came in 1960 at Sharpville near Johannesburg where police fired on crowds protesting the pass laws, killing sixty-nine people. The ANC and a more militant offshoot, the Pan-African Congress (PAC), were banned and their leaders (including Nelson Mandela, Oliver Tambo, and Govan Mbeki) were subsequently arrested and confined at Robben Island prison.

Repression did not end the civil rights struggle. In fact, the racist government faced a dilemma, which continues to make matters worse. In an industrialized country where the nonwhite majority comprises 85 percent of the population yet occupies just one-third of its poorest and segregated land area, the system is vitally dependent on the black labor force and consumer market. To solve this problem within the framework of white supremacy, the government has created a more oppressive apparatus for segregating blacks into their own townships and bogus independent homelands. To speed the process, a Group Areas Act allows the government to destroy traditional black communities — "black spots" typically earmarked for white housing development — and to move the nonwhite population to barren lands, which are now arbitrarily designated as their true homelands, far distant from the urban centers. The mythology supposes that these "independent republics" run by black puppets of apartheid will one day be separate nations, although they have no economic base or recognized authority. On the contrary, the homelands survive only on the wages that black workers earn in the mines and urban industries essential to the South African economy. The heavy burden that racism imposes on blacks and, irrationally, on the white economy appears in the problem of labor supply. Black workers must either live outside their so-called homelands in camps convenient to jobs as permitted with special passes or make long bus commutes back-and-forth each day. Indeed, to make the artificial system work, the government spends millions each year subsidizing an elaborate arrangement of forced busing in which black workers often must leave the homeland for their urban job at 3 or 4 AM and return late at night, spending as many hours on the bus as at work (Lelyveld 1985).

Resistance to apartheid from many quarters escalated in the 1970s and 1980s. Soweto, another Johannesburg township, succeeded Sharpville as a symbol of bloody repression in 1976 when police killed 600 blacks protesting forced teaching of the Afrikaans language in schools of South Africa's largest black settlement. Steve Biko, leader of a new black consciousness movement was murdered in police custody in 1977, a fate shared by countless other victims of arbitrary arrest and torture. In the 1980s, the government moved to quiet opposition with reforms, most of them meaningless, such as powerless political representatives in the homelands and a colored parliament. Perhaps in a miscalculation, however, black trade unions have been allowed. The internal opposition to apartheid now concentrates in two places, the unions relatively unified in *Cosatu* (Congress of South African Trade Unions) and in the unemployed militant youth of the townships. From abroad the banned ANC has turned to guerrilla war tactics from its base in Angola, and a number of multinational companies and conservative financiers have pulled their investments out of the country.

In the late 1980s under the government of F. W. de Klerk, the South African state finally began to respond to international pressure and entertain the prospect of black participation in democratic rule. Nelson Mandela, the imprisoned ANC leader, was released, and serious negotiations concerning the structure of an interim government and multiracial legislature were started. At the same time, however, communal violence among blacks flared, some of it inspired by the government and some of it generated by conflicts between black tribal and political factions. Mandela has not always been able to control ANC rivals, and de Klerk was under attack from the right. Democracy in South Africa is still in the future, and peace is not yet imaginable. But, like so many other struggles for freedom, including those in the United States, the South African movement is irreversible.

CONCLUSION

Rebellion is not the reverse of social order, not the world of social control turned upside down. Rather, it is an essential feature of the negotiated social order, a last-ditch response or defensively militant effort by exploited groups to change burdensome systems of social control. Rebellion therefore is a desperate extension of the more common forms of noncompliance and resistance. The several theories that endeavor to explain it capture parts of the complex pattern of causation. Hardship is important in the sense of identifying objective conditions, and relative deprivation makes it clear that group perceptions of those conditions and their alternatives are keys to mobilization. A complete explanation, however, depends on understanding aggrieved groups and social-control agencies as they interact politically over conflicting moral claims. Rebellion and dissent are mistakenly considered rare. They lurk in the moral order, popular culture, and sundry forms of deviance. They limit the reach of social control and periodically make a fight over what some people believe is rightfully theirs.

Repeatedly in previous chapters, we have encountered protest movements that challenge the prevailing tenets of social control: urban riots aimed at police, merchants, and unemployment; farm workers marching for unionism under the galvanizing symbols of ethnicity and religion; blacks and women organizing around the rightful claims of those relegated to the lower ranks of employment; populists who saw their rural way of life torn away by government-supported agribusiness; young people who read Thoreau and resisted going to an ill-conceived war; and older people who mourned for many of their children who did go.

In all of these cases, the agencies of social control did two things. First, they urged that people stay in their place: women in the home, blacks in the lower and unemployed ranks of labor, Mexicans in the fields, students in the classroom and the draft pool. Second, they labeled as deviant those people who challenged the established order: women's libbers, black revolutionaries, hippies, Communists — labels designed to belittle dissidents and presume silliness or malice in their motives. In these familiar examples it is the tenets of social control that look silly or mean-spirited today. Yet there are countless other instances of social control where the rebels failed or where few would doubt the legitimacy of social prohibitions and the depravity of their violators (e.g., agrarian radicals, welfare and convict rights advocates, predatory criminals, traitors).

Two points emphasized in this discussion help explain the difference. First, deviance is always arbitrary in the specific sense that it relies on group support and conforms to changing social definitions. What may be understood today as legitimate dissent or tragic illness was often, even recently, considered heinous perversion (criminal insanity or pacifism in World War I). Some of today's depravities are doubtless headed for reappraisal with advances in medical and political knowledge. Second, the process in which those definitions are maintained or altered is in the broadest sense a political struggle. Neither of these statements implies "moral relativism" or the notion that, because it changes, the moral order is frivolous. On the contrary, because it is socially determined and changes with emergent needs, the moral order mediates our social and material existence.

These observations summarize the interplay of social control and dissent, the social consensus and struggle over tolerable conduct. But there is a more profound lesson in the conflict over social control. Laws, courts, agencies, and the instruments of control generally are designed to protect the state and the social order of class and status on which the state rests. Yet the same instruments that support this order are avaliable to and used by disadvantaged groups to defend their beleaguered rights. The so-called tools of the establishment are sometimes converted to undoing or transforming the social order. The civil rights movement used the streets, the vulnerabilities of political parties, and the courts. Nicaragua's revolution resulted from the very greed of the dictatorship, and the South African regime is threatened because it must rely on black labor and implausible racial doctrines that support its regulation. Social change happens dialectically or, more simply, through the tendency for one day's orderly solutions to be followed by the next day's changes that contradict or undermine earlier intents while serving new ones. The comfortable racism of one era, for example, becomes the political burden and moral energizer of another. It is for this reason that social change moves fitfully in unexpected directions. Thompson's study of the Black Act concludes with an observation about law and class power that applies as well to social control and the interests of class, status, and the state. Thompson says it is not true that "law equals class power," but

> on the one hand it is true that the law did mediate existent class relations to the advantage of the rulers. . . . On the other hand, the law mediated these class relations through legal forms that imposed, again and again, inhibitions upon the actions of the rulers. . . . in certain limited areas, the law itself [becomes] a genuine forum within which certain kinds of class conflict were fought out. (1975, 264)

If the conclusion is not simple, neither is it mysterious. Social control is power, not total power but a fair assembly of society's dominant class and status group interests. At the same time, control is always limited by noncompliance, resistance, and rebellion and, ultimately, by the cultural values that support them. Sometimes social control coerces with force, and on rarer occasions it collapses under rebellious siege or sheer desertion of the faithful. More often it tolerates and negotiates within rule-governed limits. In that case, the rules shift from authoritative do's and don'ts to a forum of strategies for struggle and change. Recognizing, finally, that all these things may go on simultaneously across interrelated arenas of social control — where gains in civil rights may facilitate antiwar efforts and both reinforce the defense of class privilege — we appreciate the complexity of social change but also see clearly its locus.

At bottom, societies are therefore neither integrated nor especially orderly — except, perhaps, in their persistent struggles that transform the prevailing order to someone's advantage. Societies are forever changing, and it is within the conflicts over the direction of change that societies reveal their structure.

SELECTED BIBLIOGRAPHY

BECKER, HOWARD S. 1975. *Outsiders: Studies in the Sociology of Deviance.* Glencoe, IL: Free Press.

BELL, DANIEL. 1961. "Crime as an American Way of Life." Pp. 127–50 in *The End of Ideology: On the Exhaustion of Political Ideas in the Fifties.* New York: Collier Books.

BLACK, DONALD. 1983. "Crime as Social Control." *American Sociological Review* 48:34–45.

BLACK, GEORGE. 1981. *Triumph of the People: The Sandinista Revolution in Nicaragua.* London: Zed Press.

BLUMER, HERBERT. 1962. "Society as Symbolic Interaction." Pp. 179–92 in *Human Behavior and Social Processes: An Interactionist Approach,* edited by Arnold M. Rose. Boston: Houghton Mifflin.

BULLOCK, HENRY ALLEN. 1961. "Significance of the Racial Factor in the Length of Prison Sentences." *Journal of Criminal Law, Criminology, and Police Science* 52:411–17.

CAUTE, DAVID. 1978. *The Great Fear: The Anti-Communist Purge Under Truman and Eisenhower.* New York: Simon & Schuster.

CLINARD, MARSHALL B. and PETER C. YEAGER. 1978. "Corporate Crime." *Criminology* 16:255–72.

DAVIES, JAMES C. 1962. "Toward a Theory of Revolution." *American Sociological Review* 6:5–19.

ERIKSON, KAI T. 1966. *Wayward Puritans: A Study in the Sociology of Deviance.* New York: Wiley.

FOUCAULT, MICHEL. 1979. *Discipline and Punishment: The Birth of the Prison.* New York: Vintage Books.

KATZ, JACK. 1979. "Legality and Equality: Plea Bargaining in the Prosecution of White-Collar Crime and Common Crimes." *Law and Society Review* 13:431–59.

———. 1980. "The Social Movement Against White-Collar Crime." Pp. 161–84 in *Criminology Review Yearbook,* edited by Egon Bittner and Sheldon Messinger. Beverly Hills, CA: Sage.

KITSUSE, JOHN I. 1962. "Societal Reaction to Deviant Behavior: Problems of Theory and Method." *Social Problems* 9:247–56.

LELYVELD, JOSEPH. 1985. *Move Your Shadow: South Africa, Black and White.* New York: Viking Penguin.

MATZA, DAVID. 1961. "Subterranean Traditions of Youth." *The Annals* 338:102–18.

PETERSON, RUTH D. and JOHN HAGAN. 1984. "Changing Concepts of Race: Towards an Account of Anomalous Findings of Sentencing Research." *American Sociological Review* 49:56–70.

POPE, LISTON. 1942. *Millhands and Preachers: A Study of Gastonia.* New Haven, CT: Yale University Press.

ROBY, PAMELA A. 1969. "Politics and Criminal Law: Revision of the New York State Penal Law on Prostitution." *Social Problems* 17:83–109.

ROSZAK, THEODORE. 1969. *The Making of a Counter Culture: Reflections on the Technocratic Society and Its Youthful Opposition.* Garden City, NY: Anchor Books.

SCHUMAN, HOWARD. 1972. "Two Sources of Antiwar Sentiment in America." *American Journal of Sociology* 78:513–36.

SCOTT, JAMES C. 1985. *Weapons of the Weak: Everyday Forms of Peasant Resistance.* New Haven, CT: Yale University Press.

SNIDER, LAUREEN. 1982. "Traditional and Corporate Theft: A Comparison of Sanctions." Pp. 235–58 in *White-Collar and Economic Crime,* edited by Peter Wickman and Timothy Dailey. Lexington, MA: Heath.

SNYDER, DAVID and CHARLES TILLY. 1972. "Hardship and Collective Violence in France: 1830 to 1960." *American Sociological Review* 37:520–32.

SUTHERLAND, EDWIN H. 1940. "White Collar Criminality." *American Sociological Review* 5:1–12.

———. 1949. *White Collar Crime.* New York: Dryden Press.

———. 1956. "Crime of Corporations." Reprint. In *White Collar Criminal: The Offender in Business and the Professions,* edited by Gilbert Geis. New York: Atherton Press, 1968.

SYKES, GRESHAM. 1965. *The Society of Captives: A Study of a Maximum Security Prison.* New York: Atheneum.

THOMPSON, E. P. 1975. *Whigs and Hunters: The Origin of the Black Act.* New York: Pantheon Books.

THOREAU, HENRY DAVID. 1849. *On the Duty of Civil Disobedience.* Reprint. New York: New American Library, Signet Classics Edition, 1960.

PART IV

CONCLUSION

9

Sociology and Civil Society

Poland's free trade union movement Solidarity was established in 1980 when workers locked themselves into the Gdansk shipyard, demanding official recognition and wage increases. Although banned in 1981, Solidarity continued its struggle, and in 1989 the government restored Solidarity's legal status, granted the first free elections since World War II, and agreed to improve the standard of living. (Reuters/Bettmann newsphotos)

SOLIDARITY

In 1980 Polish workers arose in a national movement of solidarity. They rose against a stagnant economy and escalating food prices, on one side, and against an oppressive Communist party-state on the other. Specifically, they began by resisting government reforms that pushed up food prices. Their movement, however, was more than a reaction to hard times and a lack of freedom. Throughout the 1970s, those two grievances had prompted rebellion. The great difference in 1980 was that industrial workers, farmers, students and professors, miners, and state employees united under the banner of Solidarity — the organization and demand for independent trade unions. Workers in the port city of Gdansk took center stage in the uprising when they locked themselves into the Lenin shipyard and demanded recognition from the state. All over Poland, working people united behind the same objective; farmers formed a wing of Rural Solidarity, and the universities became a center for planning and even for social research on the unfolding movement. Solidarity's twenty-one demands went far beyond economic issues. They wanted independent trade unions, release of political prisoners, abolition of censorship, and free parliamentary elections.

Within eighteen months, the Polish government outlawed Solidarity, mobilized the army to suppress further strikes, and jailed thousands of activists. Yet the movement's magnetic appeal endured. In the summer of 1988, more than six years after the ban, workers were still striking and flaunting their allegiance to the people's movement. By the spring of 1989, the government was forced to reinstate Solidarity and schedule elections. On December 9, 1990, Solidarity leader Lech Walesa was elected president of Poland.

What is the meaning of Solidarity? What categories are we to use to analyze the action of people on their own behalf and authority? Adam Michnik, a Polish writer and historian twice imprisoned for his participation in Solidarity, explains its special significance:

> The essence of the spontaneously growing Independent and Self-governing Labor Union Solidarity lay in the restoration of social ties, self-organization aimed at guaranteeing the defense of labor,

> civil, and national rights. For the first time in the history of communist rule in Poland "civil society" was being restored and it was reaching a compromise with the state. (1985, 124)

Solidarity illustrates two points central to this chapter. First, it provides a case study of civil society, helping define that term. Second, it provides an introduction to the recent and monumental changes in Central Europe that have realigned today's world system. The theme of this chapter is that we need the concept civil society to understand these changes.

CIVIL SOCIETY

The concept civil society has a long history in Western philosophy, but only recently have social scientists begun to revive it as a way of formulating modern problems. Perhaps the most important stimulus for reintroducing civil society into sociological thinking is the writing of the Italian theorist Antonio Gramsci. Gramsci (1971) wrote in the 1930s when Italy was drifting into fascism, an all-intrusive corporate state. He feared the expanding power of the state, or "political society," and regarded civil society as the people's only defense against the narrow interests that controlled the state. Gramsci saw civil society as standing apart from both political society and the economy, which he described as "an ensemble of organisms commonly called private" (p. 12). Civil society's essential feature is that its authority derives from culture rather than laws of the market or government. Civil society "operates without 'sanctions' or compulsory 'obligations,' but nevertheless exerts a collective pressure and obtains objective results in the form of an evolution of customs, ways of thinking and acting, morality, etc." (p. 242). In a marvelous metaphor, Gramsci likened civil society to trenches on the battlefield:

> The superstructures of civil society are like the trench-systems of modern warfare. In war it would sometimes happen that a fierce artillery attack seemed to have destroyed the enemy's entire defense system, whereas in fact it had only destroyed the outer perimeter; and at the moment of their advance and attack the assailants would find themselves confronted by a line of defense which was still effective. (1971, 235)

Polish Solidarity is a perfect example of Gramsci's social trenches. The great working-class majority of Poland's population was attacked by the artillery of inflation, public debt, economic stagnation, an authoritarian state, and party management of their workplaces. But when the state advanced on the people with impoverishing economic reforms, it met the entrenched opposition of civil society — the independently organized forces of custom, national pride, workers' rights, and democracy.

In any society, there are realms of group life that people regard as their own — activities motivated not primarily by material ambition or the requirements of the state but by the hope of accomplishing some public good in collaboration with others. This is what goes on in churches, neighborhood associations, reform groups, and even recreational clubs where members share their concerns about public issues. If these activities are closely involved in the economy or penetrated by the state (as we saw in Chapter 6), they are nevertheless distinguished by their origins and purposes. **Civil society** is a useful concept for distinguishing voluntary, nonprofit, public action based on cultural standards rather than on principles of state obligation or market gain. Civil society is composed of organizations and actions whose primary purpose is to promote the public good, irrespective of the differences among them.

The concept civil society is not new to social theory. It was used by British philosophers in the eighteenth century to describe broadly the foundations of the state. Alexis de Tocqueville and Émile Durkheim used other terms (*associations* and *intermediary groups*) to describe the phenomenon of civil society. Tocqueville and Max Weber both toured the United States and analyzed its unique grass-roots organizational life in terms of voluntary associations and Protestant sects — two elements of civil society.

Modern sociology has neglected civil society until recently. As sociology developed a set of categories and theories with which to comprehend industrialization, urbanization, and bureaucratization including state expansion, the local world of civil society seemed less relevant to the sweep of change. Indeed, the concept civil society was superseded by more formalistic alternatives such as organizations and voluntary associations — often ones that were part of the state and economy rather than in a separate civil realm. The distinction that civil society had once drawn was dissolved. Whatever did not belong to state and economy, on one hand, or family, on the other, was residual, not worth bothering about. Particularly, it became an article of faith that collective action was generated from economic conflicts (Karl Marx) or state opportunities (Weber), and certainly not from people's associations (Tocqueville, Durkheim). The idea that major historical change could come from the actions of independent trade unions, students, farmers, and intellectuals gathered together in a civic union did not seem possible. But that was before 1989.

WHAT HAPPENED IN 1989?

The history of our time will refer to 1989 as a marker and symbol in much the same way as other histories use 1789 to stand for revolution, 1914 for world war, or 1929 for economic depression. In 1989, of course, the Soviet-backed socialist states of Central Europe fell to popular democratic movements, followed shortly in 1990 by the devolution of the Soviet Union (USSR) into separate republics also converted to democratization. The change came as a complete surprise to social scientists, foreign-area specialists, and intelligence experts. For all the effort and expense that went into analyses of the Communist bloc, few if any Western specialists understood in 1988 that the empire was about to fall. Why did it happen, why in 1989, and why all at once? Part of the answer lies in a set of historical forces that had been building for years, and part lies in the unexpectedly central role that civil society would play in the democratic revolution.

Central Europe and the USSR suffered from economic failures over the last two decades, but these were only limited failures when compared with other accomplishments. The Soviet brand of socialism succeeded at transforming a technologically backward and agrarian nation into a major industrial power in the first half of the twentieth century. From the Russian Revolution in 1917 to the world's first communications satellite (Sputnik) in 1957, the USSR created a complex of heavy industry (steel, machinery, power) capable of defeating Nazi Germany and carrying the USSR into the space age. Indeed, Soviet technology seemed to have surpassed U.S. technology. Sputnik shocked U.S. educators into a renewed emphasis on math and science and soon led to President Kennedy's promise to put a man on the moon. But Soviet socialism never succeeded at moving to the next stage of industrialization based on consumer goods and electronic and bio-technologies. The failure to develop a wide range of economical consumer goods became more troublesome as Cold War tensions eased and closer communications revealed to the socialist public that great disparities between its standard of living and Western experience existed. The socialist economies had not so much failed — certainly not by contrast to Third World conditions that Eastern Europe shared at the turn of the century — but, nevertheless, failed painfully in comparison with the West.

The key problem with Soviet socialism was that fundamental decisions about production and investment were based on political considerations rather than market pressures. Prices were fixed by administrative judgments rather than by popular preferences and availability. As a result, prices had no "real" meaning and, particularly, provided no measure of profitability with which to guide economic planning. When the system failed in agriculture, which was by nature decentralized and vulnerable to individual initiative (decollectivization), reform from the grass-roots through noncooperation and innovation was possible (Szelenyi 1987). However, to meet the limited and rapidly shifting demand, centralized industry was unable to adapt to the strategies of "flexible specialization" (Piore and Sabel 1984) employed in the West. The socialist model was least successful at producing precisely the kind of goods that had carried the West to a new plateau of industrial development and were in growing demand behind the protective barriers surrounding Eastern Europe. Like the Third World, socialist countries began to borrow heavily from Western banks and governments in the 1970s in an effort to modernize their production with the purchase of new technologies and to subsidize consumption of imported goods. From Central Europe to the USSR, countries began to run up large foreign debts.

To summarize the whole dilemma, the Stalinist systems were too rigid; managers resisted change, using political clout to force even greater investments into obsolete firms and production processes. In some cases, most notably Poland and Hungary, foreign loans started to be used simply to purchase consumer goods to make people happier, to shore up the crumbling legitimacy of regimes that had lost what youthful vigor they had once possessed and were now viewed simply as tools of a backward occupying power. This worked until the bills came due and prices had to be raised (Chirot 1991).

To pay its debt, Poland, for example, attempted to boost the volume of its exports and to earn hard currency for Polish hams and sausages, among other Western favorites. The result was a sharp decline in food supplies available for the home market and a rapid increase in prices. Worse yet, to raise money, the government diverted the most desirable goods (e.g., the best cuts of meat) to the "dollar shops," stores that accepted only foreign (hard) currencies that only privileged members of the Communist party or corrupt merchants were able to get. Shortages had the dual effect of creating want and exposing corruption. The moral economy was aroused. Food riots followed throughout Poland in 1980 and 1981, establishing the foundation of Solidarity and "the precipitant cause of the revolution" (Ash 1983, 33).

Besides these compelling economic conditions, the causes of 1989 were equally political and moral. Soviet socialism was based less on the elimination of capitalist inequality than on its replacement with a system of bureaucratic privilege maintained by the fusion of the state and Communist party — the "new class" as critics as far back as the 1950s had called it. Economic inefficiency was one side of this system, but perhaps the more objectionable side was official corruption. Government and political party elites enjoyed the key jobs, comfortable incomes, access to housing, a place in the countryside, and the prestigious opportunity to travel to the West. Corruption, of course, was key to the failure of economic reform because its many forms included overpaid managers, contract graft, smuggling, dispirited workers, and many more.

At bottom, corruption undermined state legitimacy and encouraged an opposition movement.

> The whole movement toward the creation of alternate social institutions, free of the corruption and dishonesty of the official structures, was the great ideological innovation of what began to emerge in Poland in the 1970s and 1980s as the movement toward the creation of a "civil society." Traditional revolutionary resistance . . . might be fruitless because [it] could bring down a heavy military intervention by the Soviets. But simply by beginning to turn away from the state, by refusing to take it seriously, Polish and then other Central European intellectuals exposed the shallowness of communism's claims and erased what little legitimacy Communist regimes still had. (Chirot 1991, 223)

Poland and Solidarity initiated a series of revolutionary changes that accelerated during the 1980s. When Solidarity strikes challenged the government, Soviet military intervention was threatened as the kind of decisive move that had suppressed previous rebellions: Hungary in 1956, Czechoslovakia in 1968, Poland in 1970. The late-1980s were different. The USSR was suffering its own economic trouble, trying to withdraw from its most recent military debacle in Afghanistan. Intervention in Eastern Europe was not in the cards. The satellites would be left to go their own way as attention turned to stabilizing the USSR. As Soviet Prime Minister Mikhail Gorbachev pressed for an economic and diplomatic opening to the West (*glasnost*), the first free elections in the socialist bloc were agreed to in Poland.

The promised restoration of democracy in Poland led next to a reform movement in Hungary. Communist party leaders met with dissident groups, and together they agreed to a plan for multiparty elections in October 1989. For public relations purposes during a late summer visit by President Bush, Hungarian leaders officially opened their border with Austria, although in fact the border had been crossed freely for some time. Coming at the end of the August vacation period in Europe, the public gesture encouraged thousands of East German tourists to escape by this convenient route to the West. As an estimated 100,000 East Germans fled the country, a democratic movement called New Forum was created within Hungary. New Forum organized a series of massive and successful demonstrations in Dresden, Leipzig, and East Berlin for popular elections.

Czechoslovakia was next with Civic Forum, a democratic movement mainly of intellectuals led by playwright Vaclav Havel, who became prime minister in subsequent elections. By now the Soviet-backed states had lost any claim to legitimacy or popular following. Czechoslovakia's Velvet Revolution was an overnight success.

> What was happening in Prague was that the crowd was coalescing into a political force. Unlike Poland, where Solidarity had forged programs and structures during nearly a decade of clandestine activities, Czechoslovakia had no mass dissident movement. Nor were the churches able to serve as sanctuaries for protest, as was true in both Poland and East Germany. Essentially there were only a handful of often-persecuted dissidents associated with groups. . . . And yet, when the dam of repression began to give way, these few people — a playwright, a priest, and a number of coal stokers — formed a political nucleus that very quickly attracted a mass of followers. (Gwertzman and Kaufman 1991, 305–306)

Romania was the last of the Soviet satellites to fall on Christmas Day 1989 when the dictator Nicolae Ceausescu was executed by a revolutionary crowd. The popular uprising began in the western city of Timisoara over the threatened exile of a dissident pastor. When crowds rallied to his support and the army split over orders to fire on the demonstration, the uprising spread across the country.

Finally, of course, the unified government of the fifteen Soviet republics collapsed in the summer of 1991 after seventy-four years as one of the world's superpowers. The last days of the USSR saw Prime Minister Gorbachev maneuvering to simultaneously revitalize the stagnant economy and reform the corrupt party while holding the republics in line. Gorbachev's political reform involved replacing the moribund party apparatus with new state machinery in the form of an elected Congress of People's Deputies composed of 2250 members, one-third of whom came from party-controlled organizations (Draper 1992). At bottom, the old party was too deeply rooted in all of the institutions of Soviet society. Gorbachev lacked the support or the institutional vehicle to realize his goal of creating a truly alternative government. The union simply evaporated.

Returning to the why questions posed at the beginning of this section, the democratic revolution in Eastern Europe and the USSR resulted from a combination of mounting economic failures and the mobilization of civil society around political and moral alternatives; it happened in 1989 because immediate problems of debt, exports, shortages, and protests led to reforms, which other countries adopted, in Poland and the USSR; and it happened all at once because the democratic movements correctly perceived that the USSR would not intervene to bolster illegitimate governments.

In a lucid treatment of these events, Daniel Chirot (1991) argues that the fundamental explanation for the revolutions of 1989 was moral — a mobilization of civil society in response to the corruption and lies of the party-state. Failing to appreciate the role of popular legitimacy and focusing instead on state elites and technical economic problems, social scientists missed what was happening until it was all over: "Western observers never grasped the significance of this creation of an alternative 'civil society' " (p. 231).

THE GLOBAL DEMOCRATIC TRANSITION

Often lost in the media attention to the "fall of communism," and partisan attempts to claim credit for it, is the worldwide transition from authoritarian and one-party states to multiparty democracies. The dramatic events of Eastern Europe in 1989 were part of an international movement sweeping Latin America, Africa, and parts of Asia at the same time. Although the movements in each country and continental region have their own specific causes, they are also influenced by a sea change in the deeper structure of the world system, a shift toward economic liberalization and political reform.

The extent of democratization over the last twenty years is remarkable. In the early 1970s, among nineteen important Latin American countries, only Colombia, Costa Rica, and Venezuela had competitive democratic systems. Today, virtually all do, and many of the transformations came in the mid-1980s (the return to elected civilian governments in Argentina, Brazil, and Chile and the beginnings of an effective opposition in Mexico). In Africa multiparty political reform movements appeared in more than twenty countries during the late 1980s and early 1990s. Important elections were held in Nigeria, Zambia, and Algeria (although in the latter, opposition gains were soon usurped). Promising democratic governments have taken office in Benin and Zambia, and a number of other countries such as Kenya are feeling the international pressure for change. Finally, democratic movements were established across Asia in the 1980s in South Korea, the Philippines, Thailand, Burma, and Nepal. What explains this worldwide coincidence?

Three general trends have combined at the global level to exert pressure and encouragement for democratization. First is the crisis of debt and economic stagnation in Eastern Europe and the Third World, discussed previously. On one hand, hard times have mobilized popular opposition groups. On the other hand, economic reforms urged by world organizations such as the International Monetary Fund (see Chapter 7) have reduced the resources available to authoritarian states that use patronage to purchase support.

Second, recent years have seen the acceleration of a process broadly termed "globalization," which includes international migration, cross-national investment, global cities (see Chapter 3), and greater importance of international trade. The stable operation of this international order favors democratic governments ruled by law and by market principles in politics as well as in the economy.

Third, democracy has become contagious. Just as the revolution of 1989 spread across Eastern Europe, so in Africa and Latin America opposition groups have been inspired by example — including the examples of Poland and East Germany. Contagion is more than a germ in the air. Organizations devoted to human rights have helped spread democratic aims and held states responsible for abuses in an international forum.

Turning from these very general global influences to concrete democratic movements, opposition typically emerges from civil society. The movements cited above have a common denominator; they appear in countries struggling with severe economic problems, center in urban areas, and mobilize a cross section of middle- and working-class interests. Two countries far removed from one another illustrate these propositions: Zambia in southern Africa and Mexico.

In October 1991, Frederick Chiluba was chosen to succeed Kenneth Kaunda, one of the grand old men of African political independence, in Zambia's first election in nineteen years:

> Mr. Chiluba, the 48-year-old chairman of the Zambia Council of Trade Unions, has an enthusiastic following among the large urban population — half of Zambia's people live in the cities, an unusually high proportion for Africa. City people appear to be disillusioned with low salaries, galloping inflation and an almost total breakdown in social services. . . . As the state-run economy has foundered, Mr. Kaunda has faced protests at home and condemnations from the World Bank and the International Monetary Fund. . . . The government legalized political party activity in December last year after an attempted coup and riots over increased food prices in mid-1990. But Mr. Kaunda refused to allow any new registration of voters after the Movement for Multiparty Democracy was established, thus preventing the party from capitalizing on much of its support [among new voters]. (*New York Times*, October 31, 1991)

Notice the elements here. The movement is urban-based, drawing recruits from the trade unions and youth. Austerity is a problem. Citizens are turning to electoral politics in the wake of riots. The multiparty movement is a new and independent initiative. With these factors in his favor, Mr. Chiluba won the election handily.

The opposition has not been so successful in Mexico. At the national level, the Institutional Revolutionary Party (PRI, its Spanish initials) has ruled as a one-party state since its inception in the early 1930s. During the last decade, things began to change. In 1982 Mexico became the most celebrated and defining case of the debt crisis. As a result, it received large amounts of new lending and debt relief, but those came at the price of thoroughgoing structural reforms in the economy — notably reduced government spending and sources of patronage. For years the government and PRI had manipulated urban political groups with selective awards of property titles to squatter-settlement lots, bus transportation, water, electricity, and subsidized food in government stores (Eckstein 1988). The patronage ran out, and the politics of cooptation were exhausted morally; urban groups began to mobilize in Christian based–community organizations, neighborhood associations, and human rights groups. A coalition of political groups formed the Democratic Revolution Party (PRD). Running for president as the opposition PRD candidate in 1988, Cuauhtémoc Cárdenas carried the majority of Mexico City's 18 million voters but fell short of winning the election nationwide (only because of official vote fraud, the PRD and some observers would allege). Nevertheless, in subsequent elections at the state and local levels, opposition parties have won several governorships and a growing number of city-mayor posts. Reforms in the official party and in government, in conjunction with the North American Free Trade Agreement, suggest that democratization will continue in Mexico.

In both these cases, and in Eastern Europe, democratic movements have come from the civil society: trade unions, neighborhood associations, church and human rights groups. Established political parties closely identified with the state and major interest groups in the economy have not been the centers of change. Democratic movements found their social bases and their legitimacy in civil society, in cultural principles and human aspirations. Without a sociological vocabulary that recognizes these sources of action, major events of today's world cannot be explained.

PUBLIC LIFE IN AMERICA

Although the United States has not experienced the dramatic change of Eastern Europe and the Third World in recent years, civil society plays a key role here too. In Chapter 2, we discussed the important book by Robert Bellah and associates, *Habits of the Heart: Individualism and Commitment in American Life* (1985), which argues that modern American society is adrift — reasonably secure in material ways but confused about the purpose of life beyond individual satisfactions. Americans have a vague sense that their lives lack direction, and they express this as a desire, in their words, to "reconnect," perhaps to return to the values that they think can be found in small towns. But all of this is a vague feeling of unease. We lack a discourse for thinking and talking about a good society and a morally coherent life. Our lives, in other terms, are preoccupied with work and personal ambition, overregulated by the state, but devoid of commitment to civil society.

Such, at least, is the diagnosis of modern society by many keen observers. Bellah and associates bring up to date analyses that began with sociology's classical tradition: alienation in Marx, anomie in Durkheim, and the obscured link between private troubles and public issues, which C. Wright Mills's sociological imagination aims to penetrate. At bottom, *Habits of the Heart* makes a statement about individual lives and the social condition of the 1980s, which parallels *The Grapes of Wrath* fifty years earlier.

As our discussions of the state and world system suggest, modern social life has become more complex — most people are evermore remote from the centers of power that influence their lives. If civil society stands "between the economic structure and the state with its legislation and its coercion" (Gramsci 1971, 208), it is a narrow ledge. Yet claims about the weakening of civil society require examination. We know that people still resist and rebel (see Chapter 8) when their rights are violated and opportunities appear. We know that an equality revolution swept the United States in the not distant past. What is the condition of civil society today?

The 1980s saw a shift from civic action in pursuit of status equality to class issues, many of them based on narrow group interest (e.g., tax revolt), but some based on the public good, now framed as the quality of life. The environmental movement is foremost as a broad-based citizens' effort to preserve an endangered natural realm and to protect public health with curbs on such menaces as air and water pollution, toxic wastes, and nuclear fallout. The number of environmental groups, from the Sierra Club to Friends of the Earth, and issues, from whales to energy conservation, are too familiar to require listing. The more interesting aspect of the environmental movement is the manner in which it has generated a new consciousness about our individual and collective responsibility. Not only are people more attentive to their own health (avoiding such pollutants as cigarettes or preservatives and taking exercise), they increasingly accept collective obligations to conserve energy or water, to avoid littering, and to recycle. Public gains should not be overestimated. When gasoline shortages disappear and energy price increases level off, wasteful habits return. Federal budget cuts have eliminated the Department of Energy and reduced funding for environmental cleanup. But the need for energy efficiency today is far greater than it was fifteen years ago, and a new environmental awareness emphasizing public responsibility has taken hold.

Closely akin to the environmental movement is a revitalization of neighborhood and local community organizations devoted to public issues as diverse as housing, child care, aging, city planning, consumer protection, health, and many more. Harry Boyte (1980) called it a "new citizen movement" and, in a more recent book with Sara Evans (Evans and Boyte 1988), emphasized that democratic change in the United States had to come out of those "free spaces" in the community where alternative approaches can flourish. Less focused on particular issues than broadly committed to grass-roots organization, Evans and Boyte stress local responsibility for defining problems as people understand them rather than in terms of politicians' programmed answers. Although Evans and Boyte may be overly optimistic about the results of these groups, they are certainly correct in emphasizing that the frequency with which communities organize — to fight crime or development, for example — has increased.

Other instances of active civil society could be cited. Churches, and not just fundamentalist ones, are more healthy today than in the recent past. Self-help groups have multiplied to deal with problems from addiction to child abuse. Generally, we can conclude that if the state and market have expanded their scope in the modern world, civil society has held some of its ground and responded to new challenges where the state is either ineffective (e.g., in treating addiction) or sometimes indifferent (e.g., on environmental issues).

SOCIAL SCIENCE AND THE GOOD SOCIETY

What is the connection between civil society and sociology as the kind of critical enterprise described in this book?

First, I suggest that the aims of critical sociology are shared to a greater or lesser extent by other social sciences. Second, following the distinction here between state, economy, and civil society, critical social sciences deal with all three. Finally, and this is the main point, if modern society suffers from an anemic public life, an increasingly confined space for civil society, then the social sciences can help provide an alternative.

Bellah and associates end their book *Habits of the Heart* by advocating "social science as public philosophy" (1985, 297ff.). The idea is that the social sciences should be the forum in which public choices are discussed and evaluated, rather than remote academic specialties that posture at an objectivity no science practiced by humans can have:

> Social science is not a disembodied cognitive enterprise. It is a tradition, or a set of traditions, deeply rooted in the philosophical and humanistic (and, to more than a small extent, the religious) history of the West. Social science makes assumptions about the nature of persons, the nature of society, and the relation between persons and society. It also, whether it admits it or not, makes assumptions about good persons and a good society and considers how far these conceptions are embodied in our actual society. Becoming conscious of the cultural roots of these assumptions would remind the social scientist that these assumptions are contestable and that the choice of assumptions involves controversies that lie deep in the history of Western thought. Social science as public philosophy would make the philosophical conversation concerning these matters its own. (Bellah et al. 1985, 301)

We must take care to understand this proposition. It is not true that the social sciences can prescribe morals. It is simply this: *Social science can inform moral choices.*

Some readers are likely to doubt the proposition, perhaps to interpret it as an inflated and self-serving claim. The fact is, however, that the social sciences have already been given this responsibility. In modern societies, public debate no longer relies on the wisdom of literature, philosophy, and politics for answers to moral questions. Instead, as Alan Wolfe argues,

> The gap between the need for codes of moral obligation and the reality of societies that are confused about where they can be found is filled, however uncomfortably, by the contemporary social sciences. When modern intellectuals talk about the proper balance between obligations to oneself and responsibilities to others, they do so in the language of economics, political science, and sociology. (1989, 10)

The modern public debate, for example, seldom inquires, along the lines of philosophy or art, What is virtue, or what is truth? Instead, we want to know what will promote economic growth. This is a master value shared by the public and social scientists alike, with only occasional exceptions among environmentalists, for example, who question its implications. It is today's version of the good society, whether or not its advocates recognize it as such. When, next, the practical question is raised about how we can attain economic growth, the social sciences are the authority to which the public debate turns. Will tax cuts stimulate investment and growth, promoting the general welfare in the long run even though they favor the rich immediately? Is growth the best way to include the underprivileged in society's trickle-down benefits? Can we help the have-nots without growth, or would that lead to class conflict initiated by the threatened haves? Such questions could be multiplied, but the point is clear. These are social science questions. Modern society has a clear notion of "the good," and it depends on social science for direction on how to attain that good.

For better or worse, social science is providing the moral ideas for our age. This may be for the worse to the extent that social scientists and the public are unaware of what they are doing and uncritical about their basic assumptions (e.g., that growth produces good results). Yet, it may also be for the better to the extent that everyone involved is aware of what they are doing, critical of their assumptions and the evidence on which those rest, and open to alternatives. If social science already plays this role, then it must forthrightly meet the responsibility.

REFLECTIONS

Sociology is a complex craft. But so is social life, not to mention any craft that we come to know well. In this book, I have tried to clarify, but not minimize, these complexities because they merit appreciation, reflect human inventiveness, and reveal some of what was once mysterious. We still do not know what we need, much less what we want to know. We know even less about how to put to good use what we do know. But we know a lot more and have come a long way from the cocksure common sense that Paul Lazarsfeld was moved to rebut in the defense of sociology forty-five years ago. Most important, we know better how to find out what we want to know, how to formulate why questions in ways that yield useful critical answers.

Whether our aim is to practice or simply to appreciate the sociological craft, its classical tradition is still the best starting place. Today's social issues reenact the past in certain essentials. The questions that Marx posed about the fate of the working class and Weber's vision of status group insurgency are as vital as ever — perhaps more so in an age of disappearing industrial jobs and politically organized minorities, women, and religious groups. Sometimes the classical answers, such as Durkheim's analysis of social regulation and suicide, are still the best we have. At other times, better answers are produced by evaluating classical formulations and learning, for example, that class and status interact in ways not previously understood. The present profitably reworks experience. We do not have to be Marxists or Durkheimians to come better equipped to today's problems. As historian E. P. Thompson says, if we want to understand social inequality, "Marx is on our side; we are not on the side of Marx" (1978, 192).

The classical tradition shades imperceptibly into a second, third, and fourth generation of modern sociological research classics. Some sociology is extraordinarily good reading: C. Wright Mills's distillation of the meaning of *White Collar* life, Liston Pope's empathetic account of the disrupted world of *Millhands and Preachers*, Herbert Gans's sensible description of *The Urban Villagers*, and so many more that inform this book.

Critical work goes on in the sense of research and writing that explores to the roots of social problems, whether those lie in capitalism, the welfare state, or civil society. Critical sociology is also constructive sociology by virtue of its premise that people make their own history and may make it better by understanding the consequences of social choices. It connects biography and history in ways that begin to make clear what the choices are.

Sociology has progressed in recent years. It is less parochial, more international, less deluded by grand theory, more responsible for workable theories, more crafty. Major concepts such as class and status have departed anecdotal usage in favor of rigorous application. The role of the state has assumed its appropriate (and classical) place in social analyses. It is now clear that the world economy affects the domestic social organization of all countries.

If the foregoing are mainly, and appropriately, intellectual accomplishments, sociology also affects public life. Although sociological research seldom pleases its subjects, its central virtue is that it is a public discourse and method. If it sometimes serves special interests, and it does, sociology is still relatively open to informed criticism and better argument. Among social sciences that are increasingly captured by or written for technical experts, sociology remains a forum for examining where we are headed as a people — it is a public science and a publicly accessible means for achieving an open society.

SELECTED BIBLIOGRAPHY

ASH, TIMOTHY GARTON. 1983. *The Polish Revolution.* Sevenoaks, Kent, England: Hodder and Stoughton.

BELLAH, ROBERT N., RICHARD MADSEN, WILLIAM M. SULLIVAN, ANN SWIDLER, and STEVEN M. TIPTON. 1985. *Habits of the Heart: Individualism and Commitment in American Life.* Berkeley: University of California Press.

BOYTE, HARRY C. 1980. *The Backyard Revolution: Understanding the New Citizen Movement.* Philadelphia: Temple University Press.

CHIROT, DANIEL. 1991. *The Crisis of Leninism and the Decline of the Left: The Revolutions of 1989.* Seattle: University of Washington Press.

DRAPER, THEODORE. 1992. "Who Killed Soviet Communism." *New York Review of Books* (June 11):7–14.

ECKSTEIN, SUSAN. 1988. *The Poverty of Revolution: The State and the Urban Poor in Mexico,* Revised edition. Princeton, NJ: Princeton University Press.

EVANS, SARA M., and HARRY C. BOYTE. 1988. *Free Spaces: The Sources of Democratic Change in America.* Chicago: University of Chicago Press.

GRAMSCI, ANTONIO. 1971. "State and Civil Society." Pp. 210–276 in *Selections from the Prison Notebooks of Antonio Gramsci,* edited by Quintin Hoare and Geoffrey Nowell Smith. New York: International Publishers.

GWERTZMAN, BERNARD, and MICHAEL T. KAUFMAN. 1991. *The Collapse of Communism,* revised and updated. New York: Times Books/Random House.

MICHNIK, ADAM. 1985. *Letters from Prison and Other Essays.* Berkeley: University of California Press.

PIORE, MICHAEL, and CHARLES F. SABEL. 1984. *The Second Industrial Divide.* New York: Basic Books.

SZELENYI, IVAN. 1987. *Socialist Entrepreneurs.* Madison: University of Wisconsin Press.

THOMPSON, E. P. 1978. *The Poverty of Theory and Other Essays.* New York: Monthly Review.

WOLFE, ALAN. 1989. *Whose Keeper? Social Science and Moral Obligation.* Berkeley: University of California Press.

Index

... for Studying U Soc.

...g. Urban Rock

Intro U Soc.

Mixing Passion and Thought

Ways to Understand U Space.

Two Roads from Wirth

The Language

Compromised Purposes/Goals

p. 12 The Classical Tradition in U Soc.

Impressions

Attitude is a Little Thing
that makes a big Difference